RELIABILITY OF STRUCTURES

SECOND EDITION

RELIABILITY OF STRUCTURES

SECOND EDITION

Andrzej S. Nowak

Kevin R. Collins

CRC Press is an imprint of the
Taylor & Francis Group, an **informa** business

CRC Press
Taylor & Francis Group
6000 Broken Sound Parkway NW, Suite 300
Boca Raton, FL 33487-2742

First issued in paperback 2019

ISBN-13: 978-0-415-67575-8 (hbk)
ISBN-13: 978-0-367-86627-3 (pbk)

Library of Congress Cataloging-in-Publication Data

Nowak, Andrzej S.
Reliability of structures / Andrzej Nowak, Kevin R. Collins. -- Second edition.
pages cm
Includes bibliographical references and index.
ISBN 978-0-415-67575-8 (hardcover : alk. paper)
1. Structural analysis (Engineering) 2. Reliability (Engineering) I. Collins, Kevin R.
II. Title.

TA645.N64 2013
624.1'7--dc23 2012031217

Visit the Taylor & Francis Web site at
http://www.taylorandfrancis.com

and the CRC Press Web site at
http://www.crcpress.com

Contents

Preface

The objective of this book is to provide the reader with a practical tool for reliability analysis of structures. The presented material is intended to serve as a textbook for a one-semester course for undergraduate seniors or graduate students. The material is presented assuming that the reader has some background in structural engineering and structural mechanics. Previous exposure to probability and statistics is helpful but not required; the most important aspects of probability and statistics are reviewed early in the text.

Many of the available books on reliability are written for researchers, and these texts often approach the subject from a very mathematical and theoretical perspective. The focus of this book is on practical applications of structural reliability theory. The book does not provide detailed mathematical proofs of the underlying theory; instead, the book presents the basic concepts, interpretations, and equations and then explains to the reader how to use them. The book should be useful for both students and practicing structural engineers and hopefully will broaden their perspective by considering reliability as another important dimension of structural design. In particular, the presented methodology is applicable in the development of design codes, development of more reliable designs, optimization, and rational evaluation of existing structures.

The text is divided into 10 chapters with regard to topics.

Chapter 1 provides an introduction to structural reliability analysis. The discussion deals with the objectives of the study of reliability of structures and the sources of uncertainty inherent in structural design.

Chapter 2 provides a brief review of the theory of probability and statistics. The emphasis is placed on the definitions and formulas that are needed for derivation of the reliability analysis procedures. The material covers the definition of a random variable and its parameters such as the mean, median, standard deviation, coefficient of variation, cumulative distribution function, probability density function, and probability mass function. The probability distributions commonly used in structural reliability applications are reviewed; these include the normal; lognormal; extreme Type I,

II, and III; uniform; Poisson; and gamma distributions. A brief discussion of Bayesian methods is also included.

In Chapter 3, functions of random variables are considered. Concepts and parameters such as covariance, coefficient of correlation, and covariance matrix are described. Formulas are derived for parameters of a function of random variables. Special cases considered in this chapter are the sum of uncorrelated normal random variables and the product of uncorrelated lognormal random variables.

Chapter 4 presents some simulation techniques that can be used to solve structural reliability problems. The Monte Carlo simulation technique is the focus of this chapter. Two other methods are also discussed: the Latin Hypercube sampling method and Rosenblueth's point estimate method.

The concepts of limit states and limit state functions are defined in Chapter 5. Reliability and probability of failure are considered as functions of load and resistance. The fundamental structural reliability problem is formulated. The reliability analysis methods are also presented in Chapter 5. The simple second-moment mean value formulas are derived. Then, the Hasofer–Lind reliability index is defined. An iterative procedure is shown for variables with full distributions available.

The development of a reliability-based design code is discussed in Chapter 6. The presented material includes the basic steps for finding load and resistance factors and a calibration procedure used in several recent research projects.

Load models are presented in Chapter 7. The considered load components include dead load, live load for buildings and bridges, and environmental loads (such as wind, snow, and earthquake). Some techniques for combining loads together in reliability analyses are also presented.

Resistance models are discussed in Chapter 8. Statistical parameters are presented for steel beams, columns, tension members, and connections. Noncomposite and composite sections are considered. For reinforced concrete members and prestressed concrete members, the parameters are given for flexural capacity and shear. The results are based on the available test data and simulations.

Chapter 9 deals with the important topic of system reliability. Useful formulas are presented for a series system, a parallel system, and mixed systems. The effect of correlation between structural components on the reliability of a system is evaluated. The approach to system reliability analysis is demonstrated using simple practical examples.

Models of human error in structural design and construction are reviewed in Chapter 10. The classification of errors is presented with regard to mechanism of occurrence, cause, and consequences. Error survey results are discussed. A strategy to deal with errors is considered. Special focus is placed on the sensitivity analysis. Sensitivity functions are presented for typical structural components.

Acknowledgments

Work on this book required frequent discussions and consultations with many experts in theoretical and practical aspects of structural reliability. Therefore, we would like to acknowledge the support and inspiration we received over many years from our colleagues and teachers, in particular Niels C. Lind, Palle Thoft-Christensen, Dan M. Frangopol, Mircea D. Grigoriu, Rudiger Rackwitz, Guiliano Augusti, Robert Melchers, Michel Ghosn, Fred Moses, James T.P. Yao, Ted V. Galambos, M.K. Ravindra, Brent W. Hall, Robert Sexsmith, Yozo Fujino, Hitoshi Furuta, Gerhard Schueller, Y.K. Wen, Wilson Tang, C. Allin Cornell, Bruce Ellingwood, Janusz Murzewski, John M. Kulicki, Dennis Mertz, Jozef Kwiatkowski, and Tadeusz Nawrot.

Thanks are due to many former and current doctoral students, in particular Rajeh Al-Zaid, Hassan Tantawi, Abdulrahim Arafah, Juan A. Megarejo, Jianhua Zhou, Jack R. Kayser, Shuenn Chern Ting, Sami W. Tabsh, Eui-Seung Hwang, Young-Kyun Hong, Naji Arwashan, Ahmed S. Yamani, Hani H. Nassif, Jeffrey A. Laman, Hassan H. El-Hor, Sangjin Kim, Vijay Saraf, Chan-Hee Park, Po-Tuan Chen, Juwhan Kim, Thomas Murphy, Siddhartha Ghosh, Anna Rakoczy, Krzysztof Waszczuk, and Przemyslaw Rakoczy.

This book is a revised version of the previous edition. We would like to thank current and former doctoral students at the University of Nebraska, namely, Anna M. Rakoczy, Krzysztof Waszczuk, and Przemyslaw Rakoczy, for preparation of the text, figures, and examples. Thanks are also due to Dr. Maria Szerszen, Kathleen Seavers, Tadeusz Alberski, Ahmet Sanli, Junsik Eom, Charngshiou Way, and Gustavo Parra-Montesinos who helped with the preparation of some of the text, figures, and examples in the first edition.

Finally, we would like to thank our wives, Jolanta and Karen, for their patience and support.

Acknowledgments

[illegible]

Authors

Andrzej S. Nowak has been a Robert W. Brightfelt Professor of Engineering at the University of Nebraska since 2005 after 25 years at the University of Michigan, where he was a professor of civil engineering (1979–2004). He received his MS (1970) and PhD (1975) from Politechnika Warszawska in Poland. He then worked at the University of Waterloo in Canada (1976–1978) and the State University of New York in Buffalo (1978–1979). Professor Nowak's research has led to the development of a probabilistic basis for the new generation of design codes for highway bridges, including load and resistance factors for the American Association of State Highway and Transportation Officials (AASHTO) Code, American Concrete Institute (ACI) 318 Code for Concrete Structures, Canadian Highway Bridge Design Code, and fatigue evaluation criteria for BS-5400 (United Kingdom). He has authored or coauthored more than 400 publications, including books, journal papers, and articles in conference proceedings. Professor Nowak is an active member of national and international professional organizations, and he chaired a number of committees associated with professional organizations such as the American Society of Civil Engineers (ASCE), ACI, Transportation Research Board (TRB), International Association for Bridge and Structural Engineering (IABSE), and International Association for Bridge Maintenance and Safety (IABMAS). He is an Honorary Professor of Politechnika Warszawska and Politechnika Krakowska, and a Fellow of ASCE, ACI, and IABSE. Prof. Nowak received the ASCE Moisseiff Award, Bene Merentibus Medal, and Kasimir Gzowski Medal from the Canadian Society of Civil Engineers.

Kevin R. Collins is a member of the faculty at the University of Cincinnati Blue Ash College (UCBA) in Cincinnati, OH. Prior to joining UCBA, he worked at Valley Forge Military College (Wayne, Pennsylvania), Lawrence Technological University (Southfield, Michigan), the United States Coast Guard Academy (New London, Connecticut), and the University of Michigan (Ann Arbor). He received his bachelor of civil engineering (BCE) degree from the University of Delaware in May 1988, his MS degree from

Virginia Polytechnic Institute and State University in December 1989, and his PhD degree from the University of Illinois in October 1995. Between his MS and PhD degrees, he worked for MPR Associates, Inc., in Washington, DC, for 2.5 years. Dr. Collins' research interests are in the areas of earthquake engineering, structural dynamics, and structural reliability. Dr. Collins is a member of the American Society for Engineering Education (ASEE) and the honor societies of Chi Epsilon, Tau Beta Pi, and Phi Kappa Phi.

Chapter 1

Introduction

1.1 OVERVIEW

Many sources of uncertainty are inherent in structural design. Despite what we often think, the parameters of the loading and the load-carrying capacities of structural members are not *deterministic* quantities (i.e., quantities that are perfectly known). They are *random variables*, and thus absolute safety (or zero probability of failure) cannot be achieved. Consequently, structures must be designed to serve their function with a finite probability of failure.

To illustrate the distinction between deterministic versus random quantities, consider the loads imposed on a bridge by car and truck traffic. The load on the bridge at any time depends on many factors such as the number of vehicles on the bridge and the weights of the vehicles. As we all know from daily experience, cars and trucks come in many shapes and sizes. Furthermore, the number of vehicles that pass over a bridge fluctuates, depending on the time of day. Since we do not know the specific details about each vehicle that passes over the bridge or the number of vehicles on the bridge at any time, there is some uncertainty about the total load on the bridge. Hence, the load is a random variable.

Society expects buildings and bridges to be designed with a reasonable safety level. In practice, these expectations are achieved by following code requirements specifying design values for minimum strength, maximum allowable deflection, and so on. Code requirements have evolved to include design criteria that take into account some of the sources of uncertainty in design. Such criteria are often referred to as *reliability-based design criteria*. The objective of this book is to provide the background needed to understand how these criteria were developed and to provide a basic tool for structural engineers interested in applying this approach to other situations.

The reliability of a structure is its ability to fulfill its design purpose for some specified design lifetime. Reliability is often understood to equal the probability that a structure will not fail to perform its intended function.

The term "failure" does not necessarily mean catastrophic failure but is used to indicate that the structure does not perform as desired.

1.2 OBJECTIVES OF THE BOOK

This book attempts to answer the following questions:

How can we measure the safety of structures? Safety can be measured in terms of reliability or the probability of uninterrupted operation. The complement to reliability is the probability of failure. As we discuss in later chapters, it is often convenient to measure safety in terms of a reliability index instead of probability.

How safe is safe enough? As mentioned earlier, it is impossible to have an absolutely safe structure. Every structure has a certain nonzero probability of failure. Conceptually, we can design the structure to reduce the probability of failure, but increasing the safety (or reducing the probability of failure) beyond a certain optimum level is not always economical. This optimum safety level has to be determined.

How does a designer implement the optimum safety level? Once the optimum safety level is determined, appropriate design provisions must be established so that structures will be designed accordingly. Implementation of the target reliability can be accomplished through the development of probability-based design codes.

1.3 POSSIBLE APPLICATIONS

Structural reliability concepts can be applied to the design of new structures and the evaluation of existing ones. Many modern design codes are based on probabilistic models of loads and resistances. Examples include the American Institute of Steel Construction (AISC, 2011)[1] Load and Resistance Factor Design (LRFD) code for steel buildings (AISC, 2006), American Association of State Highway and Transportation Officials LRFD code (AASHTO, 2012), Canadian Highway Bridge Design Code (2006), and the European codes (EN EUROCODES, n.d.). In general, reliability-based design codes are efficient because they make it easier to achieve either of the following goals:

- For a given cost, design a more reliable structure.
- For a given reliability, design a more economical structure.

[1] Many acronyms are used in structural engineering and structural reliability. Appendix A lists the acronyms used in this book.

The reliability of a structure can be considered as a rational evaluation criterion. It provides a good basis for decisions about repair, rehabilitation, or replacement. A structure can be condemned when the nominal value of load exceeds the nominal load-carrying capacity. However, in most cases, a structure is a system of components, and failure of one component does not necessarily mean failure of the structural system. When a component reaches its ultimate capacity, it may continue to resist the load while loads are redistributed to other components. System reliability provides a methodology to establish the relationship between the reliability of an element and the reliability of the system.

1.4 HISTORICAL PERSPECTIVE

Many of the current approaches to achieving structural safety evolved over many centuries. Even ancient societies attempted to protect the interests of their citizens through regulations. The minimum safety requirements were enforced by specifying severe penalties for builders of structures that did not perform adequately. The earliest known building code was used in Mesopotamia. It was issued by Hammurabi, the king of Babylonia, who died about 1750 BC. The "code provisions" were carved in stone, and these stone carvings are preserved in the Louvre Museum in Paris, France. The responsibilities were defined depending on the consequences of failure. If a building collapsed killing a son of the owner, then the builder's son would be put to death. If the owner's slave was killed, then the builder's slave was executed, and so on.

For centuries, the knowledge of design and construction was passed from one generation of builders to the next one. A master builder often tried to copy a successful structure. Heavy stone arches often had a considerable safety reserve. Attempts to increase the height or span were based on intuition. The procedure was essentially trial and error. If a failure occurred, that particular design was abandoned or modified.

As time passed, the laws of nature became better understood. Mathematical theories of material and structural behavior evolved, providing a more rational basis for structural design. In turn, these theories provided the necessary framework in which probabilistic methods could be applied to quantify structural safety and reliability. The first mathematical formulation of the structural safety problem can be attributed to Mayer (1926), Streletskii (1947), and Wierzbicki (1936). They recognized that load and resistance parameters are random variables, and therefore, for each structure, there is a finite probability of failure. Their concepts were further developed by Freudenthal in the 1950s (e.g., Freudenthal, 1956). The formulations involved convolution functions that were too difficult to evaluate by hand. The practical applications of reliability analysis were not possible until the pioneering work of Cornell and Lind in the late 1960s and early 1970s. Cornell proposed a second-moment reliability index in 1969. Hasofer and Lind formulated a definition of a format-invariant reliability index in 1974. An efficient numerical procedure was formulated for calculation of the reliability index by Rackwitz and Fiessler (1978). Other important contributions have been made by Ang, Veneziano, Rosenblueth, Esteva, Turkstra, Moses, Grigoriu, Der Kiuregian, Ellingwood, Corotis, Frangopol, Fujino, Furuta, Yao, Brown, Ayyub, Blockley, Stubbs, Mathieu, Melchers, Augusti, Shinozuka, and Wen. By the end of 1970s, the reliability methods reached a degree of maturity, and now they are readily available for applications. They are used primarily in the development of new design codes.

The developed theoretical work has been presented in books by Thoft-Christensen and Baker (1982), Augusti et al. (1984), Madsen et al. (2006), Ang and Tang (1984), Melchers (1999), Thoft-Christensen and Morotsu (1986), and Ayyub and McCuen (2002), to name just a few. Other books available in the area of structural reliability include Murzewski (1989) and Marek et al. (1996).

It is important to note that most reliability-based codes in current use apply reliability concepts to the design of structural members, not structural systems. In the future, one can expect a further acceleration in the development of analytical methods used to model the behavior of structural systems. It is expected that this focus on system behavior will lead to additional applications of reliability theory at the system level.

1.5 UNCERTAINTIES IN THE BUILDING PROCESS

The building process includes planning, design, construction, operation/use, and demolition. All components of the process involve various uncertainties. These uncertainties can be put into two major categories with regard to causes: natural and human.

1. *Natural causes* of uncertainty result from the unpredictability of loads such as wind, earthquake, snow, ice, water pressure, or live load. Another source of uncertainty attributable to natural causes is the mechanical behavior of the materials used to construct the structure. For example, material properties of concrete can vary from batch to batch and also within a particular batch.
2. *Human causes* include intended and unintended departures from an optimum design. Examples of these uncertainties during the design phase include approximations, calculation errors, communication problems, omissions, lack of knowledge, and greed. Similarly, during the construction phase, uncertainties arise due to the use of inadequate materials, methods of construction, bad connections, or

changes without analysis. During operation/use, the structure can be subjected to overloading, inadequate maintenance, misuse, or even an act of sabotage.

Because of these uncertainties, loads and resistances (i.e., load-carrying capacities of structural elements) are random variables. It is convenient to consider a random parameter (load or resistance) as a function of three factors:

1. *Physical Variation Factor:* This factor represents the variation of load and resistance that is inherent in the quantity being considered. Examples include a natural variation of wind pressure, earthquake, live load, and material properties.
2. *Statistical Variation Factor:* This factor represents uncertainty arising from estimating parameters based on a limited sample size. In most situations, the natural variation (physical variation factor) is unknown and it is quantified by examining limited sample data. Therefore, the larger the sample size, the smaller the uncertainty described by the statistical variation factor.
3. *Model Variation Factor:* This factor represents the uncertainty due to simplifying assumptions, unknown boundary conditions, and unknown effects of other variables. It can be considered as a ratio of the actual strength (test result) and strength predicted using the model.

How these three factors come into a reliability analysis is discussed in later chapters.

Chapter 2

Random variables

The purpose of this chapter is to review aspects of the theory of probability and statistics needed for reliability analysis of structures.

2.1 BASIC DEFINITIONS

2.1.1 Sample space and event

The concepts of sample space and event can best be demonstrated by considering an experiment. For example, the experiment might test material strength, measure the depth of a beam, or determine occurrence (or non-occurrence) of a truck on a particular bridge during a specified period of time. In these experiments, the outcomes are unpredictable. All possible outcomes of an experiment comprise a *sample space*. Combinations of one or more of the possible outcomes or ranges of outcomes can be defined as *events*.

To further illustrate these concepts, consider the following two examples.

Example 2.1

Consider an experiment in which some number (n) of standard concrete cylinders is tested to determine their compressive strength, f_c', as shown in Figure 2.1.

Assume that the test results are

$$x_1, x_2, x_3, \ldots, x_n$$

where x_i is the outcome (i.e., the experimental value of f_c') of the ith cylinder.

For this experiment, the sample space is an interval including all positive numbers because the compressive strength can be any positive value. The defined sample space for concrete cylinder tests is called a *continuous* sample space. Theoretically, even $f_c' = 0$ is possible (but unlikely) when the mix is made without any cement. The actual

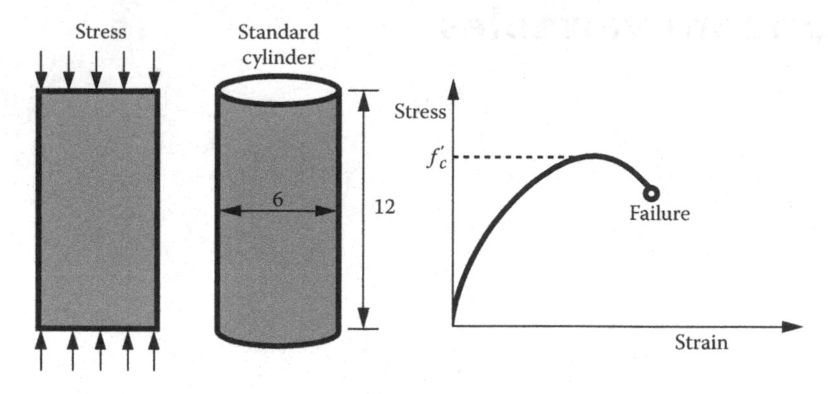

Figure 2.1 Concrete cylinder test considered in Example 2.1.

compressive strength varies randomly, and n test results supply only a limited amount of information about its variation.

Events E_1, E_2, ..., E_n can be defined as ranges of values (or intervals) of compressive strength. For example, E_1 could be defined as the event when the compressive strength is between 0 kips per square inch (ksi) and 1 ksi (1 kip = 1000 lb and 1 ksi = 6.9 MPa). Similarly, E_2 could be defined as the event when the strength is between 1 ksi and 2 ksi.

Example 2.2

Consider another experiment. A reinforced concrete beam is tested to determine one of the two possible modes of failure:

Mode 1: failure occurs by crushing of concrete
Mode 2: failure occurs by yielding of steel

In this case, the sample space consists of two discrete failure modes: mode 1 and mode 2. This sample space has a finite number of elements and it is called a *discrete* sample space. Each mode of failure can be considered an event.

Two special types of events should be mentioned. A *certain event* is defined as consisting of the entire sample space. The implication of this definition is that a certain event will *definitely* occur. In Example 2.1 above, a certain event would be when the compressive strength data are greater than or equal to zero. An *impossible event* is defined as an outcome that *cannot* occur. Again, in the context of Example 2.1, an impossible event would be when the compressive strength is less than zero.

2.1.2 Axioms of probability

The following axioms of classical probability theory are included only as a quick reference. A more comprehensive discussion of probability can be found in any introductory-level probability textbook (e.g., Miller et al., 2010; Milton and Arnold, 2002; Ross, 2009; Montgomery and Runger, 2010; Ang and Tang, 2006; Ayyub and McCuen, 2002).

Let E represent an event, and let Ω represent a sample space. The notation $P()$ is used to denote a probability function defined on events in the sample space.

Axiom 1

For any event E,

$$0 \le P(E) \le 1 \tag{2.1}$$

where $P(E)$ is the probability of event E. In words, the probability of any event must be between 0 and 1 inclusive. ■

Axiom 2

$$P(\Omega) = 1 \tag{2.2}$$

In words, this axiom states that the probability of occurrence of an event corresponding to the entire sample space (i.e., a certain event) is equal to 1. ■

Axiom 3

Consider n *mutually exclusive* events $E_1, E_2, \ldots, E_n$.

$$P\left(\bigcup_{i=1}^{n} E_i\right) = \sum_{i=1}^{n} P(E_i) \tag{2.3}$$

where $P\left(\bigcup_{i=1}^{n} E_i\right)$ represents the probability of the union of all events E_1, $E_2, \ldots, E_n$. In other words, it represents the probability of occurrence of E_1 or E_2 or … or E_n. ■

Mutually exclusive events exist when the occurrence of any one event excludes the occurrence of the others. Two or more mutually exclusive events cannot occur simultaneously. For example, returning to Example 2.1 above, if we denote compressive strength by f_c', examples of mutually exclusive events would be as follows:

$$E_1 = \{0 \le f_c' < 1000 \text{ psi}\} \tag{2.4a}$$

$$E_2 = \{1000 \text{ psi} \le f_c' < 2000 \text{ psi}\} \tag{2.4b}$$

$$E_3 = \{2000 \le f_c' < 3000 \text{ psi}\} \tag{2.4c}$$

$$E_4 = \{f_c' \ge 3000 \text{ psi}\} \tag{2.4d}$$

For these four events, the union of all events is the sample space defined earlier, i.e.,

$$\bigcup_{i=1}^{4} E_i = \{0 \le f_c' < \infty\} \tag{2.5}$$

These four mutually exclusive events are also *collectively exhaustive* because the union of all four events is the entire sample space.

2.1.3 Random variable

A random variable is defined as a function that maps events onto intervals on the axis of real numbers. This is schematically shown in Figure 2.2. In most cases, a random variable in this book is designated by a capital letter.

A probability function is defined by events. This definition can be extended using random variables. Let $X(E)$ be a function that assigns a value

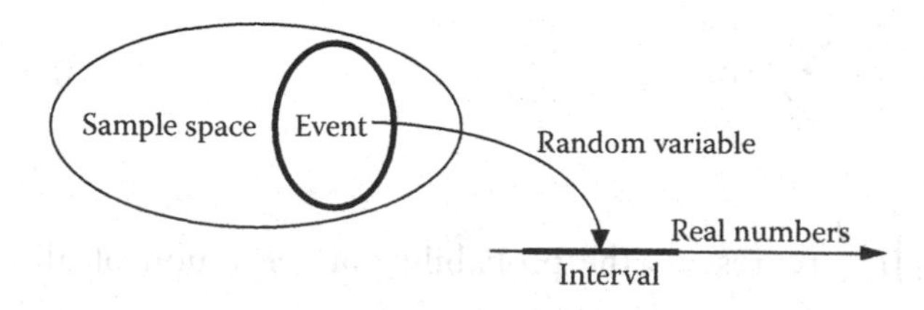

Figure 2.2 Schematic representation of a random variable as a function.

to an event E. As an example, consider the concrete strength in Example 2.1 discussed earlier. The value of the strength, f_c', has units of force per unit area. If an observed value of strength is $f_c' = 3127$ psi, then one possible definition of the random variable might simply be $X(f_c') = f_c'(\text{psi})$. In this case, the parameter that is measured is called the random variable; this is common in many engineering applications. Alternatively, we could define the random variable to be

$$X(f_c') = \frac{f_c'(\text{psi})}{1000} - 2.0\ \text{ksi} \tag{2.6}$$

then $f_c' = 3127$ psi corresponds to $X(f_c') = 1.127$ ksi.

A random variable can be either a *continuous* random variable or a *discrete* random variable. In the previous paragraph, the random variable was continuous because the variable could assume any value on the positive real axis. An example of a discrete random variable is as follows:

$$X(f_c') = \begin{Bmatrix} 1 & \text{if} & 0 & \leq & f_c' & < & 1000\,\text{psi} \\ 2 & \text{if} & 1000 & \leq & f_c' & < & 2000\,\text{psi} \\ 3 & \text{if} & 2000 & \leq & f_c' & < & 3000\,\text{psi} \\ 4 & \text{if} & & & f_c' & & 3000\,\text{psi} \end{Bmatrix} \tag{2.7}$$

In this case, the random variable can assume only four discrete integer values.

2.1.4 Basic functions

The *probability mass function* (PMF) is defined for *discrete* random variables as follows: $p_X(x)$ = probability that a discrete random variable X is equal to a specific value x where x is a real number. Note that the random variable (with an uncertain value) is denoted by a capital letter, whereas a specific value or realization of the variable is denoted by a lower case letter. Mathematically,

$$p_X(x) = P(X = x) \tag{2.8}$$

For example, if X is a discrete random variable describing concrete strength (f_c') as defined in Equation 2.7, then the values of the PMF function would be

$$p_X(1) = P(X = 1) \tag{2.9a}$$

$$p_X(2) = P(X = 2) \tag{2.9b}$$

$$p_X(3) = P(X = 3) \tag{2.9c}$$

$$p_X(4) = P(X = 4) \tag{2.9d}$$

Equations 2.9a through 2.9d are represented graphically in Figure 2.3 for a hypothetical set of values of the PMF function.

The *cumulative distribution function* (CDF) is defined for both *discrete* and *continuous* random variables as follows: $F_X(x)$ = the total sum (or integral) of all probability functions (continuous and discrete) corresponding to values less than or equal to x. Mathematically,

$$F_X(x) = P(X \leq x) \tag{2.10}$$

Consider the f_c' intervals previously defined by Equations 2.9a through 2.9d. Let X be a discrete random variable and assume the values of the probability mass function are as follows:

$$p_X(1) = 0.05 \tag{2.11a}$$

$$p_X(2) = 0.20 \tag{2.11b}$$

$$p_X(3) = 0.65 \tag{2.11c}$$

$$p_X(4) = 0.10 \tag{2.11d}$$

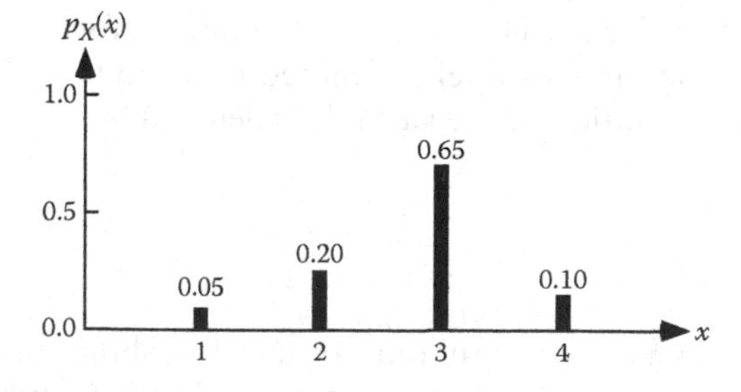

Figure 2.3 A probability mass function.

The corresponding CDF function is shown in Figure 2.4. Note that the CDF function is always a *nondecreasing* function of x.

For *continuous* random variables, the *probability density function* (PDF) is defined as the first derivative of the CDF. The PDF [$f_X(x)$] and the CDF [$F_X(x)$] for continuous random variables are related as follows:

$$f_X(x) = \frac{d}{dx} F_X(x) \tag{2.12}$$

$$F_X(x) = \int_{-\infty}^{x} f_X(\xi)\, d\xi \tag{2.13}$$

To illustrate these relationships, consider a continuous random variable X. The PDF and CDF functions might look like those shown in Figures 2.5 and 2.6, respectively. Equation 2.13 represents the shaded area under the PDF as shown in Figure 2.7 for the case $x = a$.

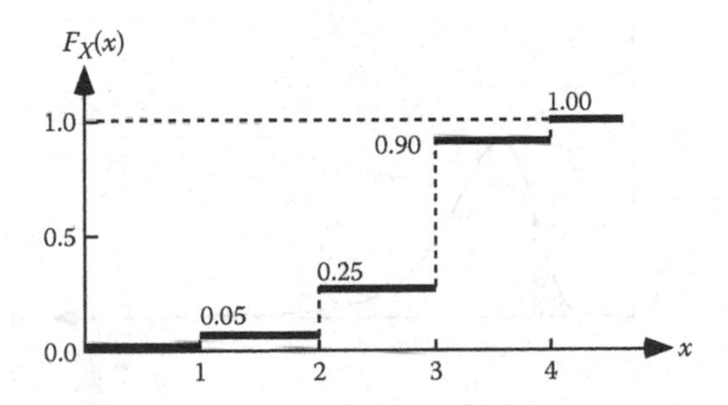

Figure 2.4 A CDF for a discrete random variable.

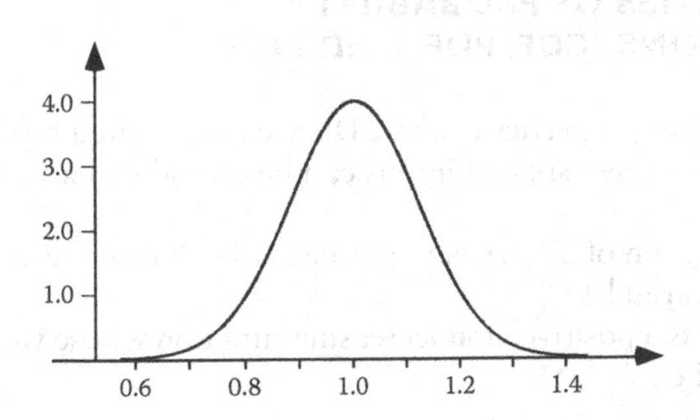

Figure 2.5 Example of a PDF.

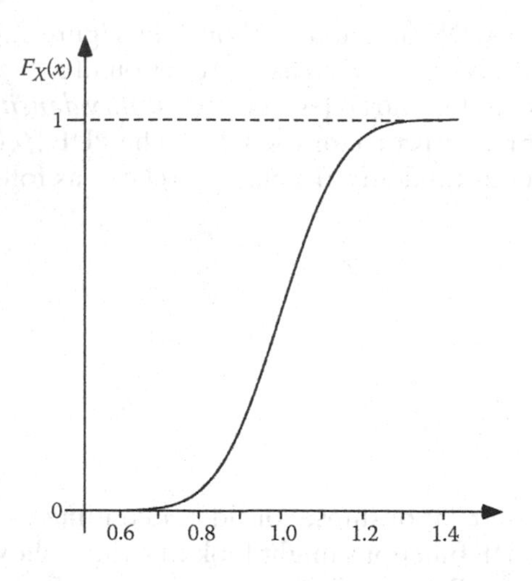

Figure 2.6 Example of a CDF.

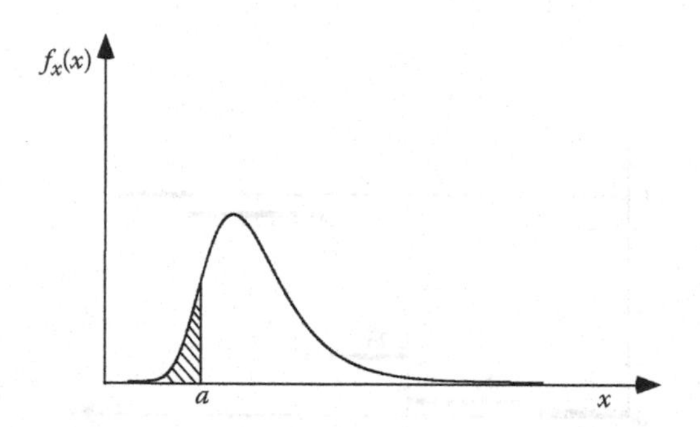

Figure 2.7 Relationship between CDF and PDF described by Equation 2.13.

2.2 PROPERTIES OF PROBABILITY FUNCTIONS (CDF, PDF, AND PMF)

Several important properties of the CDF are enumerated below. Any function that satisfies these six conditions can be considered a CDF.

1. The definition of a CDF is the same for both discrete and continuous random variables.
2. The CDF is a positive, nondecreasing function whose value is between 0 and 1, i.e.,

$$0 \leq F_x(x) \leq 1 \tag{2.14}$$

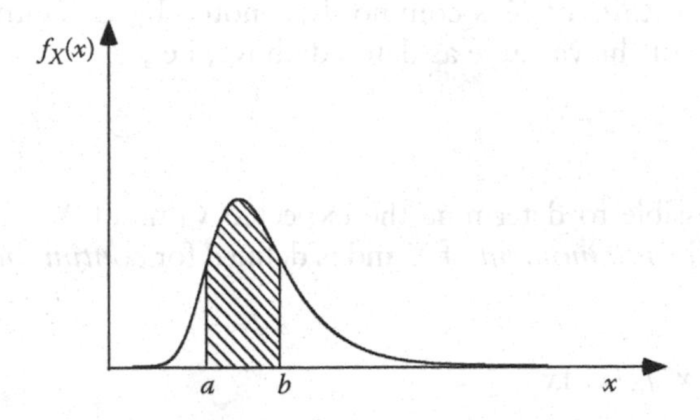

Figure 2.8 Graphical representation of $F_X(b) - F_X(a)$ in Equation 2.15.

3. If $x_1 < x_2$, then $F_X(x_1) \le F_X(x_2)$.
4. $F_X(-\infty) = 0$.
5. $F_X(+\infty) = 1$.
6. For continuous random variables,

$$P(a \le x \le b) = F_X(b) - F_X(a) = \int_a^b f_X(\xi)\,d\xi \tag{2.15}$$

Figure 2.8 provides a graphical interpretation of Equation 2.15.

2.3 PARAMETERS OF A RANDOM VARIABLE

2.3.1 Basic parameters

Consider a random variable X. Although the value of the variable is uncertain, there are certain parameters that help to mathematically describe the properties of the variable.

The *mean value* of X is denoted by μ_X. For a *continuous* random variable, the mean value is defined as

$$\mu_X = \int_{-\infty}^{+\infty} x f_X(x)\,dx \tag{2.16}$$

For a *discrete* random variable, the mean value is defined as

$$\mu_X = \sum_{\text{all } x_i} x_i\, p_X(x_i) \tag{2.17}$$

The *expected value* of X is commonly denoted by $E(X)$ and is equal to the mean value of the variable as defined above, i.e.,

$$E(X) = \mu_X \tag{2.18}$$

It is also possible to determine the expected value of X^n. This expected value is called the *nth moment* of X and is defined for *continuous* variables as

$$E(X^n) = \int_{-\infty}^{+\infty} x^n f_X(x)\,dx \tag{2.19}$$

and for *discrete* variables, the nth moment is defined as

$$E(X^n) = \sum_{\text{all } x_i} x_i^n \, p_X(x_i) \tag{2.20}$$

The *variance* of X, commonly denoted as σ_X^2, is defined as the expected value of $(X - \mu_X)^2$ and is equal to

$$\sigma_X^2 = \int_{-\infty}^{+\infty} (x - \mu_X)^2 f_X(x)\,dx \quad \text{(continuous random variable)} \tag{2.21a}$$

$$\sigma_X^2 = \sum_{\text{all } x_i} (x_i - \mu_X)^2 \, p_X(x_i) \quad \text{(discrete random variable)} \tag{2.21b}$$

An important relationship exists among the mean, variance, and second moment of a random variable X:

$$\sigma_X^2 = E(X^2) - \mu_X^2 \tag{2.22}$$

The *standard deviation* of X is defined as the positive square root of the variance:

$$\sigma_X = \sqrt{\sigma_X^2} \tag{2.23}$$

The nondimensional *coefficient of variation*, V_X or COV, is defined as the standard deviation divided by the mean:

$$V_X = \frac{\sigma_X}{\mu_X} = \text{COV} \tag{2.24}$$

This parameter is always taken to be positive by convention even though the mean may be negative.

2.3.2 Sample parameters

The parameters defined in Section 2.3.1 are the theoretical properties of the random variable because they are all calculated based on knowledge of the probability distributions of the variable. In many practical applications, we do not know the true distribution, and we need to estimate parameters using test data. If a set of n observations $\{x_1, x_2, \ldots, x_n\}$ is obtained for a particular random variable X, then the true mean μ_X can be approximated by the *sample mean* $\bar{x}$ and the true standard deviation σ_X can be approximated by the *sample standard deviation* s_X.

The sample mean is calculated as

$$\bar{x} = \frac{1}{n}\sum_{i=1}^{n} x_i \tag{2.25}$$

The sample standard deviation is calculated as

$$s_X = \sqrt{\frac{\sum_{i=1}^{n}(x_i - \bar{x})^2}{n-1}} = \sqrt{\frac{\left(\sum_{i=1}^{n} x_i^2\right) - n(\bar{x})^2}{n-1}} \tag{2.26}$$

2.3.3 Standard form

Let X be a random variable. The *standard form* of X, denoted by Z, is defined as

$$Z = \frac{X - \mu_X}{\sigma_X} \tag{2.27}$$

The mean of Z is calculated as follows. We note that the mathematical expectation (mean value) of an arbitrary function, $g(X)$, of the random variable X is defined as

$$\mu_{g(X)} = E[g(X)] = \int_{-\infty}^{+\infty} g(x) f_X(x)\,dx \tag{2.28}$$

Using this definition with $Z = g(X)$, we can show that

$$\mu_Z = E\left[\frac{X-\mu_X}{\sigma_X}\right] = \frac{1}{\sigma_X}\left[E(X) - E(\mu_X)\right] = \frac{1}{\sigma_X}\left(\mu_X - \mu_X\right) = 0 \tag{2.29}$$

$$\sigma_Z^2 = E(Z^2) - \mu_Z^2 = E\left[\left(\frac{X-\mu_X}{\sigma_X}\right)^2\right] - 0 = \frac{1}{\sigma_X^2}E\left[(X-\mu_X)^2\right] = \frac{\sigma_X^2}{\sigma_X^2} = 1 \tag{2.30}$$

Thus, the mean of the standard form of a random variable is 0 and its variance is 1.

2.4 COMMON RANDOM VARIABLES

Any random variable is defined by its CDF, $F(x)$. The PDF, $f_X(x)$, of a continuous random variable is the first derivative of $F_X(x)$. The most important random variables used in structural reliability analysis are as follows: uniform, normal, lognormal, gamma, extreme Type I, extreme Type II, extreme Type III, and Poisson. Each of these will be briefly described in the following sections.

2.4.1 Uniform random variable

For a uniform random variable, the PDF function has a constant value for all possible values of the random variable within a range $[a,b]$. This means that all numbers are equally likely to appear. Mathematically, the PDF function is defined as follows:

$$\text{PDF} = f_X(x) = \begin{cases} \dfrac{1}{b-a} & a \le x \le b \\ 0 & \text{otherwise} \end{cases} \tag{2.31}$$

where a and b define the lower and upper bounds of the random variable. The PDF and CDF for a uniform random variable are shown in Figure 2.9.

The mean and variance are as follows:

$$\mu_X = \frac{a+b}{2} \tag{2.32}$$

$$\sigma_X^2 = \frac{(b-a)^2}{12} \tag{2.33}$$

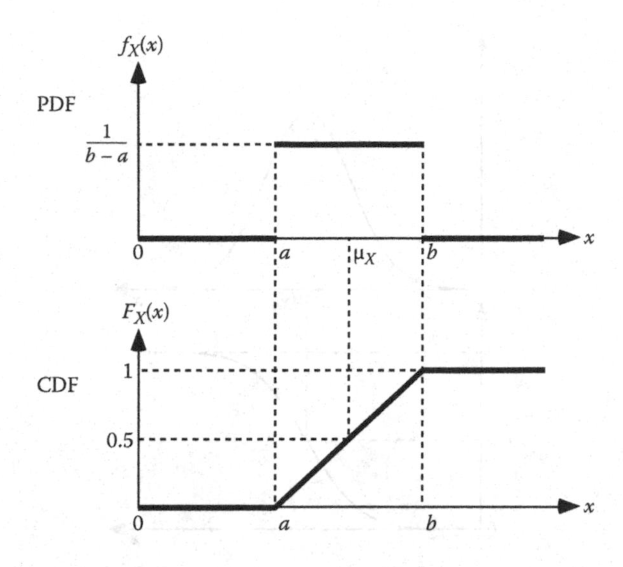

Figure 2.9 PDF and CDF of a uniform random variable.

2.4.2 Normal random variable

The normal random variable is the most important distribution in structural reliability theory. The PDF for a normal random variable X is

$$f_X(x) = \frac{1}{\sigma_X\sqrt{2\pi}} \exp\left[-\frac{1}{2}\left(\frac{x-\mu_X}{\sigma_X}\right)^2\right] \tag{2.34}$$

where μ_X and σ_X are the mean and standard deviation, respectively. Note that the term in parentheses in Equation 2.34 is in standard form as presented in Equation 2.27. Figure 2.10 shows the general shape of both the PDF and CDF of a normal random variable.

There is no closed-form solution for the CDF of a normal random variable. However, tables have been developed to provide values of the CDF for the special case in which $\mu_X = 0$ and $\sigma_X = 1$. If we substitute these values in Equation 2.34 above, we get the PDF for the *standard normal variable z*, which is often denoted by $\phi(z)$:

$$\phi(z) = \frac{1}{\sqrt{2\pi}} \exp\left[-\frac{1}{2}(z)^2\right] = f_Z(z) \tag{2.35}$$

The CDF of the standard normal variable is typically denoted by $\Phi(z)$. Many popular mathematics and spreadsheet programs have a standard

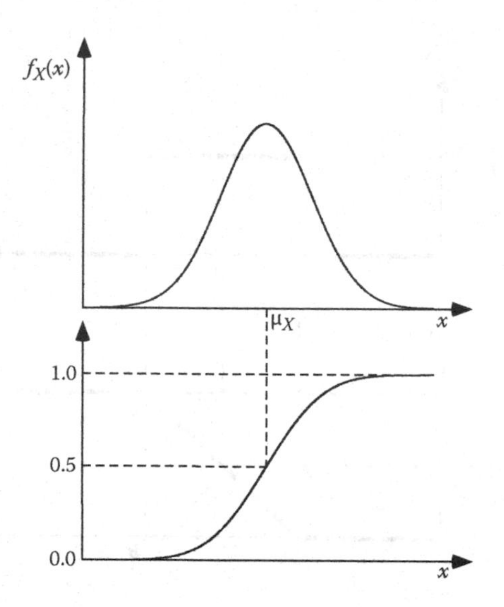

Figure 2.10 PDF and CDF of a normal random variable.

normal CDF function built in. Values of $\Phi(z)$ are listed in Table B.1 (in Appendix B) for values of z ranging from 0 to −8.9. Values of $\Phi(z)$ for $z > 0$ can also be obtained from Appendix B by applying the symmetry property of the normal distribution, i.e.,

$$\Phi(z) = 1 - \Phi(-z) \tag{2.36}$$

Figures 2.11 and 2.12 show the shapes of $\Phi(z)$ and $\Phi(z)$.

The probability information for the standard normal random variable can be used to obtain the CDF and PDF values for an arbitrary normal random variable by performing a simple coordinate transformation. Let X

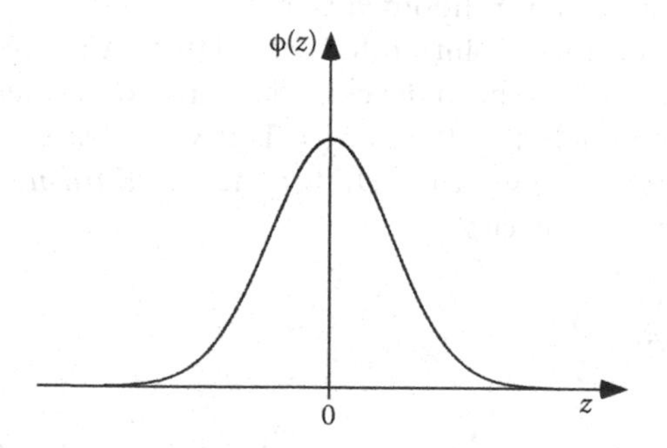

Figure 2.11 PDF $\phi(z)$ for a standard normal random variable.

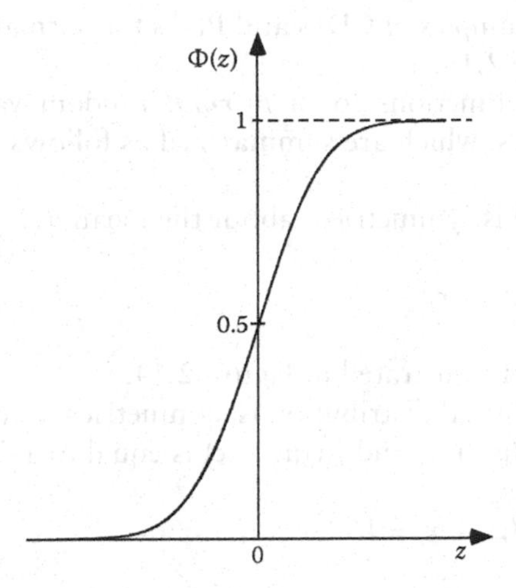

Figure 2.12 CDF $\Phi(z)$ for a standard normal random variable.

be a general normal random variable, and let Z be the standard form of X. By rearranging Equation 2.27, we can show that

$$X = \mu_X + Z\sigma_X \tag{2.37}$$

By the definition of CDF, we can write

$$F_X(x) = P(X \leq x) = P(\mu_X + Z\sigma_X \leq x) = P\left(Z \leq \frac{x - \mu_X}{\sigma_X}\right) \tag{2.38}$$

or

$$F_X(x) = \Phi\left(\frac{x - \mu_X}{\sigma_X}\right) = F_Z(z) \tag{2.39}$$

Similarly, a relationship can be derived relating the PDF of any normal random variable, $f_X(x)$, to the PDF of the standard normal variable, $\phi(x)$:

$$f_X(x) = \frac{\mathrm{d}}{\mathrm{d}x} F_X(x) = \frac{\mathrm{d}}{\mathrm{d}x} \Phi\left(\frac{x - \mu_X}{\sigma_X}\right) = \frac{1}{\sigma_X} \phi\left(\frac{x - \mu_X}{\sigma_X}\right) \tag{2.40}$$

Therefore, using the relationships in Equations 2.39 and 2.40, one can construct the distribution functions for an arbitrary normal random variable (given μ_X and σ_X) using the information provided in Table B.1

(in Appendix B). Examples of CDFs and PDFs for normal random variables are shown in Figure 2.13.

The distribution functions for a *normal* random variable have some important properties, which are summarized as follows:

1. The PDF $f_X(x)$ is symmetrical about the mean μ_X.

$$f_X(\mu_X + x) = f_X(\mu_X - x) \tag{2.41}$$

   This property is illustrated in Figure 2.14.
2. Because the normal distribution is symmetrical about its mean value, the sum of $F_X(\mu_X + x)$ and $F_X(\mu_X - x)$ is equal to 1. That is,

$$F_X(\mu_X + x) + F_X(\mu_X - x) = 1 \tag{2.42}$$

   This is a generalization of the property expressed for $\Phi(z)$ in Equation 2.36.

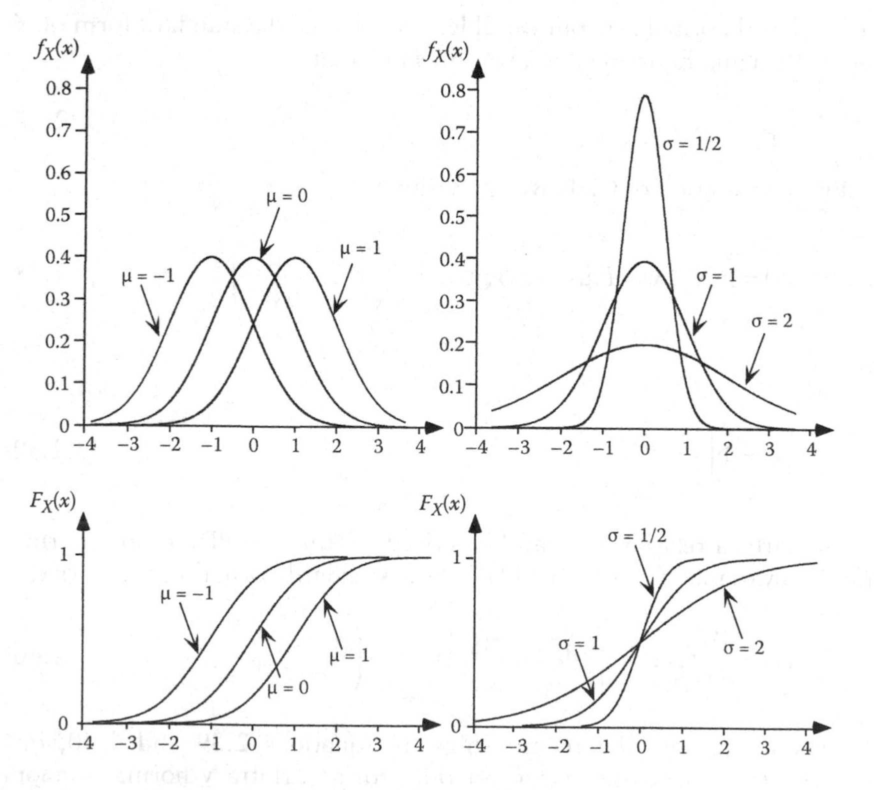

Figure 2.13 **Examples of PDFs and CDFs for normal random variables.**

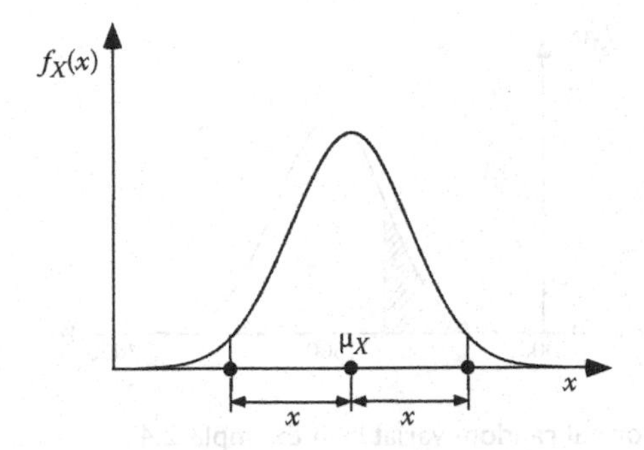

Figure 2.14 Normal random variable PDF is symmetrical about the mean.

Since the normal distribution is so important and fundamental to structural reliability theory, we illustrate its use by some simple examples.

Example 2.3

a. If Z is a standard normal random variable, and $z = -2.16$, what are the PDF and CDF values?
   From Appendix B,

$$\Phi(z = -2.16) = 0.0154.$$

From Equation 2.35,

$$\Phi(z = -2.16) = 0.0387.$$

b. If $z = +1.51$, what is $\Phi(1.51)$?
   From Appendix B, $\Phi(-1.51) = 0.0655$. Using Equation 2.36, $\Phi(1.51) = 1 - \Phi(-1.51) = 1 - 0.0655 = 0.9345$.
c. Given $\Phi(z) = 0.80 \times 10^{-4}$, what is the corresponding value of z?
   From Appendix B, we note that this value of $\Phi(z)$ does not correspond to a specific tabulated value of z. Therefore, we must use interpolation. For $z = -3.77$, $\Phi = 0.816 \times 10^{-4}$ and for $z = -3.78$, $\Phi = 0.784 \times 10^{-4}$. Interpolating between these values, the desired value of z is about -3.775.

Example 2.4

Assume X is a normal random variable with $\mu_X = 1500$ and $\sigma_X = 200$. The PDF for the variable is shown in Figure 2.15.

a. Calculate $F_X(1300)$.
   From Equation 2.39,

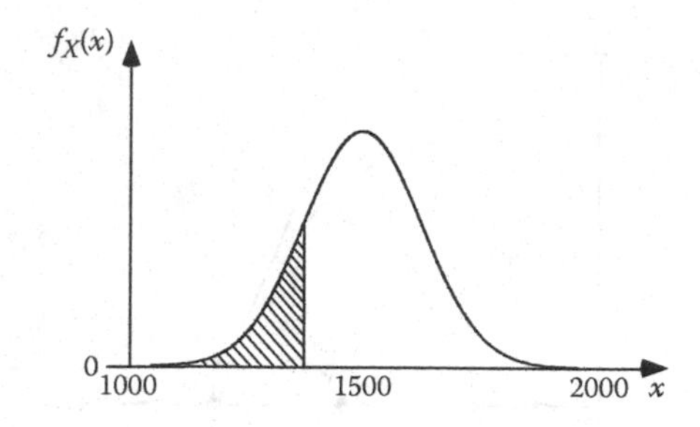

Figure 2.15 PDF of normal random variable in Example 2.4.

$$F_X(x) = \Phi\left(\frac{x-\mu_X}{\sigma_X}\right) \Rightarrow F_X(1300) = \Phi\left(\frac{1300-1500}{200}\right) = \Phi(-1)$$

From Appendix B, $\Phi(-1) = 0.159$.

b. Calculate $F_X(1900)$.
   From Equation 2.39,

$$F_X(x) = \Phi\left(\frac{x-\mu_X}{\sigma_X}\right) \Rightarrow F_X(1900) = \Phi\left(\frac{1900-1500}{200}\right) = \Phi(+2)$$

From Equation 2.36, $\Phi(2) = 1 - \Phi(-2)$.
From Appendix B, $\Phi(-2) = 0.228 \times 10^{-1}$. Therefore, $\Phi(2) = 1 - (0.228 \times 10^{-1}) = 0.977$.

c. Calculate $F_X(1700)$.
   Observe that $x = 1700$ is 200 units away from the mean value of 1500. Using Equation 2.42, we can write

$$\begin{aligned} F_X(1500+200) &= 1 - F_X(1500-200) \\ &= 1 - 0.159 \text{ (obtained in part a above)} \\ &= 0.841. \end{aligned}$$

d. Calculate $f_X(x)$ for $x = 1300$.
   From Equation 2.40,

$$f_X(x) = \frac{1}{\sigma_X}\phi\left(\frac{x-\mu_X}{\sigma_X}\right) \Rightarrow f_X(x) = \frac{1}{200}\phi\left(\frac{1300-1500}{200}\right) = \frac{1}{200}\phi(-1)$$

Using Equation 2.35, $\phi(-1) = 0.242$. Therefore, $f_X(1300) = 0.00121$.

e. Calculate $f_X(1500)$.
Using Equation 2.40,

$$f_X(x) = \frac{1}{\sigma_X}\phi\left(\frac{x-\mu_X}{\sigma_X}\right) \Rightarrow f_X(x) = \frac{1}{200}\phi\left(\frac{1500-1500}{200}\right) = \frac{1}{200}\phi(0)$$

Using Equation 2.35, $\phi(0) = 0.399$. Therefore, $f_X(1500) = 0.00199$.

As we will see later, it is often necessary to calculate the inverse of the CDF of the standard normal distribution function $\Phi(z)$. Although the inverse does not exist in closed form, an approximate formula for the inverse does exist, and it gives reasonable results over a wide range of probability values. Let $p = \Phi(z)$. The inverse problem would be to find $z = \Phi^{-1}(p)$. The following formula can be used if p is less than or equal to 0.5:

$$z = \Phi^{-1}(p) \approx -t + \frac{c_0 + c_1 t + c_2 t^2}{1 + d_1 t + d_2 t^2 + d_3 t^3} \quad \text{for } p \le 0.5 \tag{2.43}$$

where

$c_0 = 2.515517$
$c_1 = 0.802853$
$c_2 = 0.010328$
$d_1 = 1.432788$
$d_2 = 0.189269$
$d_3 = 0.001308$

and

$$t = \sqrt{-\ln(p^2)} \tag{2.44}$$

For $p > 0.5$, Φ^{-1} is calculated for $p^* = (1 - p)$ and then we use the following relationship:

$$z = \Phi^{-1}(p) = -\Phi^{-1}(p^*) \tag{2.45}$$

2.4.3 Lognormal random variable

The random variable X is a *lognormal* random variable if $Y = \ln(X)$ is normally distributed. A lognormal random variable is defined for positive values only ($x > 0$). The PDF and CDF can be calculated using distributions $\phi(z)$ and $\Phi(z)$ for the standard normal random variable Z as follows:

$$F_X(x) = P(X \le x) = P(\ln X \le \ln x) = P(Y \le y) = F_Y(y) \tag{2.46}$$

Since Y is normally distributed, we can use the standard normal functions as discussed in Section 2.4.2. Specifically,

$$F_X(x) = F_Y(y) = \Phi\left(\frac{y-\mu_Y}{\sigma_Y}\right) \tag{2.47}$$

where $y = \ln(x)$, $\mu_Y = \mu_{\ln(X)}$ = mean value of $\ln(X)$, and $\sigma_Y = \sigma_{\ln(X)}$ = standard deviation of $\ln(X)$. These quantities can be expressed as functions of μ_X, σ_X, and V_X using the following formulas:

$$\sigma_{\ln(X)}^2 = \ln\left(V_X^2+1\right) \tag{2.48}$$

$$\mu_{\ln(X)} = \ln(\mu_X) - \frac{1}{2}\sigma_{\ln(X)}^2 \tag{2.49}$$

If V_X is less than 0.2, the following approximations can be used to find $\sigma_{\ln(X)}^2$ and $\mu_{\ln(X)}$:

$$\sigma_{\ln(X)}^2 \approx V_X^2 \tag{2.50}$$

$$\mu_{\ln(X)} \approx \ln(\mu_X) \tag{2.51}$$

For the PDF function, using Equation 2.12, we can show that

$$f_X(x) = \frac{d}{dx}F_X(x) = \frac{d}{dx}\Phi\left(\frac{\ln(x)-\mu_{\ln(X)}}{\sigma_{\ln(X)}}\right) = \frac{1}{x\sigma_{\ln(X)}}\phi\left(\frac{\ln(x)-\mu_{\ln(X)}}{\sigma_{\ln(X)}}\right) \tag{2.52}$$

The general shape of the PDF function for a lognormal variable is shown in Figure 2.16.

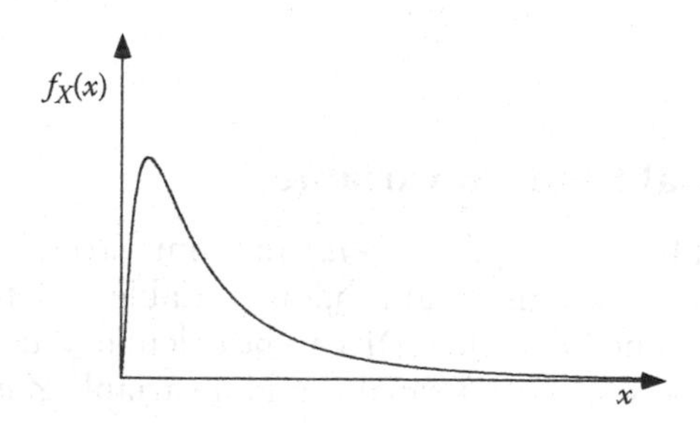

Figure 2.16 PDF of a lognormal random variable.

The lognormal distribution is widely used in structural reliability analyses. The following example illustrates its use.

Example 2.5

Let X be a lognormal random variable with a mean value of 250 and a standard deviation of 30. Calculate $F_X(200)$ and $f_X(200)$.

$$V_X = \frac{\sigma_X}{\mu_X} = \frac{30}{250} = 0.12$$

$$\sigma^2_{\ln(X)} = \ln\left(V_X^2 + 1\right) = 0.0143; \quad \sigma_{\ln(X)} = 0.1196$$

$$\mu_{\ln(X)} = \ln(\mu_X) - \frac{1}{2}\sigma^2_{\ln(X)} = \ln(250) - 0.5(0.0143) = 5.51$$

$$F_X(200) = \Phi\left(\frac{\ln(x) - \mu_{\ln(X)}}{\sigma_{\ln(X)}}\right) = \Phi\left(\frac{\ln(200) - 5.51}{0.1196}\right) = \Phi(-1.77) = 0.0384$$

$$f_X(200) = \frac{1}{x\sigma_{\ln(X)}}\phi\left(\frac{\ln(x) - \mu_{\ln(X)}}{\sigma_{\ln(X)}}\right) = \frac{1}{(200)(0.1196)}\phi(-1.77)$$

$$= \frac{0.0833}{23.92} = 0.00348$$

2.4.4 Gamma distribution

The PDF of a gamma random variable is useful for modeling sustained live load, such as in buildings. It is defined by

$$f_X(x) = \frac{\lambda(\lambda x)^{k-1} e^{-\lambda x}}{\Gamma(k)} \quad \text{for } x > 0 \tag{2.53}$$

where λ and k are distribution parameters. The function $\Gamma(k)$ is the gamma function, which is defined as

$$\Gamma(k) = \int_0^\infty e^{-u} u^{k-1}\, du \tag{2.54}$$

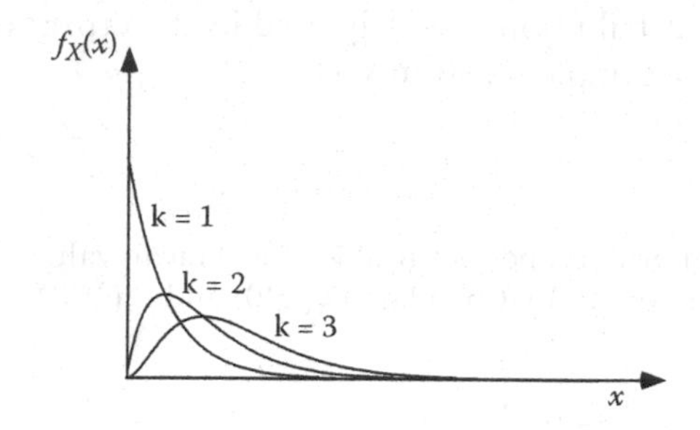

Figure 2.17 PDFs of gamma random variables.

For integer values of k, the gamma function has these properties:

$$\Gamma(k) = (k-1)(k-2)\ldots(2)(1) = (k-1)! \tag{2.55a}$$

$$\Gamma(k+1) = \Gamma(k)\,k \tag{2.55b}$$

Values of $\Gamma(k)$ for $1 \le k \le 2$ are tabulated in Table C.1 (in Appendix C). Some PDF functions for various values of k are shown in Figure 2.17.

The mean and variance can be calculated as follows:

$$\mu_X = \frac{k}{\lambda} \tag{2.56}$$

$$\sigma_X^2 = \frac{k}{\lambda^2} \tag{2.57}$$

2.4.5 Extreme Type I (Gumbel distribution, Fisher–Tippett Type I)

Extreme value distributions, as the name implies, are useful to characterize the probabilistic nature of the extreme values (largest or smallest values) of some phenomenon over time. For example, consider n time intervals. Each interval might be 1 year. During each year, there will be a maximum value of some phenomenon (such as wind speed). Suppose we want to determine the probability distribution for those largest annual wind speeds. Let $W_1, \ldots, W_n$ be the largest wind speeds in years 1 through n. Then, $X = \max(W_1, W_2, \ldots, W_n)$ might be characterized as an extreme Type I random variable. The CDF and PDF for this random variable are

$$F_X(x) = \exp(-\exp(-\alpha(x-u))) \quad \text{for} \quad -\infty < x < \infty \tag{2.58}$$

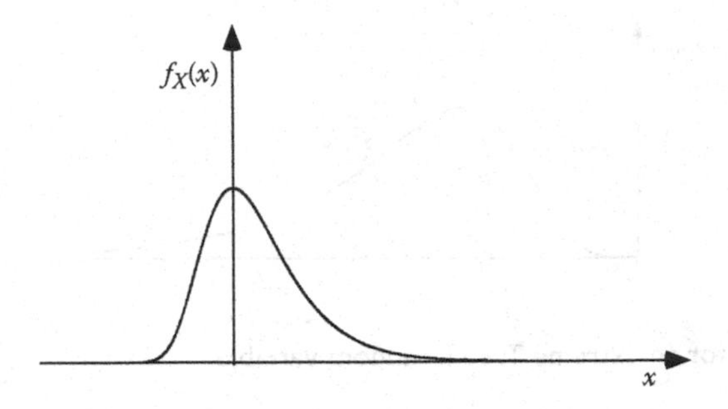

Figure 2.18 PDF of an extreme Type I random variable.

$$f_X(x) = \alpha \exp(-\exp(-\alpha(x - u)))\exp(-\alpha(x - u)) \tag{2.59}$$

where u and α are distribution parameters and $\exp(x) = e^x$. The basic shape of the Type I PDF function is shown in Figure 2.18.

The mean and standard deviation can be calculated using the following approximations (Benjamin and Cornell, 1970):

$$\mu_X \approx u + \frac{0.577}{\alpha} \tag{2.60}$$

$$\sigma_X \approx \frac{1.282}{\alpha} \tag{2.61}$$

Thus, if the mean and standard deviation are known, Equations 2.60 and 2.61 can be rearranged and solved for the corresponding values of the distribution parameters as follows:

$$\alpha \approx \frac{1.282}{\sigma_X} \tag{2.62}$$

$$u \approx \mu_X - 0.45\sigma_X \tag{2.63}$$

2.4.6 Extreme Type II

An extreme Type II distribution sometimes gives the best approximation of the distribution of the maximum seismic load applied to a structure. The CDF and PDF are

$$F_X(x) = e^{-(u/x)^k} \quad \text{for } 0 < x < \infty \tag{2.64}$$

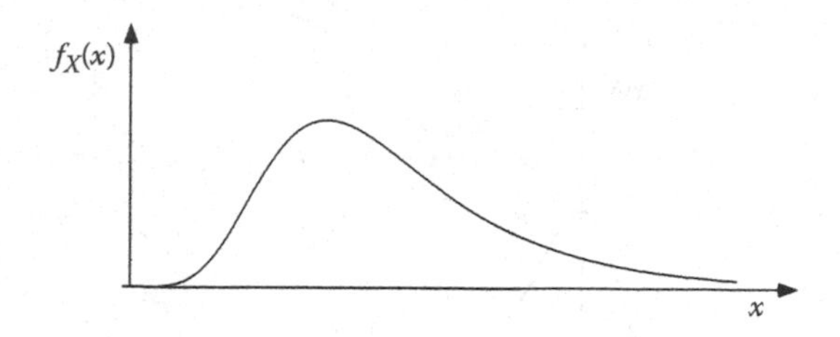

Figure 2.19 PDF for an extreme Type II random variable.

$$f_X(x) = \frac{k}{u}\left(\frac{u}{x}\right)^{k+1} e^{-(u/x)^k} \tag{2.65}$$

where u and k are distribution parameters. The PDF for an extreme Type II variable has the general shape shown in Figure 2.19.

The mean and standard deviation can be calculated as follows:

$$\mu_X = u\,\Gamma\left(1-\frac{1}{k}\right) \quad \text{for } k>1 \tag{2.66}$$

$$\sigma_X^2 = u^2\left[\Gamma\left(1-\frac{2}{k}\right)-\Gamma^2\left(1-\frac{1}{k}\right)\right] \quad \text{for } k>2 \tag{2.67}$$

Note that the COV, V_X, is a function of k only. Graphs exist to calculate V_X for any k. (See, for example, Ang and Tang, 1984.)

2.4.7 Extreme Type III (Weibull distribution)

The extreme Type III distribution is defined by three parameters. There is a different function for the largest and the smallest values.

For the largest values, the CDF is defined by

$$F_X(x) = e^{-\left(\frac{w-x}{w-u}\right)^k} \quad \text{for } x \le w \tag{2.68}$$

where w, u, and k are parameters. The mean and standard deviation are

$$\mu_X = w-(w-u)\Gamma\left(1+\frac{1}{k}\right) \tag{2.69a}$$

$$\sigma_X^2 = (w-u)^2\left[\Gamma\left(1+\frac{2}{k}\right)-\Gamma^2\left(1+\frac{1}{k}\right)\right] \tag{2.69b}$$

For the smallest values, the CDF is defined by

$$F_X(x) = 1 - e^{-\left(\frac{x-\varepsilon}{u-\varepsilon}\right)^k} \quad \text{for } x \geq \varepsilon \tag{2.70}$$

where u, ε, and k are the parameters. For the smallest value case, the mean and variance can be calculated using the following formulas:

$$\mu_X = \varepsilon + (u+\varepsilon)\Gamma\left(1+\frac{1}{k}\right) \tag{2.71a}$$

$$\sigma_X^2 = (u-\varepsilon)^2\left[\Gamma\left(1+\frac{2}{k}\right)-\Gamma^2\left(1+\frac{1}{k}\right)\right] \tag{2.71b}$$

2.4.8 Poisson distribution

The Poisson distribution is a discrete probability distribution that can be used to calculate the PMF for the number of occurrences of a particular event in a time or space interval $(0,t)$. For example, the Poisson distribution can be used to represent the number of earthquakes that occur within a certain time interval or the number of defects in a certain length of rod.

The following important underlying assumptions behind the Poisson distribution must be considered before it is used in a probabilistic analysis:

- The occurrences of events are independent of each other. In other words, the occurrence or nonoccurrence of events in a prior time interval has no effect on the occurrence of events in the time interval being considered.
- Two or more events cannot occur simultaneously.

Let N be a discrete random variable representing the number of occurrences of an event within a prescribed time (or space) interval $(0,t)$. Let ν represent the mean occurrence rate of the event. This is usually obtained from statistical data. The Poisson PMF function is defined as

$$P(N = n \text{ in time } t) = \frac{(\nu t)^n}{n!}e^{-\nu t} \quad n = 0,1,2,\ldots,\infty \tag{2.72}$$

The mean and standard deviation of the random variable N are

$$\mu_N = \nu t; \quad \sigma_N = \sqrt{\nu t} \tag{2.73}$$

An alternate parameter that is often used with the Poisson distribution is the return period (or interval) τ. The return period is simply the reciprocal of the mean occurrence rate ν:

$$\tau = \frac{1}{\nu} \tag{2.74}$$

The return period is a *deterministic* number representing the *average* time interval between occurrences of events. The actual time interval between events is a random number.

Example 2.6

Suppose that the average occurrence rate of earthquakes (with magnitudes between 5 and 8) in a region surrounding Los Angeles, California, has been determined to be 2.14 earthquakes/year. Determine

A. The return period for earthquakes in this magnitude range
B. The probability of exactly three earthquakes (magnitude between 5 and 8) in the next year
C. The annual probability of an earthquake with magnitude between 5 and 8

Solution:

A. The return period is calculated using Equation 2.74:

$$\tau = \frac{1}{\nu} = \frac{1}{2.14} = 0.47 \text{ year}$$

In other words, *on average*, there is one earthquake (in the defined magnitude range) about every 6 months.

B. The probability of exactly three earthquakes in the next year is determined using Equation 2.72 with $t = 1$ and $n = 3$:

$$P(N = 3 \text{ in } 1 \text{ year}) = \frac{\left[(2.14)(1)\right]^3}{3!} e^{-[(2.14)(1)]}$$
$$= 0.192$$

C. To find the annual probability of an earthquake, it is helpful to interpret the question as follows. The annual probability of an earthquake is the annual probability of *at least one* earthquake. Therefore,

$$P(\text{at least one earthquake}) = 1 - P(\text{no earthquakes})$$

or

$$P(N \geq 1) = 1 - P(N = 0).$$

Therefore,

$$P(N \geq 1) = 1 - \frac{\left[(2.14)(1)\right]^0}{0!} e^{-\left[(2.14)(1)\right]}$$
$$= 1 - e^{-\left[(2.14)(1)\right]}$$
$$= 0.88$$

2.5 PROBABILITY PAPER

Probability paper can be used to graphically determine whether or not a set of experimental data follows a particular probability distribution. Probability paper for the normal distribution is the most common, and it is commercially available. However, it is possible to construct probability paper for many common distributions using ordinary graph paper. In this section, we will discuss how to construct and use normal probability paper.

All CDFs are nondecreasing functions of the random variable. For example, the CDF of the normal distribution has an "S-shape" as shown in Figures 2.13 and 2.20. The basic idea behind normal probability paper is to redefine the vertical scale so that the normal CDF will plot as a straight line. Conversely, if a set of data plotted on normal probability paper plots as a straight line, then it is reasonable to model the data using a normal CDF. The slope and y-intercept of the graph can be used to determine the mean and standard deviation of the distribution.

Consider a normal random variable X with mean value μ_X and standard deviation σ_X. Now, imagine a transformation in which the S-shaped CDF is "straightened" as shown in Figure 2.20. The transformation is such that each point on the original CDF can only move vertically up or down. In commercial normal probability paper, this transformation is accomplished by altering the scale of the vertical axis as shown in Figure 2.21. Observe that the values on the left vertical axis are not evenly spaced. If the coordinate pairs $(x, F_X(x))$ for a normal random variable X are plotted on normal probability paper, the graph will be a straight line.

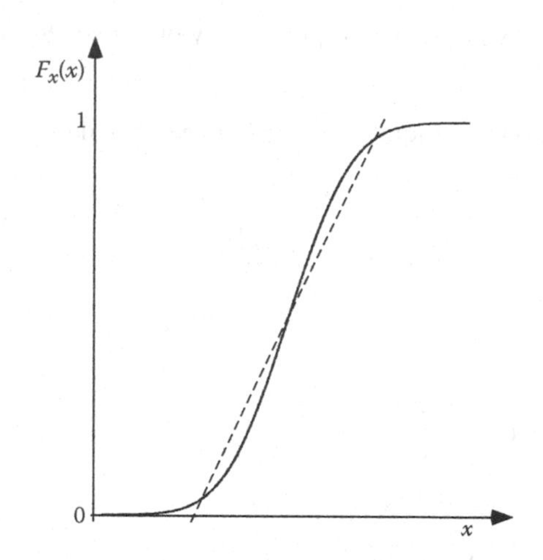

Figure 2.20 The S-shaped CDF for a normal random variable.

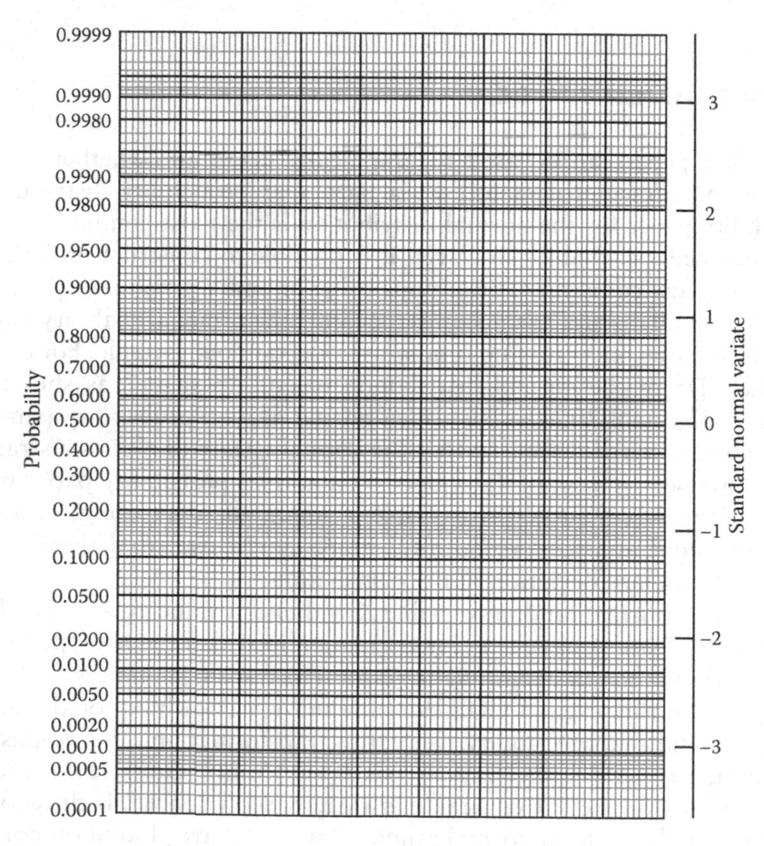

Figure 2.21 Example of normal probability paper.

Today, with the availability of spreadsheet programs and computers, it is very easy to achieve the same effect of commercial normal probability paper by performing a simple mathematical transformation and plotting a standard linear (xy) graph. Recall that the standardized form Z of a normal random variable X is

$$Z = \frac{X - \mu_X}{\sigma_X} = \left(\frac{1}{\sigma_X}\right) X + \left(\frac{-\mu_X}{\sigma_X}\right) \tag{2.75}$$

For any realization x of the normal random variable X, the corresponding standardized value is

$$z = \frac{x - \mu_X}{\sigma_X} = \left(\frac{1}{\sigma_X}\right) x + \left(\frac{-\mu_X}{\sigma_X}\right) \tag{2.76}$$

The corresponding probability based on the normal CDF would be

$$F_X(x) = p = \Phi\left(\frac{x - \mu_X}{\sigma_X}\right) \tag{2.77}$$

If we take the inverse of Equation 2.77, we get

$$\Phi^{-1}(p) = z = \left(\frac{1}{\sigma_X}\right) x + \left(\frac{-\mu_X}{\sigma_X}\right) \tag{2.78}$$

Equation 2.78 represents a linear relationship between $z = \Phi^{-1}(p)$ and x, and this provides the rationale behind normal probability paper. The vertical axis on the right side of Figure 2.21 was obtained by transforming the probability values on the left scale using Equation 2.78. Observe that the values on this scale are evenly spaced. If $\Phi^{-1}(p)$ versus x is plotted on standard (linear) graph paper, a straight line plot will result.

The relationship expressed in Equation 2.78 is further illustrated in Figure 2.22. In this figure, the uneven probability scale and the corresponding linear scale are both shown on the left side of the plot. Data points from a general normal distribution are plotted, and a straight line is obtained. Observe that the value of x corresponding to $F_X(x) = 0.5$ (or $z = \Phi^{-1}(0.5) = 0$) is the mean value μ_X. From Equation 2.78, we note that the slope of the straight line is the reciprocal of the standard deviation. If we move away from the mean value by an amount $n\sigma_X$ where n is an integer and σ_X is the standard deviation, the corresponding value of z is equal to n. This is shown by the dashed lines in Figure 2.22.

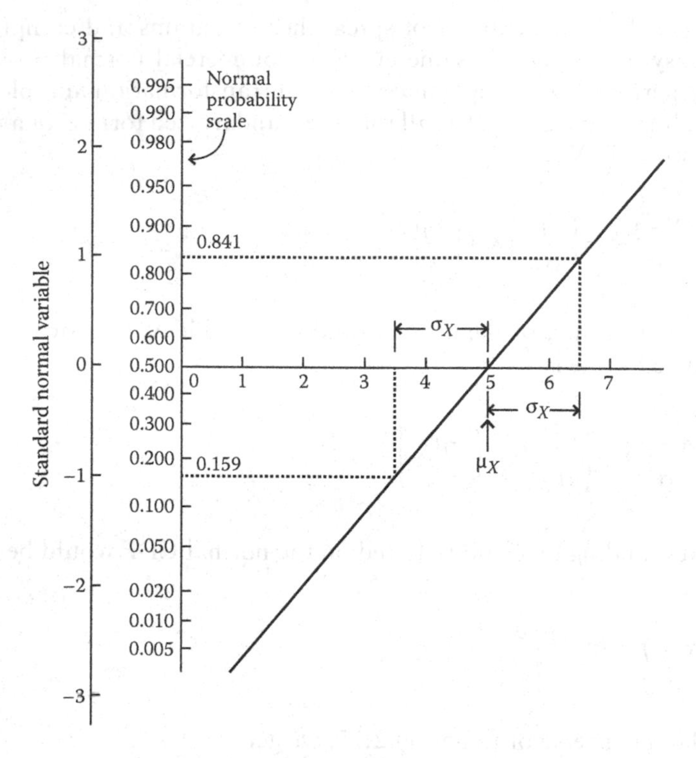

Figure 2.22 Interpretation of a straight line plot on normal probability paper in terms of the mean and standard deviation of the normal random variable.

Now consider the practical application of normal probability paper to evaluate experimental data. Consider an experiment or test in which N values of some random variable X are obtained. These values will be denoted $\{x\}$. To be able to use normal probability paper (commercial or computed), it is necessary to associate a probability value with each x value. The procedure is as follows:

1. Arrange the data values $\{x\}$ in *increasing* order. Once ordered, the first (lowest) value of x will be denoted as x_1, the next value as x_2, and so on, up to the last (largest) value x_N. Do not discard repeated values.
2. Associate with each x_i a cumulative probability p_i equal to (Gumbel, 1954)

$$p_i = \frac{i}{N+1} \tag{2.79}$$

3. If commercial normal probability paper is being used, then plot the (x_i, p_i) and go to Step 6. Otherwise, go to Step 4.
4. For each p_i, determine $z_i = \Phi^{-1}(p_i)$. Equation 2.43 can be useful in this step.
5. Plot the coordinates (x_i, z_i) on standard linear graph paper by hand or using a computer.
6. If the plot appears to follow a straight line, then it is reasonable to conclude that the data can be modeled using a normal distribution. Sketch a "best fit" line for the data. The slope of the line will be equal to $1/\sigma_X$, and the value of x at which the probability is 0.5 (or $z = 0$) will be equal to μ_X. (Alternatively, you can plot a reference line using the sample mean $\bar{x}$ and sample standard deviation s_X obtained using Equations 2.25 and 2.26.) If the data do not appear to follow a straight line, then a normal distribution is probably not appropriate. However, the plot can still provide some useful information, as discussed in later chapters.

Example 2.7

Consider the following set of data points: $\{x\} = \{6.5, 5.3, 5.5, 5.9, 6.5, 6.8, 7.2, 5.9, 6.4\} (N = 9)$. Plot the data on normal probability paper.

It is convenient to carry out steps 1–2 above by setting up a table as seen in Table 2.1. The values of (x_i, z_i) are plotted on linear xy axes as shown in Figure 2.23. The data appear to follow (at least approximately) a straight line, and thus we might conclude that the data follow a normal distribution. For comparison, a "reference" straight line is plotted based on the sample statistics $\bar{x} = 6.2$ and $s_X = 0.62$.

Table 2.1 Data table for Example 2.7

i = index value	*x_i (in increasing order)*	*Probability, $p_i = i/(N + 1)$*	$z_i = \Phi^{-1}(p_i)$
1	5.3	0.1	−1.282
2	5.5	0.2	−0.842
3	5.9	0.3	−0.524
4	5.9	0.4	−0.253
5	6.4	0.5	0
6	6.5	0.6	0.253
7	6.5	0.7	0.524
8	6.8	0.8	0.842
9	7.2	0.9	1.282

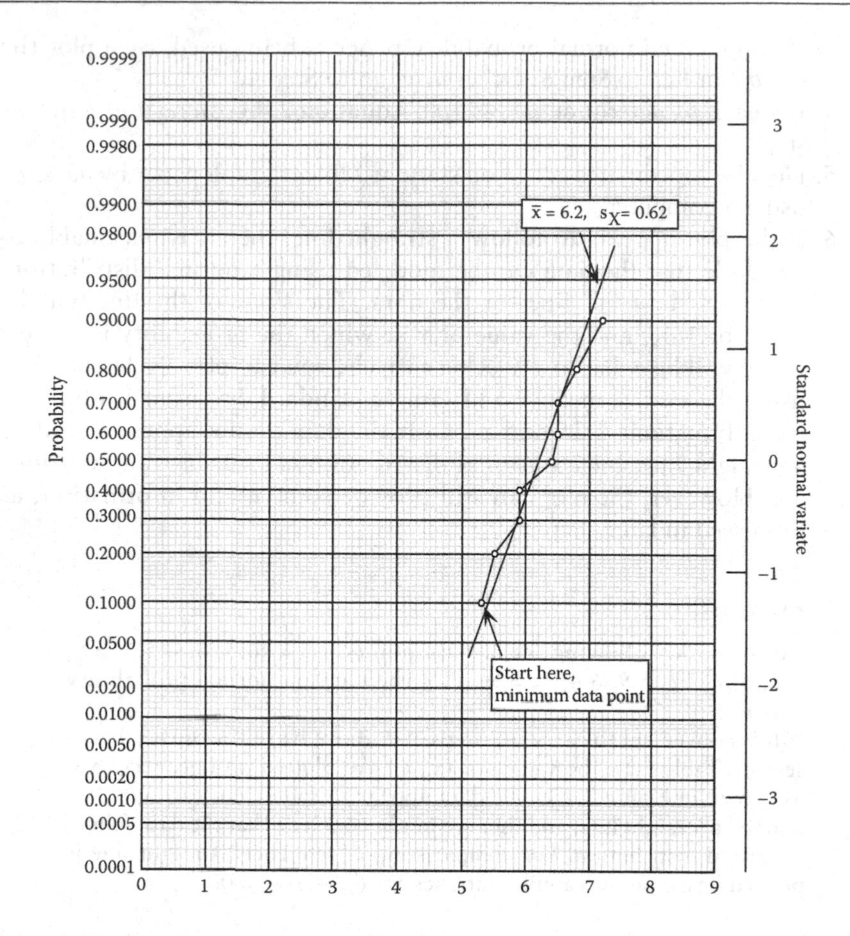

Figure 2.23 Data from Example 2.7 plotted on normal probability paper.

Example 2.8

Consider the results of a truck weight survey. The recorded values of gross vehicle weight (GVW) are shown in Table 2.2. Evaluate the data using normal probability paper.

First, the data are entered into a spreadsheet table. Then, the data are sorted and ranked in increasing order. Each value of GVW is assigned a probability using Equation 2.79. The coordinates (x_i, p_i) are then plotted on normal probability paper as shown in Figure 2.24. Alternatively, $z_i = \Phi^{-1}(p_i)$ could be calculated for each p_i value, and then the points (x_i, z_i) could be plotted on typical linear graph paper. The same plot would result.

Table 2.2 Observed GVWs used in Example 2.8

GVW (kips) [1 kip = 4.448 kN]						
106.8	23.5	88.9	22.9	113.9	29.0	65.4
18.2	25.3	36.9	14.2	18.5	21.6	26.4
63.0	50.6	41.4	39.2	74.3	44.4	29.9
34.0	22.1	39.5	27.6	29.1	30.5	39.1
42.1	47.7	49.5	33.6	40.6	32.6	47.9
31.3	42.7	44.7	20.6	10.5	108.3	10.8
21.6	177.7	34.8	22.3	22.9	30.7	15.3
15.4	32.4	42.6	21.7	21.4	43.5	12.8
24.7	17.4	42.0	77.5	20.8	40.6	17.3
58.5	23.8	88.2	31.3	23.9	70.8	15.0
19.0	33.2	17.2	29.7	42.4	15.3	92.5
48.1	30.0	31.4	77.8	30.2	52.7	19.7
14.7	45.3	30.5	28.5	13.2	55.7	41.0
49.3	25.0	37.4	35.9	20.5	24.1	13.7
16.2	49.7	60.9	42.7	11.6	23.5	92.5
37.2	57.0	22.4	32.3	17.9	20.8	27.5
41.7	85.4	45.8	77.7	21.2	24.0	48.9
47.6	50.6	30.6	32.2	73.1	49.1	70.9
30.6	15.1	39.4	49.7	24.9	46.9	42.7
24.7	11.5	36.4	36.6	74.3	42.2	25.1
88.5	11.6	63.8	17.8	50.6	28.6	22.0
14.8	36.8	37.7	89.8	153.6	99.3	33.6
36.0	31.8	34.6	73.5	17.9	14.1	89.8
40.9	18.1	40.8	28.3	19.6	10.2	40.0
38.5	20.0	40.9	15.5	38.9	63.5	19.5
71.3	22.4	11.0	85.6	56.2	18.0	14.1
27.2	38.9	30.8	15.5	53.7	57.5	36.5
52.9	75.8	146.4	89.3	52.9	24.1	12.6
58.0	12.8	12.9	67.4	41.4	33.2	35.4
80.8	13.8	21.0	28.9	50.7	31.2	37.3
40.7	24.6	31.2	27.5	19.6	17.8	97.0
23.7	61.8	52.7	110.9	34.2	135.8	47.0
30.4	20.4	52.7	28.8	12.3	39.5	21.6
20.8	20.9	111.6	53.3	29.9	25.7	24.1
44.0	11.3	23.0	46.2	51.6	12.7	33.4
22.4	63.6	22.4	30.1	45.3	40.4	41.7
22.0	105.6	21.0	23.2	71.9	36.6	27.8
44.5	53.7	34.3	41.6	10.4	150.6	44.6

(continued)

Table 2.2 Observed GVWs used in Example 2.8 (Continued)

GVW (kips) [1 kip = 4.448 kN]						
12.6	30.2	55.7	21.5	39.7	17.4	47.2
72.9	61.7	39.5	30.5	40.1	77.7	43.1
17.9	33.6	41.7	26.4	20.2	79.3	23.1
73.9	74.5	41.7	43.7	23.2	23.7	72.9
29.2	27.1	23.5	39.1	27.5	79.3	56.1
75.0	25.4	11.1	15.7	53.9	34.3	
54.9	23.8	39.1	37.7	18.9	30.4	

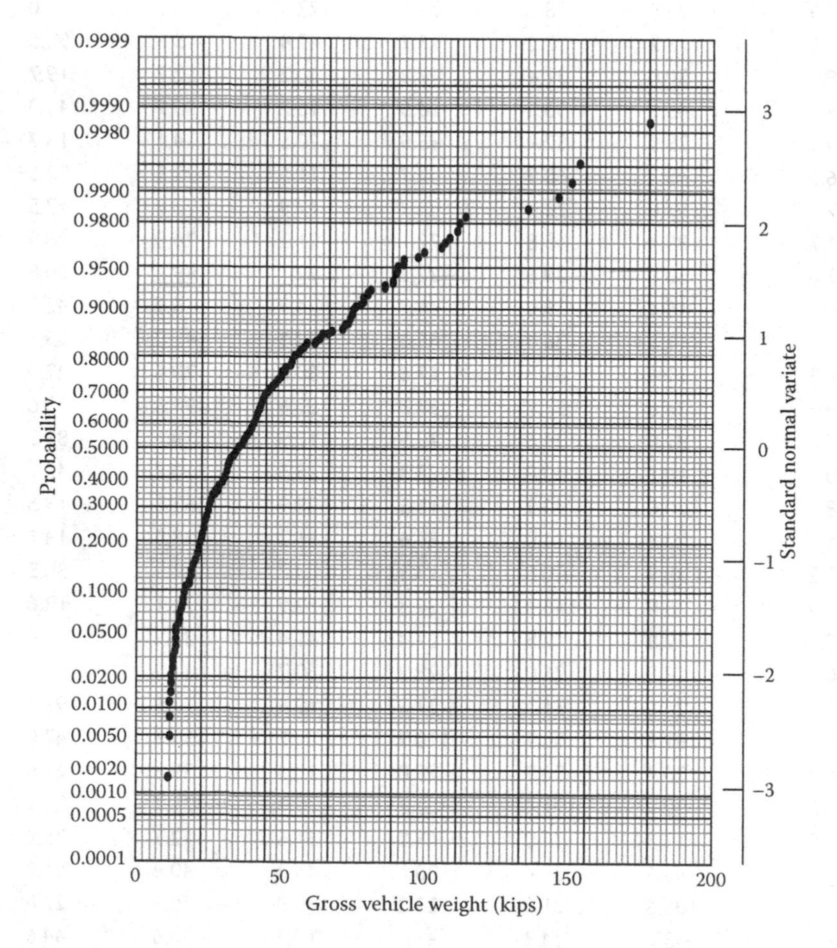

Figure 2.24 Observed data on GVW plotted on normal probability paper (1 kip = 4.448 kN).

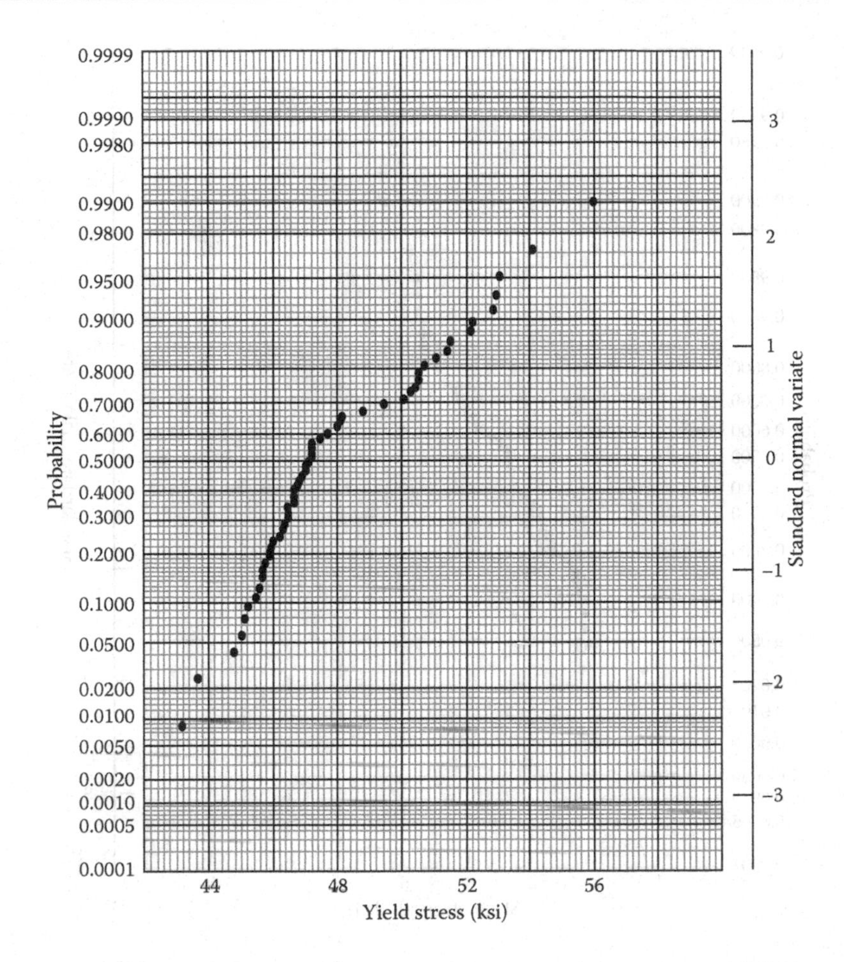

Figure 2.25 **Test results of yield stress of steel on commercial normal probability paper.**

Other examples of actual data plotted on normal probability paper are shown in Figures 2.25 and 2.26.

2.6 INTERPRETATION OF TEST DATA USING STATISTICS

In Section 2.3.2, we discussed how to calculate the sample mean and sample standard deviation for a set of observed data. In Section 2.5, we discussed how to plot the data on normal probability paper to see if the data follow a normal distribution. Another graphical technique, known as the

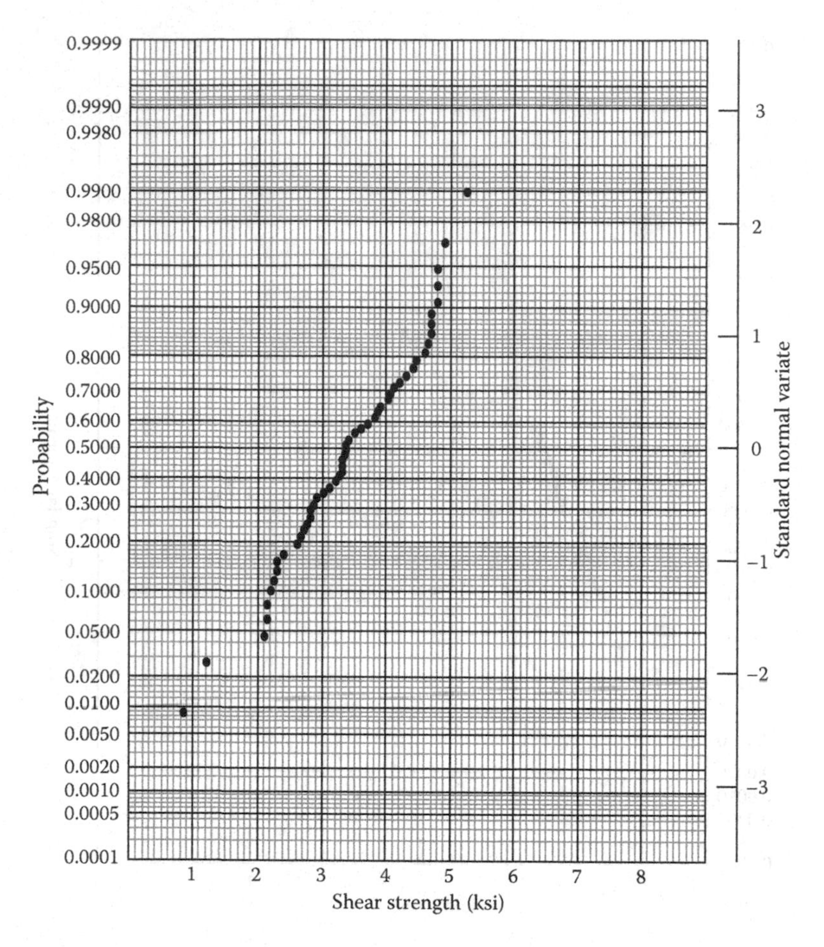

Figure 2.26 Test results of shear strength of spot welds on normal probability paper.

histogram, is sometimes useful. The basic idea is to count the number of data points that fall into predefined intervals and then make a bar graph. By looking at the bar graph, you can observe trends in the data and visually determine the "distribution" of the data. This idea is best explained by example. For more theoretical details, see the book by Spiegel and Stephens (2011).

Example 2.9

Suppose we test 100 concrete cylinders and experimentally determine the compressive strength for each specimen. We then establish intervals of values and count the number of observed values that fall in each interval. This is shown in Table 2.3. Then, for each interval, we calculate the relative frequency of occurrence, which is the number of observations for the interval divided by the total number of all observations. This corresponds to the percentage of all observations that fall in a particular interval. This has been calculated in the third column of Table 2.3. If, for each interval, we add up the frequency value for that interval and all intervals below it, we get a cumulative frequency value as shown in the last column of Table 2.3.

If we plot the values in column 3 of Table 2.3 versus the interval values in column 1, we get a *relative frequency histogram* as shown in Figure 2.27. If we plot the values in column 4 versus the interval values in column 1, we get a *cumulative frequency histogram* as shown in Figure 2.28. Figure 2.29 shows how the interval size can drastically influence the overall appearance of the histograms.

Table 2.3 Data for Example 2.9

Concrete strength f_c' *(150 psi interval)*	*Number of observations in interval*	*Frequency of occurrence*	*Cumulative frequency*
Below 1500 psi	0	0.00	0.00
1500–1650	1	0.01	0.01
1650–1800	1	0.01	0.02
1800–1950	3	0.03	0.05
1950–2100	3	0.03	0.08
2100–2250	8	0.08	0.16
2250–2400	12	0.12	0.28
2400–2550	11	0.11	0.39
2550–2700	10	0.10	0.49
2700–2850	13	0.13	0.62
2850–3000	9	0.09	0.71
3000–3150	8	0.08	0.79
3150–3300	6	0.06	0.85
3300–3450	3	0.03	0.88
3450–3600	4	0.04	0.92
3600–3750	4	0.04	0.96
3750–3900	2	0.02	0.98
3900–4050	1	0.01	0.99
4050–4200	1	0.01	1.00
4200–4350	0	0.00	1.00
4350–4500	0	0.00	1.00
Above 4500 psi	0	0.00	1.00

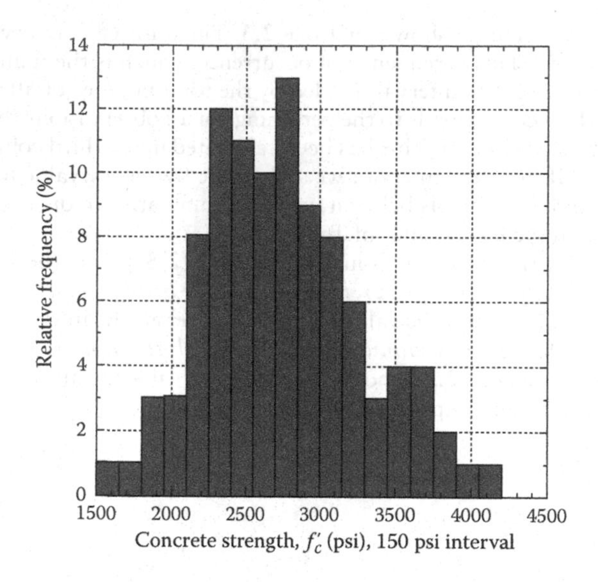

Figure 2.27 Relative frequency histogram for concrete strength (Example 2.9).

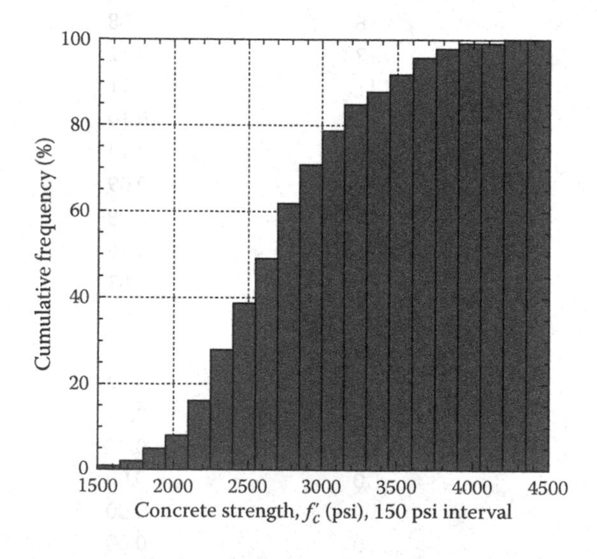

Figure 2.28 Cumulative frequency histogram for concrete strength (Example 2.9).

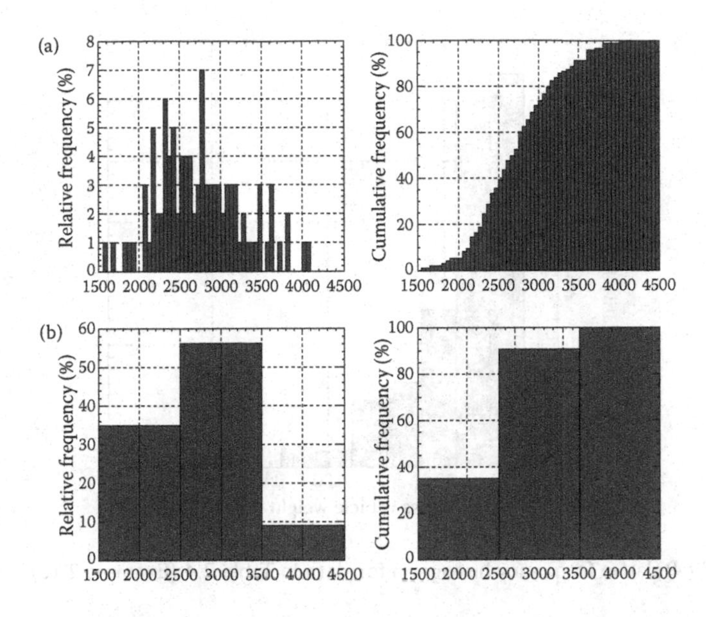

Figure 2.29 Influence of interval size on appearance of histogram (Example 2.9) (a) for interval of 50 psi and (b) for interval of 1000 psi.

Example 2.10

Consider the set of values of GVW shown in Table 2.4. Figures 2.30 and 2.31 can be plotted from calculations of relative and cumulative frequency values for the intervals defined in the table.

Table 2.4 Observed data on GVW

GVW (15 kips interval)	*Number of observations*	*Frequency of occurrence*	*Cumulative frequency*
Below 15 kips	25	0.08	0.08
15.0–30.0	101	0.32	0.40
30.0–45.0	94	0.30	0.70
45.0–60.0	40	0.13	0.83
60.0–75.0	21	0.07	0.90
75.0–90.0	17	0.05	0.95
90.0–105.0	4	0.01	0.96
105.0–120.0	6	0.02	0.98
120.0–135.0	0	0.00	0.98
135.0–150.0	2	0.01	0.99
150.0–165.0	2	0.01	1.00
165.0–180.0	1	0.00	1.00
Above 180.0 kips	0	0.00	1.00

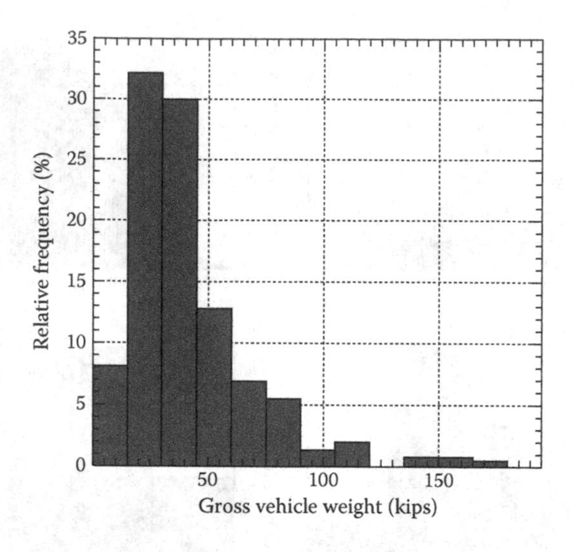

Figure 2.30 Relative frequency histogram for data in Table 2.4 (Example 2.10).

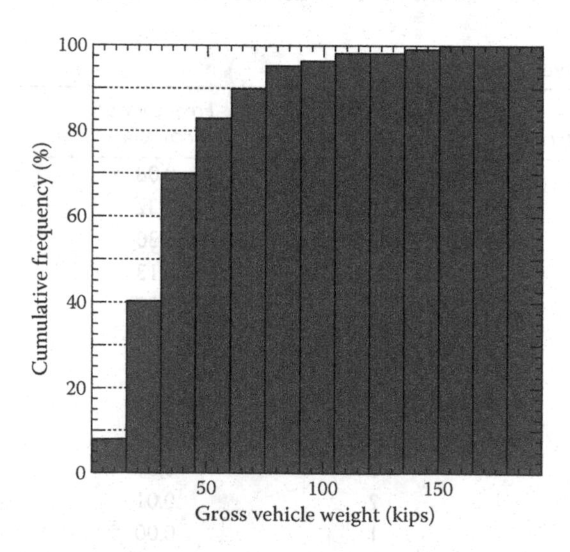

Figure 2.31 Cumulative frequency histogram for data in Table 2.4 (Example 2.10).

2.7 CONDITIONAL PROBABILITY

An important description of how two uncertain events are related is provided by the conditional probability of one event occurring given the fact that the other event has or has not occurred. Given two events E_1 and E_2, the conditional probability of E_1 occurring if E_2 has already occurred is defined as

$$P(E_1|E_2) = \frac{P(E_1 \cap E_2)}{P(E_2)} \tag{2.80}$$

where $E_1|E_2$ is the notation commonly used to denote the case in which event E_1 occurs given event E_2 has occurred. The quantity $P(E_1 \cap E_2)$ represents the probability of the *intersection* (∩) of events E_1 and E_2. The term "intersection" means that both events E_1 and E_2 occur simultaneously. This should be distinguished from the concept of union, which was introduced in Section 2.1.2. Figure 2.32 schematically shows the difference between intersection and union of events.

To illustrate the concept of conditional probability, consider the following example.

Example 2.11

Consider tests of concrete beams. Two parameters are observed: cracking moment and ultimate moment. Let M_u and M_{cr} denote the ultimate bending moment and the cracking moment, respectively. Define event E_1 as the case when $M_u \geq 150$ k-ft, and define event E_2 as the case when $M_{cr} \geq 100$ k-ft (1 k-ft = 1 kip-feet = 1.356 kN-m).

A conditional probability that the ultimate moment will be reached given that the cracking moment has been reached would be written as follows:

$$P(E_1|E_2) = P(M_u \geq 150 \text{ given } M_{cr} \geq 100) = \frac{P(E_1 \cap E_2)}{P(E_2)}$$

$$= \frac{P(M_u \geq 150 \text{ and } M_{cr} \geq 100)}{P(M_{cr} \geq 100)}$$

If two events are *statistically independent*, then the occurrence of one event has no effect on the probability of occurrence of the other event. In

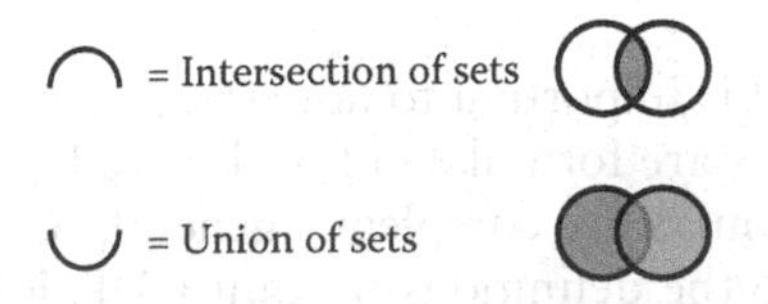

Figure 2.32 A Venn diagram showing the difference between the intersection and union of events.

terms of the conditional probability definition given by Equation 2.80, we can state that for two statistically independent events E_1 and E_2,

$$P(E_1|E_2) = P(E_1) \quad \text{and} \quad P(E_2|E_1) = P(E_2) \tag{2.81}$$

The concept of statistical independence has important implications on structural reliability analysis, as discussed later.

2.8 RANDOM VECTORS

A *random vector* is defined as a vector (or set) of random variables $\{X_1, X_2,\ldots, X_n\}$. When we deal with multiple random variables in a random vector, we can define distribution functions and density functions similar to those defined in Section 2.1.4 for single random variables. The *joint CDF*, analogous to the CDF defined in Section 2.1.4, is defined as

$$F_{X_1,X_2\ldots,X_n}\left(x_1,x_2\ldots,x_n\right) = P\left(X_1 \le x_1, X_2 \le x_2,\ldots,X_n \le x_n\right) \tag{2.82}$$

In Equation 2.82, the right-hand side of the equation should be read as the probability of the intersection of the events $X_1 \le x_1$ and $X_2 \le x_2$ and ... and $X_n \le x_n$. This function is defined for both discrete and continuous random variables. For continuous random variables, the *joint PDF* is defined as

$$f_{X_1,X_2\ldots,X_n}\left(x_1,x_2\ldots,x_n\right) = \frac{\partial^n F}{\partial_{X_1}\ldots\partial_{X_n}}\left(x_1,x_2\ldots,x_n\right) \tag{2.83}$$

For discrete random variables, the *joint probability mass function* is defined as

$$p_{X_1,X_2,\ldots,X_n}(x_1,x_2,\ldots,x_n) = P(X_1 = x_1, X_2 = x_2,\ldots,X_n = x_n) \tag{2.84}$$

For continuous random variables, we can define a *marginal density function* for each X_i as

$$f_{X_i}(x_i) = \int_{-\infty}^{\infty}\int_{-\infty}^{\infty} f_{X_1\ldots X_n}\left(x_1\ldots x_n\right)\mathrm{d}x_1\,\mathrm{d}x_2\ldots\mathrm{d}x_{i-1}\,\mathrm{d}x_{i+1}\ldots\mathrm{d}x_n \tag{2.85}$$

In Equation 2.85, it is important to note that there are $n - 1$ integrations involved. The integrals are formulated for all variables except X_i.

The preceding formulas are completely general, but they can be confusing. To help illustrate the definitions of joint CDF, joint density function, and marginal density functions, consider the case of two continuous random variables X and Y. The *joint CDF* for X and Y is defined as

$$F_{XY}(x,y) = P(X \leq x, Y \leq y) \tag{2.86}$$

The *joint PDF* is defined as

$$f_{XY}(x,y) = \frac{\partial^2 F_{XY}}{\partial x \partial y}(x,y) \tag{2.87}$$

The marginal density functions are

$$f_X(x) = \int_{-\infty}^{+\infty} f_{XY}(x,y)\,dy \tag{2.88}$$

$$f_Y(y) = \int_{-\infty}^{+\infty} f_{XY}(x,y)\,dx \tag{2.89}$$

In Section 2.7, we introduced the concept of conditional probability. This concept can be extended to define a *conditional distribution function* for a random vector. Again, consider the case of two continuous random variables X and Y. The conditional distribution function is defined as

$$f_{X|Y}(X \mid Y) = \frac{f_{XY}(x,y)}{f_y(y)} = \frac{\text{joint density}}{\text{marginal density}} \tag{2.90}$$

If the random variables X and Y are *statistically independent* (see Section 2.7), then

$$f_{X|Y}(x|y) = f_X(x) \tag{2.91a}$$

and

$$f_{Y|X}(y|x) = f_Y(y) \tag{2.91b}$$

which implies, based on Equation 2.90, that

$$f_{XY}(x,y) = f_X(x)f_Y(y) \tag{2.92}$$

Example 2.12

Consider a set of tests in which two quantities are measured: modulus of elasticity, X_1, and compressive strength, X_2. Since the values of these variables vary from test to test, as seen in Table 2.5, it is appropriate to treat them as random variables.

Table 2.5 Values of modulus of elasticity and compressive strength used in Example 2.12 (1 psi = 6.895 kPa)

Sample	f'_c *(psi)*	*E (psi)*	*Sample*	f'_c *(psi)*	*E (psi)*	*Sample*	f'_c *(psi)*	*E (psi)*
1	3059	3,335,000	34	2266	2,797,000	67	2634	3,216,000
2	3397	3,280,000	35	3414	3,087,000	68	2309	2,725,000
3	2575	3,117,000	36	2973	3,122,000	69	3062	3,317,000
4	3803	3,271,000	37	2397	2,933,000	70	2336	2,995,000
5	2887	3,201,000	38	3629	3,522,000	71	2325	2,512,000
6	2774	3,067,000	39	2797	3,042,000	72	2600	2,840,000
7	3187	3,252,000	40	3164	2,890,000	73	2197	2,636,000
8	2804	2,814,000	41	2063	2,421,000	74	3635	3,304,000
9	1563	2,354,000	42	2521	3,074,000	75	1938	2,483,000
10	2258	2,606,000	43	2643	2,962,000	76	2557	2,618,000
11	2753	3,233,000	44	4072	3,814,000	77	3566	3,990,000
12	2156	2,854,000	45	2249	2,920,000	78	2432	3,112,000
13	2752	3,020,000	46	3107	3,485,000	79	2903	3,408,000
14	2933	3,080,000	47	3009	2,942,000	80	2776	2,963,000
15	2821	3,455,000	48	2452	2,901,000	81	3239	3,497,000
16	2209	2,464,000	49	2361	2,917,000	82	2393	2,960,000
17	2774	2,853,000	50	2780	3,010,000	83	3459	3,545,000

18	2391	2,685,000	51	3113	3,454,000	84	2423	3,097,000
19	3251	2,931,000	52	3071	3,182,000	85	2330	2,697,000
20	2933	2,841,000	53	2577	2,962,000	86	3199	3,318,000
21	3049	3,034,000	54	2421	2,803,000	87	3101	3,188,000
22	2079	2,473,000	55	1878	2,534,000	88	2509	2,516,000
23	3615	3,895,000	56	3470	3,377,000	89	3306	2,823,000
24	2724	2,937,000	57	2977	3,342,000	90	2402	2,935,000
25	2690	2,999,000	58	2140	2,635,000	91	2524	2,856,000
26	2722	2,880,000	59	2087	2,208,000	92	2318	2,214,000
27	2170	2,985,000	60	2551	2,810,000	93	2884	3,089,000
28	2509	2,790,000	61	4025	3,977,000	94	2803	3,014,000
29	2172	2,663,000	62	2303	2,362,000	95	2983	3,308,000
30	3450	3,236,000	63	1650	2,335,000	96	2877	2,965,000
31	3729	3,201,000	64	2683	2,823,000	97	2192	2,553,000
32	1807	2,344,000	65	3280	3,214,000	98	2631	3,179,000
33	2438	3,144,000	66	3801	3,060,000	99	2456	2,904,000
						100	2725	3,150,000

Using the concept of histograms discussed in Section 2.6, we can get an idea of the general shape of the PDF for each individual variable and the joint PDF and joint probability distribution function. For each individual variable, we define appropriate intervals of values and then count the number of observations within each interval. The resulting relative frequency histogram for each variable is shown in Figure 2.33. To consider the joint histogram, we need to define "two-dimensional intervals." For example, one "interval" would be for values of $X_1(E)$ between 3.0×10^6 psi and 3.25×10^6 psi *and* values of $X_2(f_c')$ between 2.5×10^3 psi and 3.0×10^3 psi. Looking at Table 2.5, we see that there are 15 samples that satisfy both requirements simultaneously; these samples are highlighted in the table. Therefore, we have 15 observations in this interval out of 100 total observations, and the relative frequency value is 15/100 = 0.15. This value is indicated as the shaded block in Figure 2.34, the relative frequency histogram. A cumulative frequency histogram can also be constructed as shown in Figure 2.35. For example, to find the cumulative value of the number of times that X_1 is less than or equal to 3.0×10^6 psi and X_2 is less than or equal to

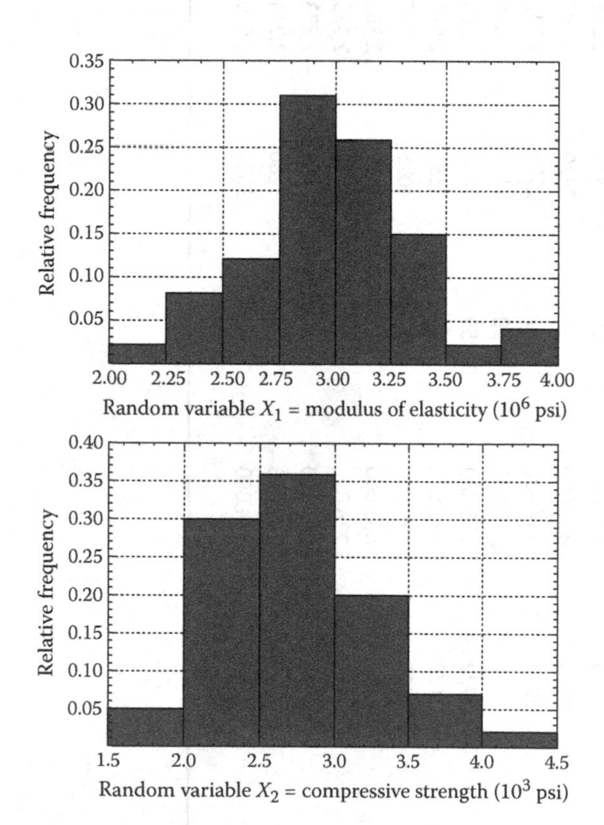

Figure 2.33 Relative frequency histograms for X_1 and X_2 considered independently.

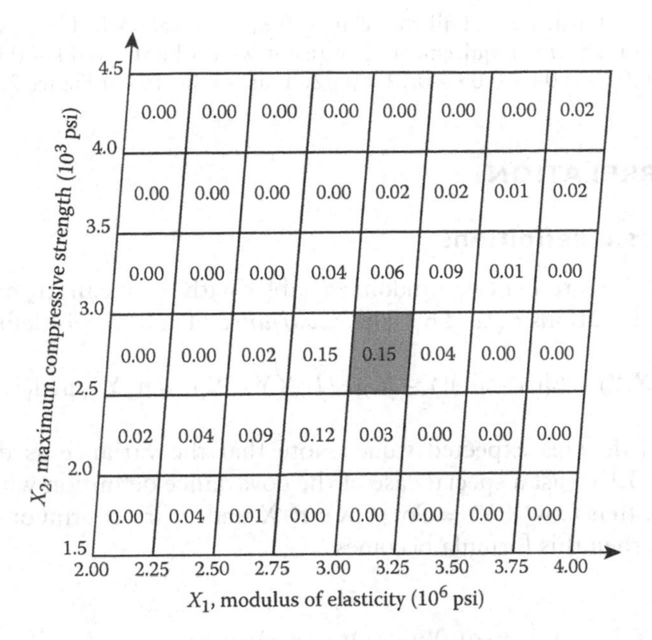

Figure 2.34 Relative frequency histogram for both X_1 and X_2.

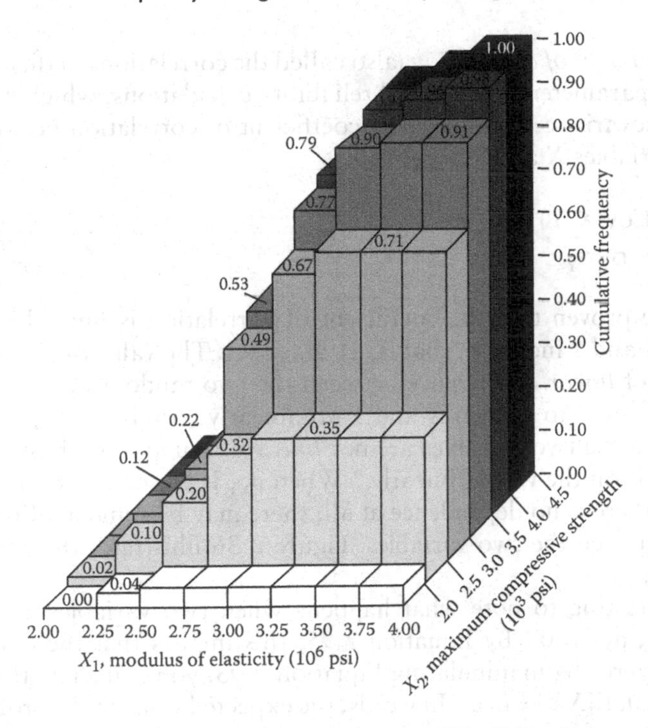

Figure 2.35 Cumulative frequency histogram for both X_1 and X_2.

2.5×10^3 psi, we add all the relative frequency values in Figure 2.34 that satisfy this requirement. The result would be $0 + 0.04 + 0.01 + 0 + 0.02 + 0.04 + 0.09 + 0.12 = 0.32$. This is reflected in Figure 2.35.

2.9 CORRELATION

2.9.1 Basic definitions

Let X and Y represent two random variables with the means μ_X and μ_Y and standard deviations σ_X and σ_Y. The *covariance* of X and Y is defined as

$$\text{Cov}(X,Y) = E[(X - \mu_X)(Y - \mu_Y)] = E[XY - X\mu_Y - \mu_X Y + \mu_X\mu_Y] \tag{2.93}$$

where $E[]$ denotes expected value. Note that the variance as defined in Section 2.3.1 is just a special case of the covariance definition when $X = Y$. Also note that $\text{Cov}(X,Y) = \text{Cov}(Y,X)$. If X and Y are continuous random variables, then this formula becomes

$$\text{Cov}(X,Y) = \int_{-\infty}^{+\infty}\int_{-\infty}^{+\infty} (x - \mu_X)(y - \mu_Y) f_{XY}(x,y)\, dx\, dy \tag{2.94}$$

The *coefficient of correlation* (also called the correlation coefficient) is an important parameter in structural reliability calculations, which is defined using the covariance formula. The coefficient of correlation between two random variables X and Y is defined as

$$\rho_{XY} = \frac{\text{Cov}(X,Y)}{\sigma_X \sigma_Y} \tag{2.95}$$

It can be proven that the coefficient of correlation is limited to values between –1 and 1 inclusive; that is, $-1 \le \rho_{XY} \le 1$. The value of ρ_{XY} indicates the degree of *linear* dependence between the two random variables X and Y. If $|\rho_{XY}|$ is close to 1, then X and Y are linearly correlated. If ρ_{XY} is close to zero, then the two variables are not *linearly* related to each other. Note the emphasis on the word "linearly." When ρ_{XY} is close to zero, it does not mean that there is no dependence at all; there may be some nonlinear relationship between the two variables. Figure 2.36 illustrates the concept of correlation.

It is interesting to note what happens when two variables are uncorrelated (i.e., $\rho_{XY} = 0$). By Equation 2.95, this implies that the covariance is equal to zero. By manipulating Equation 2.93, you can show that when $\text{Cov}(X,Y) = 0$, $E[XY] = \mu_X\mu_Y$. In words, the expected value of the product XY is the product of the expected values when the variables are uncorrelated.

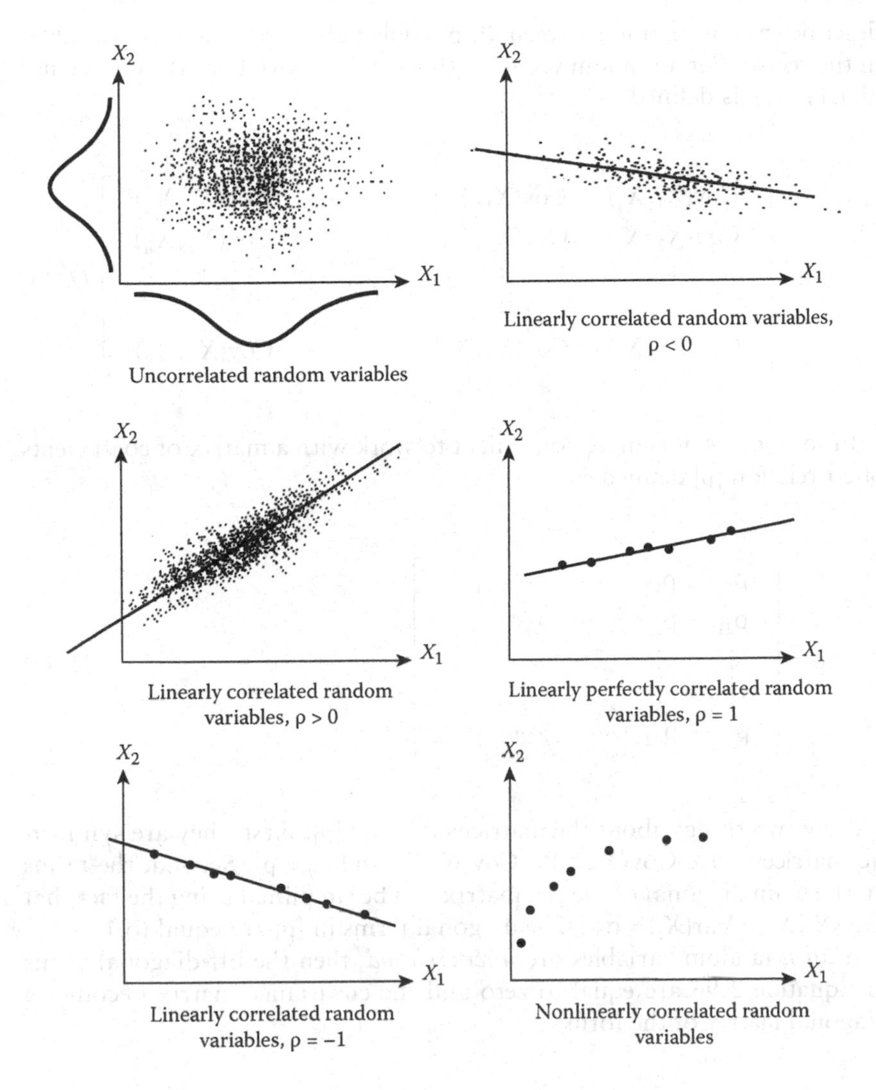

Figure 2.36 **Examples of correlated and uncorrelated random variables.**

It is important to emphasize that the terms "statistically independent" and "uncorrelated" are not always synonymous. Statistically independent is a much stronger statement than uncorrelated. If two variables are statistically independent, then they must also be uncorrelated. However, the converse is not true in general. If two variables are uncorrelated, they are not necessarily statistically independent.

The foregoing comments on correlation pertain to two random variables. When dealing with a random vector, a *covariance matrix* is used to describe the correlation between all possible pairs of the random variables in the vector. For a random vector with n random variables, the covariance matrix, $[C]$, is defined as

$$[C]=\begin{bmatrix} \text{Cov}(X_1,X_1) & \text{Cov}(X_1,X_2) & \cdots & \cdots & \text{Cov}(X_1,X_n) \\ \text{Cov}(X_2,X_1) & \text{Cov}(X_2,X_2) & & & \text{Cov}(X_2,X_n) \\ \vdots & \vdots & \ddots & & \vdots \\ \vdots & \vdots & & \ddots & \vdots \\ \text{Cov}(X_n,X_1) & \text{Cov}(X_n,X_2) & \cdots & \cdots & \text{Cov}(X_n,X_n) \end{bmatrix} \tag{2.96}$$

In some cases, it is more convenient to work with a matrix of coefficients of correlation $[\rho]$ defined as

$$[\rho]=\begin{bmatrix} \rho_{11} & \rho_{12} & \cdots & \cdots & \rho_{1n} \\ \rho_{21} & \rho_{22} & \cdots & \cdots & \rho_{2n} \\ \vdots & \vdots & \ddots & & \vdots \\ \vdots & \vdots & & \ddots & \vdots \\ \rho_{n1} & \rho_{n2} & \cdots & \cdots & \rho_{nn} \end{bmatrix} \tag{2.97}$$

Note two things about the matrices $[C]$ and $[\rho]$. First, they are symmetric matrices since $\text{Cov}(X_i,X_j) = \text{Cov}(X_j,X_i)$ and $\rho_{ij} = \rho_{ji}$. Second, the terms on the main diagonal of the $[C]$ matrix can be simplified using the fact that $\text{Cov}(X_i,X_i) = \text{Var}(X_i) = \sigma_{X_i}^2$. The diagonal terms in $[\rho]$ are equal to 1.

If *all n* random variables are *uncorrelated*, then the off-diagonal terms in Equation 2.96 are equal to zero and the covariance matrix becomes a diagonal matrix of the form

$$[C]=\begin{bmatrix} \sigma_{X_1}^2 & 0 & \cdots & \cdots & 0 \\ 0 & \sigma_{X_2}^2 & & & 0 \\ \vdots & \vdots & \ddots & & \vdots \\ \vdots & \vdots & & \ddots & \vdots \\ 0 & 0 & \cdots & \cdots & \sigma_{X_n}^2 \end{bmatrix} \tag{2.98}$$

The matrix $[\rho]$ in Equation 2.97 becomes a diagonal matrix with 1's on the diagonal.

2.9.2 Statistical estimate of the correlation coefficient

As noted in Section 2.3.2, in practice, we often do not know the underlying distributions of the variables we are observing, and thus, we have to rely on test data and observations to estimate parameters. When we have observed data for two random variables X and Y, we can estimate the correlation coefficient as follows.

Assume that there are n observations $\{x_1, x_2, ..., x_n\}$ of variable X and n observations $\{y_1, y_2, ..., y_n\}$ of variable Y. The sample mean and standard deviation for each variable can be calculated using Equations 2.25 and 2.26. Once the sample means $\bar{x}$ and $\bar{y}$ and sample standard deviations s_X and s_Y are determined, the sample estimate of the correlation coefficient can be calculated using this formula:

$$\hat{\rho}_{XY} = \frac{1}{n-1} \frac{\sum_{i=1}^{n} (x_i - \bar{x})(y_i - \bar{y})}{s_X s_Y} = \frac{1}{n-1} \frac{\left(\sum_{i=1}^{n} x_i y_i \right) - n\bar{x}\bar{y}}{s_X s_Y} \tag{2.99}$$

2.10 BAYESIAN UPDATING

2.10.1 Bayes' Theorem

Consider a set of n events $\{A_1, A_2, ..., A_n\}$ in which the events A_i are mutually exclusive and collectively exhaustive. Based on Equations 2.2 and 2.3, these conditions require that

$$A_1 \cup A_2 \cup ...... \cup A_n = \Omega \tag{2.100a}$$

$$P(A_1 \cup A_2 \cup ...... \cup A_n) = P(\Omega) = 1 \tag{2.100b}$$

where Ω is the sample space. Now consider another event E defined in the sample space Ω. By necessity, E must "overlap" with one or more of the events A_i. In other words, if E occurs, then one or more of the A_i events must occur also. We can determine the probability of E by using the total probability theorem as follows:

$$P(E) = \sum_{i=1}^{n} P(E|A_i)P(A_i) \tag{2.101}$$

Observe that the concepts of conditional probability, mutually exclusive events, and collectively exhaustive events are all used in the theorem of total probability.

Now consider the following question: if event E occurs, what is the probability of occurrence of a particular event A_i? Bayes' Theorem provides a way to answer this question. According to Bayes' Theorem, if event E occurs, the probability of occurrence of A_i can be found using

$$P(A_i|E) = \frac{P(E|A_i)\,P(A_i)}{\sum_{i=1}^{n} P(E|A_i)\,P(A_i)} \tag{2.102}$$

2.10.2 Applications of Bayes' Theorem

Bayes' Theorem is useful for combining statistical and judgmental information and for updating probabilities based on observed outcomes. For example, consider the determination of the strength of a structural component in the context of the discussion in Section 2.10.1. Define a random variable A representing the strength of the component. For simplicity of explanation, assume that the strength can only assume one of n discrete values $\{a_1, a_2, \ldots, a_n\}$. The probability of each value a_i is estimated to be $P(A = a_i) = p_i$. These probabilities are the prior probabilities and are based on past experience and judgment. Now, assume that some field tests are conducted, and we wish to update the probabilities in light of this additional information. Define p_i' as the posterior or updated probability. Let event E represent a possible test result. The test result must give one of the possible values $\{a_1', a_2', \ldots, a_n'\}$ where the primes have been added to indicate that these are possible test results. The updated probability can be found using Bayes' Theorem provided that we have some information on the conditional probabilities used in Equation 2.102. If we rewrite Equation 2.102 using the notation presented above, we get

$$P(A = a_i|E = a_j') = \frac{P(E = a_j'|A = a_i)\,P(A = a_i)}{\sum_{i=1}^{n} P(E = a_j'|A = a_i)\,P(A = a_i)} \tag{2.103}$$

or

$$p_i' = \frac{P(E = a_j'|A = a_i)\,p_i}{\sum_{i=1}^{n} P(E = a_j'|A = a_i)\,p_i} \tag{2.104}$$

The conditional probabilities reflect any uncertainty in the tests themselves. The probability $P\left(E = a'_j \middle| A = a_i\right)$ is the probability that the test result will indicate a'_j given that the true value is a_i. In other words, the conditional probability gives some idea of the confidence in the test.

Example 2.13

Consider a steel beam where some corrosion is observed. An engineer has been called in to determine the actual shear strength of the web. For evaluation purposes, it is sufficiently accurate to assume that the strength can take one of the following five values: R_V, $0.9R_V$, $0.8R_V$, $0.7R_V$, and $0.6R_V$. From past experience with the corrosion problem, an engineer estimates that the probabilities of these values are 0, 0.15, 0.30, 0.40, and 0.15, respectively.

A field test is conducted, and it is found that the strength is equal to $0.8R_V$. The reliability of the test is reasonably good but not perfect. Table 2.6 is a conditional probability matrix providing an indication of the confidence in the test. Update the probabilities in light of this new information.

The updated probabilities are calculated using Equation 2.104. First, it is helpful to understand how to interpret the matrix. Each entry is a probability $P\left(E = a'_j \middle| A = a_i\right)$. Each column can be interpreted as its own sample space in which the actual beam strength is known. The probability values in each column are the probabilities associated with a mutually exclusive and collectively exhaustive set of experimental outcomes for the particular beam strength being considered. Therefore, the sum of the probabilities in each column must add up to 1. Observe that the same property does not apply to each row.

The updated posterior probabilities are calculated using Equation 2.104:

$$p'_i = \frac{P(E = a'_j | A = a_i)p_i}{\sum_{i=1}^{n} P(E = a'_j | A = a_i)p_i}$$

Table 2.6 Test information for Example 2.13

		Actual beam strength				
		R_V	$0.9R_V$	$0.8R_V$	$0.7R_V$	$0.6R_V$
	$0.6R_V$	0	0	0.05	0.15	0.70
Test	$0.7R_V$	0	0.05	0.20	0.75	0.25
values	$0.8R_V$	0.10	0.25	0.70	0.10	0.05
	$0.9R_V$	0.30	0.65	0.05	0	0
	R_V	0.60	0.05	0	0	0
	Column sum	1.0	1.0	1.0	1.0	1.0

For this example, we need the conditional probabilities contained in the highlighted row. Each of the updated probabilities requires the same denominator in the above formula, so this quantity is calculated first:

$$\sum_{i=1}^{n} P(E=0.8R_V \mid A=a_i)p_i = (0.10)(0) + (0.25)(0.15) + (0.70)(0.30) + (0.10)(0.40) + (0.05)(0.15) = 0.295$$

Now, the updated probabilities can be calculated as follows:

$$P(A=0.6R_V \mid E=0.8R_V) = \frac{P(E=0.8R_V \mid A=0.6R_V)P(0.6R_V)}{\sum_{i=1}^{n} P(E=a'_j \mid A=a_i)p_i} = \frac{(0.05)(0.15)}{0.295} = 0.025$$

$$P(A=0.7R_V \mid E=0.8R_V) = \frac{P(E=0.8R_V \mid A=0.7R_V)P(0.7R_V)}{\sum_{i=1}^{n} P(E=a'_j \mid A=a_i)p_i} = \frac{(0.10)(0.40)}{0.295} = 0.136$$

$$P(A=0.8R_V \mid E=0.8R_V) = \frac{P(E=0.8R_V \mid A=0.8R_V)P(0.8R_V)}{\sum_{i=1}^{n} P(E=a'_j \mid A=a_i)p_i} = \frac{(0.70)(0.30)}{0.295} = 0.712$$

$$P(A=0.9R_V \mid E=0.8R_V) = \frac{P(E=0.8R_V \mid A=0.9R_V)P(0.9R_V)}{\sum_{i=1}^{n} P(E=a'_j \mid A=a_i)p_i} = \frac{(0.25)(0.15)}{0.295} = 0.127$$

Table 2.7 Comparison of prior and posterior probabilities

	Posterior probability	*Prior probability*
$P(0.6R_V)$	0.025	0.15
$P(0.7R_V)$	0.136	0.40
$P(0.8R_V)$	0.712	0.30
$P(0.9R_V)$	0.127	0.15
$P(R_V)$	0	0
Sum	1	1

$$P(A=R_V|E=0.8R_V)=\frac{P(E=0.8R_V|A=R_V)P(R_V)}{\sum_{i=1}^{n}P(E=a'_j|A=a_i)p_i}=\frac{(0.10)(0)}{0.295}=0.0$$

For comparison, the prior and posterior probabilities are listed in Table 2.7.

2.10.3 Continuous case

The examples and discussion presented in Sections 2.10.1 and 2.10.2 have focused on applications of Bayesian updating to discrete cases. However, Bayesian updating can also be applied to problems in which the random variables involved are continuous. For the continuous case, the updating is applied to the prior PDF function for the random variable. Let $f'_A(a)$ represent the prior PDF function for the continuous random variable A, and let $f''_A(a)$ represent the updated PDF function. The conditional probability expression shown in Equation 2.102 must be expressed as a continuous function to apply Bayesian updating. Let this continuous function be represented by a function $L(E|a)$. This function is sometimes referred to as a "likelihood function" (Ang and Tang, 1975). Finally, the summation in Equation 2.102 must be replaced by an integral. With these modifications, the expression for Bayesian updating in the continuous case becomes

$$f''_A(a)=\frac{L(E|a)f'_A(a)}{\int_{-\infty}^{+\infty}L(E|a)f'_A(a)\,da} \tag{2.105}$$

PROBLEMS

2.1 The results of tests to determine the modulus of rupture (MOR) for a set of timber beams is shown in Table P2.1.
   A. Plot the relative frequency and cumulative frequency histograms.
   B. Calculate the sample mean, standard deviation, and COV.
   C. Plot the data on normal probability paper.

2.2 A set of test data for the load-carrying capacity of a member is shown in Table P2.2.
   A. Plot the test data on normal probability paper.
   B. Plot a normal distribution on the same probability paper. Use the sample mean and standard deviation as estimates of the true mean and standard deviation.
   C. Plot a lognormal distribution on the same normal probability paper. Use the sample mean and standard deviation as estimates of the true mean and standard deviation.
   D. Plot the relative frequency and cumulative frequency histograms.

2.3 For the data in Table 2.5, calculate the statistical estimate of the correlation coefficient using Equation 2.99.

2.4 Variable X is to be modeled using a uniform distribution. The lower bound value is 5, and the upper bound value is 36.
   A. Calculate the mean and standard deviation of X.
   B. What is the probability that the value of X is between 10 and 20?
   C. What is the probability that the value of X is greater than 31?
   D. Plot the CDF on normal probability paper.

2.5 The dead load D on a structure is to be modeled as a normal random variable with a mean value of 100 and a COV of 8%.
   A. Plot the PDF and CDF on standard graph paper.
   B. Plot the CDF on normal probability paper.

Table P2.1 Test data for MOR of timber beams [units are pounds per square inch (psi); 1 psi = 6.895 kPa]

4464	5974	6044	5288	5866	4627	5332	3521
5001	5023	3600	4830	5133	4930	3973	5184
4880	5085	7145	5130	5802	5772	5331	5437
6486	4764	5072	3966	7022	4724	4209	
5897	3508	5693	5702	4590	6408	4434	
3758	4705	3845	4310	5289	4667	3982	

Table P2.2 Test data for load-carrying capacity of a member

3.95	3.99	4.21	4.28	4.26	4.07	4.36
4.07	4.21	3.90	4.15	4.41	3.97	
4.14	4.39	3.74	4.04	4.22	4.05	

C. Determine the probability that D is less than or equal to 95.
D. Determine the probability that D is between 95 and 105.

2.6 The ground snow load q (in pounds per square foot, psf) is to be modeled as a lognormal random variable. The mean value of the ground snow load is 8.85 psf, and the standard deviation is 5.83 psf (1 psf = 47.88 Pa).
A. Plot the PDF and CDF on standard graph paper.
B. Plot the CDF on normal probability paper.
C. Determine the probability that q is less than or equal to 7.39 psf.
D. Determine the probability that q is between 6 and 8 psf.

2.7 The yield stress of A36 steel is to be modeled as a lognormal random variable with a mean value of 36 ksi and a COV of 10% (1 ksi = 6.895 MPa).
A. Plot the PDF and CDF on standard graph paper.
B. Plot the CDF on normal probability paper.
C. Determine the probability that the yield stress is greater than 40 ksi.
D. Determine the probability that the yield stress is between 34 and 38 ksi.

2.8 The annual extreme wind speed at a particular location is to be modeled as an extreme Type I random variable. The mean value of the extreme wind is 50 miles per hour (mph) and the COV is 15% (1 mph = 1.609 km/h).
A. Plot the PDF and CDF on standard graph paper.
B. Plot the CDF on normal probability paper.
C. Determine the probability that the annual maximum wind speed is greater than 50 mph.
D. Determine the probability that the annual maximum wind speed is less than 50 mph.
E. Determine the probability that the wind speed will be between 40 and 60 ksi.

2.9 The peak ground acceleration A that is expected at a site in a 50-year time period is modeled as an extreme Type II random variable with u = 0.2g (g is the acceleration of gravity) and k = 3.
A. Plot the PDF and CDF on standard graph paper.
B. Plot the CDF on normal probability paper.
C. Determine the probability that the ground acceleration will be between 0.15g and 0.2g.
D. Determine the probability that the ground acceleration will be greater than 0.3g.

2.10 The occurrence of major snow storms (snowfall greater than 1 ft = 0.3048 m) for a particular location is assumed to follow a Poisson distribution. Over 33 years of snowfall records, such storms have occurred 94 times.

A. What is the probability of two or more major storms in the next year?
B. What is the probability of exactly two storms occurring in the next 3 years?
C. What is the return period of these storms?
D. What is the probability of more than three storms in the next 2 years?

Chapter 3

Functions of random variables

In this chapter, the concept of random variables is extended to cover functions of random variables. If a set of variables in a function is random, then the function itself will be random, and the value of the function can be described as a random variable.

3.1 LINEAR FUNCTIONS OF RANDOM VARIABLES

Let Y be a linear function of random variables $X_1, X_2, \ldots, X_n$:

$$Y = a_0 + a_1X_1 + a_2X_2 + \ldots + a_nX_n = a_0 + \sum_{i=1}^{n} a_iX_i \tag{3.1}$$

where the a_i ($i = 0, 1, \ldots, n$) are constants. By using the concepts of expected value of a function, variance, covariance, correlation coefficient, and marginal density functions presented in Chapter 2, it can be shown that the mean (expected value) of Y is

$$\mu_Y = a_0 + a_1\mu_{X_1} + a_2\mu_{X_2} + \ldots + a_n\mu_{X_n} = a_0 + \sum_{i=1}^{n} a_i\mu_{X_i} \tag{3.2}$$

and the variance of Y is

$$\begin{aligned}\sigma_Y^2 &= E\left[\left(Y - \mu_Y\right)^2\right] \\ &= E\left[Y^2\right] - \mu_Y^2 \\ &= \sum_{i=1}^{n}\sum_{j=1}^{n} a_ia_j\text{Cov}\left(X_i, X_j\right) \\ &= \sum_{i=1}^{n}\sum_{j=1}^{n} a_ia_j\rho_{X_iX_j}\sigma_{X_i}\sigma_{X_j}\end{aligned} \tag{3.3}$$

Note that the constant a_0 does not affect the variance, but it does affect the mean value. If the n random variables are uncorrelated with each other, then $\text{Cov}(X_i, X_j) = 0$ for $i \neq j$ and thus Equation 3.3 simplifies to

$$\sigma_Y^2 = \sum_{i=1}^{n} a_i^2 \sigma_{X_i}^2 \tag{3.4}$$

It is important to note that these results are valid regardless of the probability distributions of the random variables $X_1, X_2, \ldots, X_n$. However, we have not addressed the issue of determining the distribution of the linear function Y. This, in general, is difficult to determine. Some special cases are discussed in Section 3.2.

Example 3.1

Let R = load-carrying capacity (resistance) and Q = the demand or load effect. Define a "performance function," Y, in terms of R and Q as

$$Y = R - Q$$

As we will discuss later, the value of the function Y is a measure of structural safety. If $Y > 0$, the capacity is greater than the demand and thus the structure is safe. If $Y < 0$, the capacity is less than the demand, and the structure fails. Calculate the following: μ_Y, σ_Y^2, V_Y.

Mean value μ_Y: For this case, let R correspond to X_1 and Q correspond to X_2. Then, $a_0 = 0$, $a_1 = 1$, and $a_2 = -1$. Substituting into Equation 3.2 gives

$$\mu_Y = a_0 + \sum_{i=1}^{n} a_i \mu_{X_i} = 0 + (1)\mu_R + (-1)\mu_Q = \mu_R - \mu_Q$$

Variance σ_Y^2: Using Equation 3.3,

$$\begin{aligned}
\sigma_Y^2 &= \sum_{i=1}^{n}\sum_{j=1}^{n} a_i a_j \text{Cov}(X_i, X_j) \\
&= a_1^2 \text{Cov}(X_1, X_1) + 2a_1 a_2 \text{Cov}(X_1, X_2) + a_2^2 \text{Cov}(X_2, X_2) \\
&= (1)^2 \sigma_R^2 + 2(1)(-1)\text{Cov}(R, Q) + (-1)^2 \sigma_Q^2 \\
&= \sigma_R^2 + \sigma_Q^2 - 2\text{Cov}(R, Q) \\
&= \sigma_R^2 + \sigma_Q^2 - 2\rho_{RQ}\sigma_R\sigma_Q
\end{aligned}$$

If R and Q are uncorrelated, then $\rho_{RQ} = 0$ and hence $\sigma_Y^2 = \sigma_R^2 + \sigma_Q^2$.

Coefficient of variation V_Y: Using the definition from Chapter 2, the coefficient of variation is

$$V_Y = \frac{\sigma_Y}{\mu_Y} = \frac{\sqrt{\sigma_R^2 + \sigma_Q^2 - 2\text{Cov}(R,Q)}}{\mu_R - \mu_Q}$$

The reciprocal of V_Y is called a *reliability index*, and it is typically denoted by

$$\beta = \frac{1}{V_Y} = \frac{\mu_Y}{\sigma_Y}$$

The reliability index is discussed more in later chapters.

3.2 LINEAR FUNCTIONS OF NORMAL VARIABLES

In the previous section, we discussed how to determine the mean and variance for a linear function of random variables, and the distributions of the random variables were arbitrary. If the random variables are all normally distributed, then we can determine additional information about the linear function.

Let Y be a linear function of *normally distributed* random variables X_i, $i = 1, 2, \ldots, n$, as shown in Equation 3.1. The mean value μ_Y can be found using Equation 3.2. If the random variables X_i are all *uncorrelated*, then Equation 3.4 can be used to determine the variance σ_Y^2. Furthermore, it can be shown that the linear function Y of uncorrelated normal random variables is a *normal* random variable with distribution parameters μ_Y and σ_Y.

Example 3.2

Consider the simply supported beam shown in Figure 3.1.

Assume the distributed load w, concentrated load P, and the moment capacity M_R are uncorrelated normal random variables. Calculate the reliability index, β, as defined in Example 3.1. What is the probability that the beam will fail? The distribution parameters for the random variables are shown below (1 kip = 4.448 kN; 1 ft = 0.3048 m).

$$\mu_W = 1 \text{ k/ft} \qquad V_W = 10\% \qquad \Rightarrow \sigma_W = \mu_W V_W = 0.1 \text{ k/ft}$$

$$\mu_P = 12 \text{ k} \qquad V_P = 15\% \qquad \Rightarrow \sigma_P = \mu_P V_P = 1.8 \text{ k}$$

$$\mu_{M_R} = 200 \text{ k-ft} \qquad V_{M_R} = 12\% \qquad \Rightarrow \sigma_{M_R} = \mu_{M_R} V_{M_R} = 24 \text{ k-ft}$$

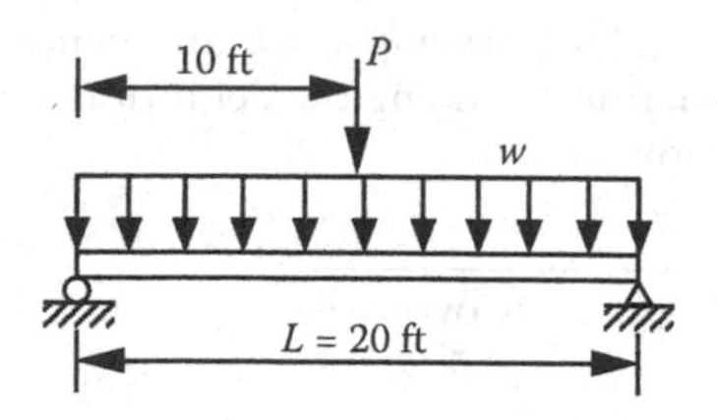

Figure 3.1 Beam considered in Example Problem 3.2.

The maximum moment due to the distributed dead load is

$$M_D = \frac{wL^2}{8} = 50w$$

The moment M_D is a normal random variable since it is a linear function of the normal random variable w. Using Equations 3.2 and 3.4, the mean value is 50 μ_W and the variance is $(50\,\sigma_W)^2$. The maximum moment due to the concentrated live load is

$$M_L = \frac{PL}{4} = 5P$$

and it is also a normal random variable because it is a linear function of the normal random variable P. Its mean value is 5 μ_P and its variance is $(5\,\sigma_P)^2$. The total maximum moment is thus $M_D + M_L$, which represents the total load effect (demand) on the beam. We can define a performance function in terms of capacity and demand as

$$Y = M_R - (M_D + M_L) = M_R - M_D - M_L$$

Thus, the beam is satisfactory if $M_R \geq (M_D + M_L)$ and unsatisfactory if $M_R < (M_D + M_L)$. Since M_R, M_D, and M_L are all uncorrelated normal random variables, the variable Y is also normal. Therefore,

$$\begin{aligned}\mu_Y = \sum a_i \mu_{M_i} &= (1)\mu_{M_R} + (-1)\mu_{M_D} + (-1)\mu_{M_L} \\ &= 200 - 50\mu_W - 5\mu_P \\ &= 200 - 50(1) - 5(12) \\ &= 90 \text{ (k-ft.)}\end{aligned}$$

$$\sigma_Y^2 = \sum a_i^2 \sigma_{M_i}^2 = (1)^2 \sigma_{M_R}^2 + (-1)^2 \sigma_{M_D}^2 + (-1)^2 \sigma_{M_L}^2$$
$$= (24)^2 + (50\sigma_W)^2 + (5\sigma_P)^2$$
$$= (24)^2 + (5)^2 + (9)^2$$
$$= 682 \text{ (k-ft.)}^2$$

$$\sigma_Y = \sqrt{\sigma_Y^2} = 26.1 \text{ (k-ft.)}$$

$$V_Y = \frac{\sigma_Y}{\mu_Y} = \frac{26.1}{90} = 0.29 = 29\%$$

$$\beta = \frac{1}{V_Y} = \frac{1}{0.29} = 3.45$$

To calculate the probability of failure, we need to understand the meaning of failure for this problem. Failure will occur when the demand exceeds the capacity. Referring to the performance function above, this can be expressed mathematically as the event when $Y < 0$. Hence,

$$P(failure) = P(Y < 0).$$

Since Y is a normal random variable, we can calculate this probability directly using the procedures described in Chapter 2 and Table B.1 in Appendix B:

$$P(Y < 0) = \Phi\left(\frac{0 - \mu_Y}{\sigma_Y}\right) = \Phi(-\beta) = \Phi(-3.45) = 0.280 \times 10^{-3} = 0.000280$$

Figure 3.2 schematically shows the interpretation of the probability of failure in terms of the probability density function for Y.

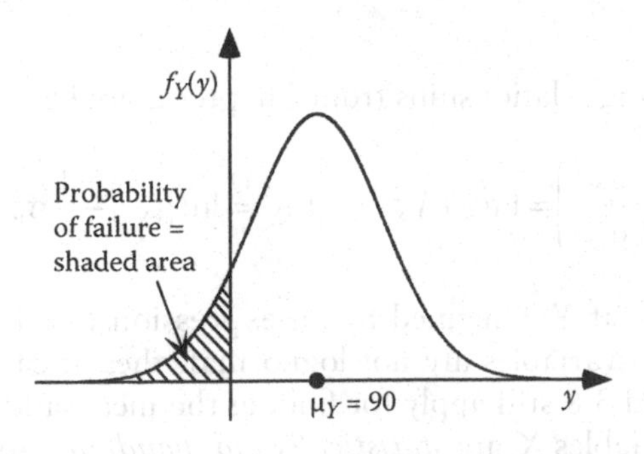

Figure 3.2 Graphical interpretation of the probability of failure for Example 3.2.

3.3 PRODUCT OF LOGNORMAL RANDOM VARIABLES

In the previous section, it was shown that the sum of uncorrelated normal random variables is also a normal random variable. In this section, we will show that another useful result occurs for the product or quotient of *statistically independent* lognormal random variables.

Let Y be a function involving the products or quotients of several random variables X_i. For example,

$$Y = K\frac{X_1 X_3}{X_2} \tag{3.5}$$

where K is a constant. Assume that these random variables are statistically independent, lognormal random variables. If we take the natural logarithm of both sides of Equation 3.5, we get

$$\begin{aligned} \ln Y &= \ln K + \ln X_1 + \ln X_3 - \ln X_2 \\ &= (\text{constant}) + \sum (\pm 1)\ (\text{normal random variables}) \end{aligned} \tag{3.6}$$

Note that Equation 3.6 is identical in form to Equation 3.1. Therefore, since Equation 3.6 represents the sum of normally distributed random variables $\ln X_i$, the quantity $\ln Y$ is a normally distributed random variable and hence Y is a *lognormally distributed random variable.*

By combining the results of Chapter 2 and Section 3.2, and generalizing Equation 3.6 for n random variables, it can be shown that the distribution parameters for the lognormal variable Y are as follows:

$$\mu_{\ln Y} = \ln K + \sum_{i=1}^{n} (\pm 1 \text{ as appropriate})\ \mu_{\ln X_i} \tag{3.7}$$

$$\sigma_{\ln Y}^2 = \sum_{i=1}^{n} \sigma_{\ln X_i}^2 \tag{3.8}$$

where the following relationships from Chapter 2 apply:

$$\sigma_{\ln X_i}^2 = \ln\left(1 + \frac{\sigma_{X_i}^2}{\mu_{X_i}^2}\right) = \ln\left(1 + V_{X_i}^2\right); \quad \mu_{\ln X_i} = \ln\left(\mu_{X_i}\right) - \frac{1}{2}\sigma_{\ln X_i}^2 \tag{3.9}$$

As a final note, if Y is defined by an expression like Equation 3.5 but all of the random variables are not lognormal, then it can be shown that Equations 3.7 and 3.8 still apply for finding the mean and variance of $\ln Y$ if the random variables X_i are *statistically independent*. However, you *cannot* say that the probability distribution of Y is lognormal.

Example 3.3

Consider a steel beam for which the cross section is classified as a compact section. This implies that its moment-carrying capacity is the plastic moment calculated as $M_p = F_y Z$, where Z = plastic section modulus and F_y is the yield stress.

Let the total load effect (demand) be denoted by Q, which is the maximum moment demand on the beam due to the applied loading. Assume F_y, Z, and Q are statistically independent lognormal random variables with means and coefficients of variation defined as follows (1 ksi = 6.895 MPa; 1 in = 25.40 mm; 1 k-ft = 1.356 kN-m):

Variable	*Mean*	*Coefficient of variation*
F_y	40 ksi	10%
Z	54 in^3	5%
Q	120 k-ft	12%

Calculate the probability of failure.

In Examples 3.1 and 3.2, the concept of a performance function was introduced, and it was formulated as a linear combination of random variables. Sometimes, it is more convenient to formulate a performance function as the product or quotient of random variables. For instance, consider the ratio M_p/Q in this example. M_p is a measure of the capacity of the member, and Q is the measure of demand. Failure occurs when capacity is less than demand, and this corresponds to the condition when the ratio M_p/Q is less than 1. If we define Y as

$$Y = \frac{M_p}{Q} = \frac{F_y Z}{Q}$$

then the failure probability corresponds to

$$P(failure) = P(Y < 1).$$

Since Y is a combination of products and quotients of independent lognormal random variables with $K = 1$, Y is also lognormal with

$$\begin{aligned}
\mu_{\ln Y} &= \ln K + \sum_{i=1}^{n} (\pm 1 \text{ as appropriate}) \mu_{\ln X_i} \\
&= \ln(1) + (1)\mu_{\ln F_y} + (1)\mu_{\ln Z} + (-1)\mu_{\ln Q} \\
&= \mu_{\ln F_y} + \mu_{\ln Z} - \mu_{\ln Q}
\end{aligned}$$

$$\sigma_{\ln Y}^2 = \sum_{i=1}^{n} \sigma_{\ln X_i}^2 = \sigma_{\ln F_y}^2 + \sigma_{\ln Z}^2 + \sigma_{\ln Q}^2$$

To use these relationships, we must first determine the values of the parameters using Equation 3.9.

$$\sigma_{\ln F_y}^2 = \ln\left(1+V_{F_{yi}}^2\right) = \ln\left[1+(0.1)^2\right] = 9.95\times 10^{-3}$$

$$\mu_{\ln F_y} = \ln\left(\mu_{F_y}\right) - \frac{1}{2}\sigma_{\ln F_y}^2 = \ln(40) - 0.5\left(9.95\times 10^{-3}\right) = 3.68$$

$$\sigma_{\ln Z}^2 = \ln\left(1+V_Z^2\right) = \ln\left[1+(0.05)^2\right] = 2.50\times 10^{-3}$$

$$\mu_{\ln Z} = \ln\left(\mu_Z\right) - \frac{1}{2}\sigma_{\ln Z}^2 = \ln(54) - 0.5\left(2.50\times 10^{-3}\right) = 3.99$$

$$\sigma_{\ln Q}^2 = \ln\left(1+V_Q^2\right) = \ln\left[1+(0.12)^2\right] = 1.43\times 10^{-2}$$

$$\mu_{\ln Q} = \ln\left(\mu_Q\right) - \frac{1}{2}\sigma_{\ln Q}^2 = \ln(120\times 12) - 0.5\left(1.43\times 10^{-2}\right) = 7.27$$

With this information, we can now calculate $\mu_{\ln Y}$ and $\sigma_{\ln Y}^2$ as follows:

$$\begin{aligned}\mu_{\ln Y} &= \mu_{\ln F_y} + \mu_{\ln Z} - \mu_{\ln Q}\\ &= 3.68 + 3.99 - 7.27 = 0.4\end{aligned}$$

$$\begin{aligned}\sigma_{\ln Y}^2 &= \sigma_{\ln F_y}^2 + \sigma_{\ln Z}^2 + \sigma_{\ln Q}^2\\ &= 9.95\times 10^{-3} + 2.50\times 10^{-3} + 1.43\times 10^{-2} = 0.0268\end{aligned}$$

$$\sigma_{\ln Y} = \sqrt{0.0268} = 0.164$$

Since Y is lognormal, the probability of $Y < 1$ can be determined using the standard normal CDF as follows:

$$P(Y<1) = \Phi\left(\frac{\ln(1) - \mu_{\ln Y}}{\sigma_{\ln Y}}\right) = \Phi\left(\frac{0-0.4}{0.164}\right) = \Phi(-2.4) = 0.82\times 10^{-2} = 0.0082$$

A graphical interpretation of the probability of failure is shown in Figure 3.3.

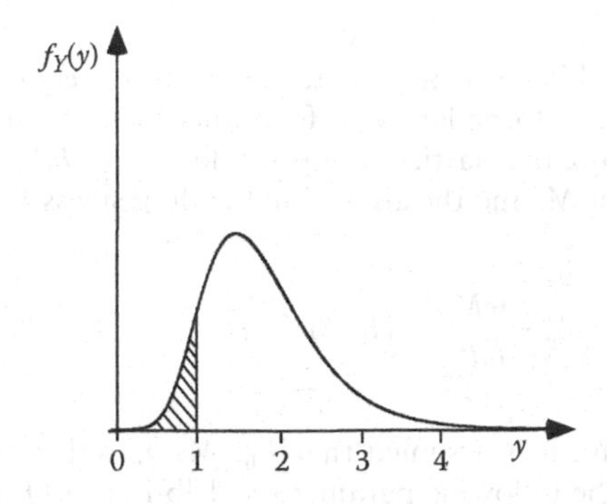

Figure 3.3 Graphical interpretation of the probability of failure in Example 3.3.

3.4 NONLINEAR FUNCTION OF RANDOM VARIABLES

In the previous sections, we discussed how to estimate the mean and variance of some relatively simple functional forms. In this section, we consider an approximate way to estimate the mean and variance of a general function.

Let the performance function Y be a general nonlinear function of the random variables X_i, $i = 1, 2, \ldots, n$. Mathematically,

$$Y = f(X_1, X_2, \ldots, X_n) \tag{3.10}$$

To calculate the mean and variance of Y, we can use a Taylor series expansion of Y to linearize the performance function, and then we can estimate the mean and variance of the linearized version of the function. The Taylor series expansion is

$$Y = f\left(x_1^*, x_2^*, \ldots, x_n^*\right) + \sum_{i=1}^{n} \left(X_i - x_i^*\right) \frac{\partial f}{\partial X_i}\bigg|_{\text{evaluated at } \left(x_1^*, x_2^*, \ldots, x_n^*\right)} + \text{(higher-order terms)} \tag{3.11}$$

where the x_i^* values (deterministic values) are "design point values" of the random variables X_i, that is, the values about which the function Y is linearized. The higher-order terms are neglected when using the linearized version of Y. The choice of the design point values of the random vector $\{X\}$ is very important in structural reliability analysis. For the present, we assume that the design point values are the mean values of the random variables.

Example 3.4

Consider a wood beam with a rectangular cross section as shown in Figure 3.4. We can formulate a performance function for beam bending by considering the elastic section modulus S (= $bd^2/6$), the applied bending moment M, and the allowable bending stress F_b as follows:

$$Y = F_b - \frac{M}{S} = F_b - \frac{6M}{bd^2} = f(F_b, M, b, d)$$

For this problem, it is assumed that F_b, M, b, and d are all random variables with the following parameters (1 lb-in = 0.1130 Nm; 1 psi = 6.895 kPa; 1 in = 25.40 mm):

$$\begin{aligned}
&\mu_M = 100{,}000 \text{ lb-in} && V_M = 0.12 && \Rightarrow \sigma_M = V_M\mu_M = 12{,}000 \text{ lb-in}\\
&\mu_{F_b} = 1600 \text{ psi} && V_{F_b} = 0.32 && \Rightarrow \sigma_{F_b} = V_{F_b}\mu_{F_b} = 512 \text{ psi}\\
&\mu_B = 5.6 \text{ in} && V_B = 0.04 && \Rightarrow \sigma_B = V_B\mu_B = 0.224 \text{ in}\\
&\mu_D = 11.4 \text{ in} && V_D = 0.03 && \Rightarrow \sigma_D = V_D\mu_D = 0.342 \text{ in}
\end{aligned}$$

The goal is to calculate the mean and variance of Y and the reliability index as defined in Example 3.1.

Since Y is a nonlinear function, we must linearize the function about the design point values. We will use the mean values as the design point values in this example. Based on Equation 3.11, the linearized form of Y will look like

$$Y \approx \left(\mu_{F_b} - \frac{6\mu_M}{\mu_B(\mu_D)^2}\right) + (F_b - \mu_{F_b})\frac{\partial f}{\partial F_b}\bigg|_* + (M - \mu_M)\frac{\partial f}{\partial M}\bigg|_*$$
$$+ (b - \mu_B)\frac{\partial f}{\partial B}\bigg|_* + (d - \mu_D)\frac{\partial f}{\partial D}\bigg|_*$$

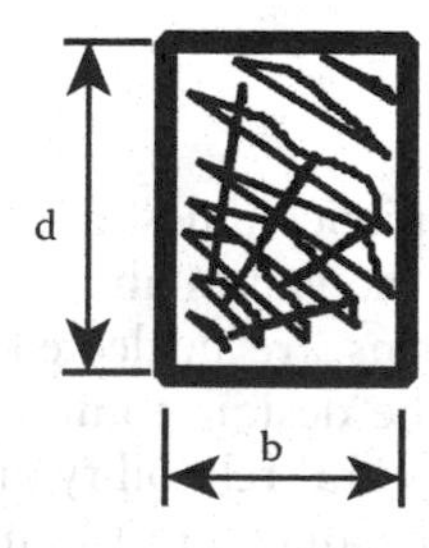

Figure 3.4 Cross section of beam considered in Example 3.4.

where $|_*$ denotes that the partial derivatives are evaluated at the design point values (mean values) of the random variables. The partial derivatives are

$$\frac{\partial f}{\partial F_b} = 1 \qquad \Rightarrow \qquad \left.\frac{\partial f}{\partial F_b}\right|_* = 1$$

$$\frac{\partial f}{\partial M} = -\frac{6}{bd^2} \qquad \Rightarrow \qquad \left.\frac{\partial f}{\partial M}\right|_* = -\frac{6}{\mu_B \mu_D^2}$$

$$\frac{\partial f}{\partial B} = \frac{6M}{b^2 d^2} \qquad \Rightarrow \qquad \left.\frac{\partial f}{\partial B}\right|_* = \frac{6\,\mu_M}{\mu_B^2 \mu_D^2}$$

$$\frac{\partial f}{\partial D} = \frac{12M}{bd^3} \qquad \Rightarrow \qquad \left.\frac{\partial f}{\partial D}\right|_* = \frac{12\,\mu_M}{\mu_B \mu_D^3}$$

Substituting these derivatives into the linearized equation, plugging in the mean values of the variables, and rearranging give the following linearized form for Y:

$$Y = F_b - 0.008244M + 147.2b + 144.6d - 2473$$

Since this is a linear function of the random variables, we can apply the results of Section 3.1 to calculate the mean and variance of Y. Assuming the variables are all uncorrelated, we can use Equations 3.2 and 3.4 to get

$$\begin{aligned}\mu_Y &= \mu_{F_b} - 0.008244\,\mu_M + 147.2\,\mu_B + 144.6\,\mu_D - 2473 \\ &= 1600 - 0.008244(100{,}000) + 147.2(5.6) + 144.6(11.4) - 2473 \\ &= 775.4 \text{ (psi)}\end{aligned}$$

$$\begin{aligned}\sigma_Y^2 &= (1)^2 \sigma_{F_b}^2 + (-0.00824)^2 \sigma_M^2 + (147.2)^2 \sigma_B^2 + (144.26)^2 \sigma_D^2 \\ &= (1)^2 (512)^2 + (-0.008244)^2 (12{,}000)^2 + (147.2)^2 (0.224)^2 + (144.6)^2 (0.342)^2 \\ &= 275{,}464 \text{ psi}^2\end{aligned}$$

$$\sigma_Y = \sqrt{275{,}464} = 524.8 \text{ psi}$$

Thus, referring to Example 3.1, the reliability index, β, is

$$\beta = \frac{\mu_Y}{\sigma_Y} = \frac{775.4}{524.8} = 1.48$$

3.5 CENTRAL LIMIT THEOREM

3.5.1 Sum of random variables

Let the function Y be the sum of n random variables X_i ($i = 1, 2, ..., n$). Furthermore, assume the X_i are statistically independent, and their probability distributions are arbitrary. The *Central Limit Theorem* states that, as n approaches infinity, the sum of these independent random variables approaches a normal probability distribution if none of the random variables tends to dominate the sum. Thus, if we have a function defined as the sum of a large number of random variables, then we would expect the sum to be approximately normally distributed.

The sum of variables is often used to model the total load on a structure. Therefore, the total load can be approximated as a normal variable.

Example 3.5

Let Q be the total load representing the effects of dead load D, live load L, and snow load S. Mathematically, Q is defined as

$$Q = D + L + S$$

Then, the Central Limit Theorem can be used to say that Q is *approximately* normal even if D, L, and S are not normal. The mean and variance of Q are, using Equations 3.2 and 3.4,

$$\mu_Q = \mu_D + \mu_L + \mu_S$$

$$\sigma_Q^2 = \sigma_D^2 + \sigma_L^2 + \sigma_S^2$$

3.5.2 Product of random variables

If we have a product of independent variables, then we can do a transformation that allows us to apply the Central Limit Theorem discussed above. Let Y be a product of statistically independent random variables of the form

$$Y = X_1 X_2 ... X_n \tag{3.12}$$

The equation could also include quotients of random variables. If we take the natural logarithm of both sides of Equation 3.12, we get

$$\ln Y = \ln X_1 + \ln X_2 + ... + \ln X_n \tag{3.13}$$

This sum can be interpreted as the sum of a series of random variables $\ln X_i$; hence, using the Central Limit Theorem, we can conclude that $\ln Y$

approaches a normal distribution as the number of random variables approaches infinity. If $\ln Y$ is normal, then Y must be lognormal. Thus, if we have a product (or quotient) of many independent random variables, the product (or quotient) approaches a lognormal distribution.

The product of variables is often used to model the resistance (or capacity) of a structure or structural element. Therefore, the resistance can be approximated as a lognormal variable.

PROBLEMS

3.1 Derive Equation 3.2.

3.2 Derive Equation 3.3.

3.3 Suppose the function Y is defined as

$$Y = 3X_1 - 7X_2$$

and X_1 and X_2 are both random variables. X_1 is a uniformly distributed random variable between 2 and 4, and X_2 is normally distributed with a mean value of 0.6 and a standard deviation of 0.1.

A. Calculate the mean and standard deviation of Y assuming that X_1 and X_2 are uncorrelated.

B. If the variables are correlated, how will this affect the mean and standard deviation of Y?

3.4 The load on a cantilever beam of length L consists of a load P at the tip (free end) and a distributed load w acting over the entire length. Both P and w are random variables. The length is assumed to be deterministic, and its value is 6 ft. The statistical parameters of P and w are as follows (1 lb = 4.448 N; 1 ft = 0.3048 m):

$$\mu_P = 4000 \text{ lb} \qquad \mu_w = 50 \text{ lb/ft}$$

$$V_P = 10\% \qquad V_w = 10\%$$

A. Calculate the mean and standard deviation of the internal bending moment at the fixed end.

B. Calculate the mean and standard deviation of the internal shear force at the fixed end.

C. If both P and w are normal random variables, what is the probability that the bending moment at the fixed end will exceed 29 k-ft (1 k-ft = 1.356 kN-m)?

D. If both P and w are normal random variables, what is the probability that the shear force at the fixed end will exceed 5000 lb?

3.5 A simply supported steel beam (compact section) is subjected to a point load P in the middle. The length of the beam is 18 ft. The plastic modulus Z of the beam about the axis of bending is 80 in^3. The load P is a random variable with a mean value of 10.2 kips and a coefficient of variation of 11%. The yield stress of the steel, F_y, is also a random variable with a mean value of 40.3 ksi and a coefficient of variation of 11.5% (1 ft = 0.3048 m; 1 in = 25.40 mm; 1 kip = 4.448 kN; 1 ksi = 6.895 MPa).

A. Formulate a performance function Y that can be used to describe the failure condition of the beam. Consider bending only. [Hint: Consider Y = (capacity) – (demand). Failure corresponds to the case when $Y < 0$.]
B. Formulate an alternative performance function Y involving the product and quotient of random variables. Consider bending only. (Hint: see Example 3.3.)
C. If P and F_y are both normal random variables, calculate the probability of failure using the results of part A.
D. If P and F_y are both lognormal random variables, calculate the probability of failure using the results of part B.

3.6 Consider a three-span continuous beam. All supports are pinned supports. Each span has a length of L. The beam has a modulus of elasticity E and a moment of inertia I. All spans are subjected to a uniformly distributed load w. The maximum deflection of the beam occurs in the outer spans and is equal to

$$0.0069\frac{wL^4}{EI}$$

Your job is to evaluate the probability that the deflection will exceed the code-specified limit of $L/360$ given the following information:

- $L = 5$ m (deterministic).
- w is lognormal with a mean value of 10 kN/m and a coefficient of variation of 0.4.
- E is lognormal with a mean value of 2×10^7 kN/m^2 and a coefficient of variation of 0.25.
- I is lognormal with a mean value of 8×10^{-4} m^4 and a standard deviation of 1.5×10^{-4} m^4.

Calculate the probability of failing the deflection criterion.

3.7 Consider the performance function

$$Y = 3X_1 - 2X_2$$

where X_1 and X_2 are both normally distributed random variables with

$$\mu_{X_1} = 16.6 \qquad \mu_{X_2} = 18.8$$
$$\sigma_{X_1} = 2.45 \qquad \sigma_{X_2} = 2.83$$

The two variables are correlated, and the covariance is equal to 2.0. Determine the probability of failure if failure is defined as the state when $Y < 0$.

3.8 The resistance (or capacity) R of a member is to be modeled using

$$R = R_n MPF$$

where R_n is the nominal value of the capacity determined using code procedures and M, P, and F are random variables that account for various uncertainties in the capacity. If M, P, and F are all lognormal random variables, determine the mean and variance of R in terms of the means and variances of M, P, and F.

Chapter 4

Simulation techniques

Several techniques may be used to solve structural reliability problems. In this chapter, simulation techniques are presented as one possible way to solve such problems. The basic idea behind simulation is, as the name implies, to numerically simulate some phenomenon and then observe the number of times some event of interest occurs. The basic concept behind simulation is relatively straightforward, but the procedure can become computationally intensive.

4.1 MONTE CARLO METHODS

4.1.1 Basic concept

Assume that we have information from N tests, and assume that we put all N test results in a bag as shown in Figure 4.1. Now suppose that we need a sample of n test results. Instead of doing n additional tests, we could randomly select n of the N test results from the bag. In Figure 4.1, this sampling technique is referred to as the "special technique."

The Monte Carlo method is a special technique that we can use to generate some results numerically without actually doing any physical testing. We can use results of previous tests (or other information) to establish the probability distributions of the important parameters in our problem. Then, we use this distribution information to generate samples of numerical data. To illustrate the basic idea, consider the following example.

Example 4.1

Consider a series of actual tests of concrete cylinders to determine the compressive strength f'_c. Assume that a relative frequency histogram has been drawn using the data, and a lognormal probability distribution as shown in Figure 4.2 appears to fit the data reasonably well.

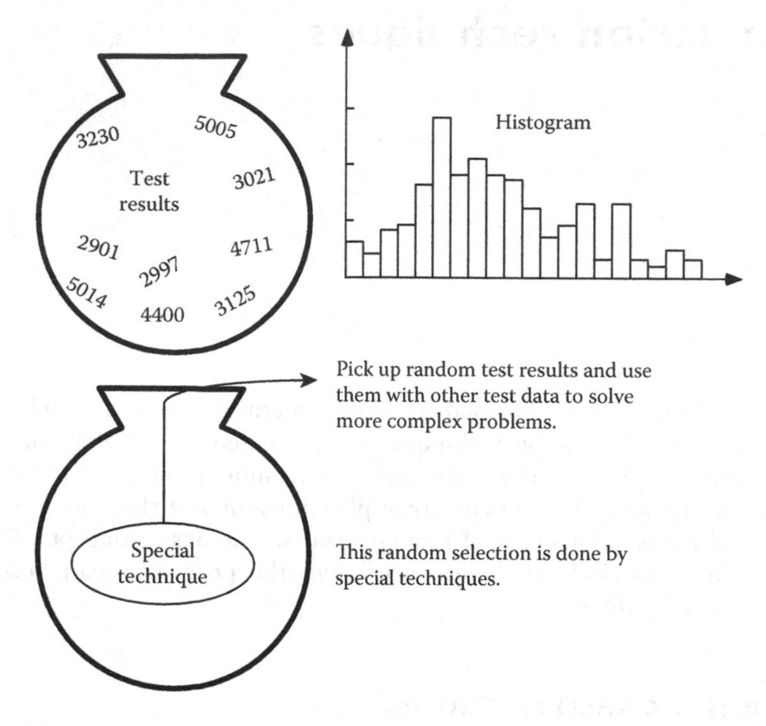

Figure 4.1 Schematic of the Monte Carlo method.

Now consider a concrete column. The compressive load-carrying capacity is $0.85f_c'A_c$ where A_c is the cross-section area of the column and is assumed to be deterministic. Assume that the applied load, Q, is normally distributed with a mean value μ_Q and a coefficient of variation V_Q. What is probability of failure, p_F?

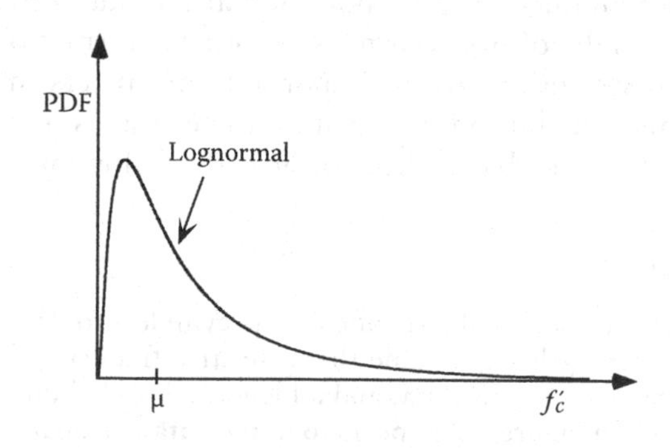

Figure 4.2 Lognormal probability density function (PDF) for Example 4.1.

The performance function for this case could be written as

$$Y = R - Q$$

where $R = 0.85f_c'A_c$. The probability of failure is the probability that $R < Q$ or

$$p_F = P(Y < 0) = P(R - Q < 0)$$

Unfortunately, in this case, the distribution of R is not normal (it is lognormal) even though Q is normal. Therefore, although we can calculate a mean and variance using the methods presented in Chapter 3, we do not have a closed-form solution for the probability distribution of Y. However, the problem can be solved using Monte Carlo simulation. The basic procedure is as follows:

1. Randomly generate a value of f_c' (using the probability distribution information given above) and calculate $R = 0.85f_c'A_c$.
2. Randomly generate a value of Q using its probability distribution.
3. Calculate $Y = R - Q$.
4. Store the calculated value of Y.
5. Repeat steps 1–4 until a sufficient number of Y values have been generated.
6. Plot the simulated data on normal probability paper (see Chapter 2) and use the graph to estimate the probability of failure. Alternatively, if a sufficient number of simulated values are available, estimate the probability of failure as

$$\bar{P} = \frac{\text{number of times that } Y < 0}{\text{total number of simulated } Y \text{ values}}$$

We have skipped some details, but the basic procedure is just like this. The details will be discussed in Sections 4.1.2 through 4.1.6.

The Monte Carlo method is often applied in three situations:

1. It is used to solve complex problems for which closed-form solutions are either not possible or extremely difficult. For example, probabilistic problems involving complicated nonlinear finite element models can be solved by Monte Carlo simulation provided that the necessary computing power is available and the required input information is known.
2. It is used to solve complex problems that can be solved (at least approximately) in closed form if many simplifying assumptions are made. By using Monte Carlo simulation, the "original" problem can be studied without these assumptions, and more realistic results can be obtained.
3. It is used to check the results of other solution techniques.

This chapter provides only a brief introduction to the topic of simulation. For additional details and applications, see Marek et al. (1996), Rubinstein (1981), Ross (1997), Ayyub and McCuen (2002), and Ang and Tang (1984).

4.1.2 Generation of uniformly distributed random numbers

The basis of all Monte Carlo simulation procedures is the generation of random numbers that are uniformly distributed between 0 and 1. Tables of randomly generated numbers are available (e.g., RAND Corporation, 1955), as are computer subroutines, and many popular mathematical programs have such subroutines built in. Table 4.1 is an example of a table of uniform random variables that was generated using a standard spreadsheet program.

Table 4.1 Simulated values of a uniformly distributed random variable (values between 0 and 1)

0.050203	0.269082	0.442000	0.390912	0.084078	0.597430	0.249519	0.892361
0.619129	0.472640	0.833705	0.876064	0.821741	0.149907	0.653035	0.908841
0.872402	0.422864	0.412275	0.462844	0.444990	0.774895	0.345225	0.834681
0.376568	0.467299	0.145451	0.926969	0.337626	0.648000	0.323649	0.656117
0.139927	0.415784	0.849178	0.307840	0.059633	0.498886	0.433912	0.320231
0.318491	0.523667	0.598193	0.005036	0.132786	0.892575	0.835353	0.666829
0.987671	0.243629	0.561388	0.414869	0.378796	0.301706	0.328349	0.937925
0.033265	0.741569	0.169408	0.118229	0.195227	0.515915	0.575213	0.709037
0.234626	0.408673	0.967040	0.398450	0.730552	0.411115	0.703421	0.591021
0.623157	0.021790	0.864834	0.319895	0.726890	0.023835	0.073214	0.012818
0.957884	0.547472	0.332286	0.970153	0.762535	0.055788	0.118198	0.082675
0.518906	0.749779	0.849239	0.173711	0.586932	0.554094	0.767510	0.751549
0.442305	0.681600	0.834803	0.406201	0.803613	0.162908	0.224067	0.956908
0.445845	0.140538	0.885769	0.760491	0.132450	0.866573	0.869015	0.663656
0.834284	0.888607	0.359783	0.297708	0.564196	0.141392	0.706259	0.199316
0.811213	0.105136	0.607227	0.119755	0.449202	0.731407	0.930631	0.250893
0.935728	0.067202	0.705435	0.150792	0.448347	0.502579	0.392346	0.740989
0.450423	0.864772	0.841975	0.671010	0.403912	0.711509	0.072970	0.974548
0.579058	0.363628	0.781091	0.455458	0.431471	0.579272	0.923032	0.748009
0.662648	0.863948	0.491226	0.011902	0.905759	0.806421	0.929014	0.749077
0.039918	0.858242	0.557054	0.278726	0.593005	0.191778	0.047029	0.050478
0.414075	0.432234	0.934263	0.173040	0.358684	0.660817	0.372265	0.925138
0.103214	0.412091	0.087985	0.395398	0.606067	0.401349	0.596393	0.385754
0.112308	0.749199	0.242988	0.543260	0.649922	0.168065	0.111393	0.986175
0.821833	0.260933	0.012574	0.562700	0.402142	0.109256	0.144383	0.030641

Once we have some realizations (u) of a uniformly distributed random number (U) between 0 and 1, we can generate realizations (x) of a uniformly distributed random number (X) between any two values a and b using the following formula:

$$x = a + (b - a)u \tag{4.1}$$

We can also generate sample values (i) for a uniformly distributed random *integer* (I) between two integer values a and b (including the values a and b) using the following formula:

$$i = a + \text{TRUNC}[(b - a + 1)u] \tag{4.2}$$

where TRUNC[ ] denotes a function that truncates its argument (i.e., removes the fractional part and returns the integer part of a real number).

Before leaving the topic of generating uniform random variables, two comments are in order. First, many random number generators require the user to input a "seed" value. This number is an integer that is used by the routine to start the simulation algorithm. By choosing a different seed, you can generate a different set of uniformly distributed numbers. In general, if you use the same seed over and over again, you will generate the same set of uniformly distributed random numbers over and over again. Second, the built-in generators found in many software packages should be used with caution. Some random number generation algorithms work better than others. A discussion of this issue is beyond the scope of this text. For a concise summary of possible problems and a listing of various random number generation algorithms, see Press et al. (1992, Chapter 7).

4.1.3 Generation of standard normal random numbers

Since the normal probability distribution plays such an important role in structural reliability analysis, the capability to simulate normally distributed random variables is important. To begin, consider a standard normal distribution. To generate a set of standard normal random numbers z_1, z_2, …, z_n, we first need to generate a corresponding set of uniformly distributed random variables u_1, u_2, …, u_n, between 0 and 1. Then, for each u_i, we can generate a value z_i using

$$z_i = \Phi^{-1}(u_i) \tag{4.3}$$

where Φ^{-1} is the inverse of the standard normal cumulative distribution function (CDF). Figure 4.3 shows this relationship graphically.

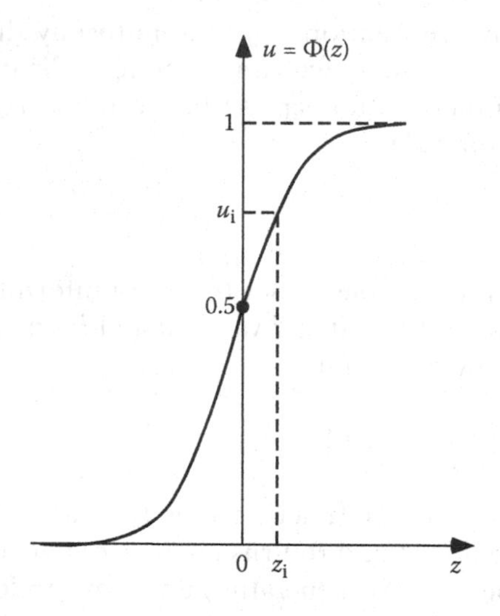

Figure 4.3 Generation of standard normal random variables.

Equation 2.43 in Chapter 2 can be used to estimate the inverse function in Equation 4.3. Many computer programs have a standard normal random number generator built in.

4.1.4 Generation of normal random numbers

In the previous section, we discussed how to generate sample values of a normally distributed random variable following a *standard* normal distribution. What if we need to generate sample values from an arbitrary normal distribution? To do this, we use the relationship discussed in Chapter 2. Assume we have a normally distributed random variable X with mean μ_X and standard deviation σ_X. The basic relationship between X and the standard normal variate Z is

$$X = \mu_X + Z\sigma_X \tag{4.4}$$

Thus, given a sample value z_i generated using the approach discussed in Section 4.1.3, the corresponding x_i value can be calculated using

$$x_i = \mu_X + z_i\sigma_X \tag{4.5}$$

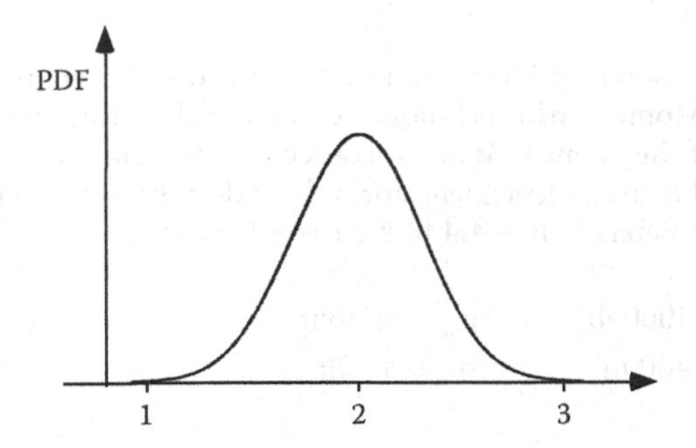

Figure 4.4 Probability density function used in Example Problem 4.2.

Example 4.2

Assume that the dead load D on a structure is a normally distributed random variable. The mean μ_D is 2.0 k/ft, and coefficient of variation V_D is 10%. The PDF for dead load is shown in Figure 4.4. Generate 10 values of the random variable D (1 kip = 4.448 kN; 1 ft = 0.3048 m).

To begin, we need to generate some uniformly distributed random variables. For this example, the first 10 values in column 1 of Table 4.1 are used. Then, Equation 4.3 is used to generate corresponding z_i values. Finally, Equation 4.5 is used to convert the z_i values to D_i values. Note that $\sigma_D = V_D\mu_D = 0.2$ k/ft. The numerical results are tabulated in Table 4.2.

Table 4.2 Summary of calculations for Example 4.2

u_i	z_i	D_i
0.050203	−1.64289	1.67
0.619129	0.303194	2.06
0.872402	1.13782	2.23
0.376568	−0.314507	1.94
0.139927	−1.08065	1.78
0.318491	−0.471923	1.91
0.987671	2.24672	2.45
0.033265	−1.83483	1.63
0.234626	−0.723697	1.86
0.623157	0.313783	2.06

Example 4.3

Consider a wood cantilever beam such as the one shown in Figure 4.5. Using the Monte Carlo technique, calculate the mean and standard deviation of the moment M at a distance of 6 ft from the free end. The loads P and w are independent normal random variables with the following parameters (1 lb = 4.448 N; 1 ft = 0.3048 m):

$$\mu_P = 4000 \text{ lb} \qquad \mu_W = 50 \text{ lb/ft}$$
$$\sigma_P = 400 \text{ lb} \qquad \sigma_W = 5 \text{ lb/ft}$$

Using statics, we know that the bending moment M at the location of interest is

$$M = 6P - 18w$$

Since P and w are both independent normal random variables and M is a linear function of P and w, we can say that M is also a normal random variable. The mean and standard deviation of M can be calculated *exactly* using Equations 3.2 and 3.4 as follows (1 k-ft = 1.356 kN m):

$$\mu_M = 6\mu_P - 18\mu_W = 23{,}100 \text{ lb-ft} = 23.1 \text{ k-ft}$$

$$\sigma_M = \sqrt{(6\sigma_P)^2 + (18\sigma_W)^2} = 2400 \text{ lb-ft} = 2.40 \text{ k-ft}$$

As we will demonstrate, we can also use Monte Carlo simulation. After the simulation, we will be able to see how the simulation results compare with the theoretically correct results.

For this example, we will calculate five values of the bending moment M using Monte Carlo simulation. Therefore, we will need to simulate five values of the load P and five values of the load w. This will require us to determine 10 values (2 simulated variables times 5 values of each variable) of a uniformly distributed random variable between 0 and 1. The uniform random numbers for this example are taken from Table 4.1. We will take the first 10 values in column 1. The first 5 will be used to calculate 5 values of P, and the last 5 will be used to calculate 5 values of w. The simulated values of P and w are calculated

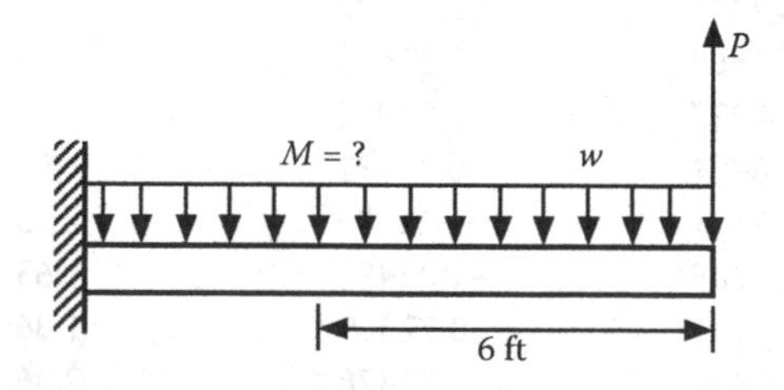

Figure 4.5 Example of a wood cantilever beam.

Table 4.3 Tabulation of calculations for Example 4.3

u_i	z_i	$P_i = 4000 + z_i\,(400)$
0.050203	−1.64289	3343
0.619129	0.303194	4121
0.872402	1.13782	4455
0.376568	−0.314507	3874
0.139927	−1.08065	3568
u_i	z_i	$w_i = 50 + z_i\,(5)$
0.318491	−0.471923	47.64
0.987671	2.24672	61.23
0.033265	−1.83483	40.83
0.234626	−0.723697	46.38
0.623157	0.313783	51.57

using Equation 4.5. The calculations are summarized in Table 4.3 for convenience.

Next, five data sets $\{P_i, w_i\}$ are formed and used to generate five values for the moment M using

$$M_i = 6P_i - 18w_i$$

The following results are generated for M_i (in lb-ft):

19,200
23,620
25,990
22,410
20,480

Hence, the sample mean and sample standard deviation can be calculated using Equations 2.25 and 2.26 from Chapter 2 as follows:

$$\bar{M} = \frac{1}{5}\sum_{i=1}^{5} M_i = 22{,}340 \text{ lb-ft} = 22.34 \text{ k-ft}$$

$$s_M = \sqrt{\frac{\left(\sum_{i=1}^{5} M_i^2\right) - 5(\bar{M})^2}{5-1}} = 2659 \text{ lb-ft} = 2.659 \text{ k-ft}$$

Comparing these values with the theoretically exact values obtained earlier, we see that the values are comparable but not as accurate as we might like. The accuracy of the Monte Carlo simulation approach increases as the number of sample values increases. If more simulated values of M were generated, we would expect our estimates of mean and standard deviation to improve. To illustrate this, the preceding calculation was repeated with the following input:

- The first 50 values of uniform random variables found in the first two columns of Table 4.1 were used to simulate values of P.
- The next 50 values of uniform random variables found in columns 3 and 4 of Table 4.1 were used to simulate values of w.
- Using the simulated values of P and w, 50 values of M were generated, and the mean and standard deviation were calculated.

Using these 50 simulated values of M, the sample mean and sample standard deviation were found to be 22.99 k-ft and 2.332 k-ft, respectively. These values are much closer to the theoretically correct values of 23.1 k-ft and 2.40 k-ft.

4.1.5 Generation of lognormal random numbers

Let X be a lognormal random variable with mean μ_X and standard deviation σ_X. To generate a sample value x_i, we begin as before by generating a sample value u_i of a uniformly distributed random number between 0 and 1. Then, a sample value z_i from a standard normal distribution is calculated using Equation 4.3. Finally, using the relationship between normal and lognormal variables, we obtain x_i using

$$x_i = \exp[\mu_{\ln X} + z_i \sigma_{\ln X}] \tag{4.6}$$

where

$$\begin{aligned} \sigma_{\ln X}^2 &= \ln\left(V_X^2 + 1\right) \\ &\approx V_X^2 \quad (\text{for } V_X < 0.20) \end{aligned} \tag{4.7}$$

$$\begin{aligned} \mu_{\ln X} &= \ln\left(\mu_X\right) - \frac{1}{2}\sigma_{\ln X}^2 \\ &\approx \ln\left(\mu_X\right) \quad (\text{for } V_X < 0.20) \end{aligned} \tag{4.8}$$

If the approximate relationships in Equations 4.7 and 4.8 are used, Equation 4.6 can be rewritten as

$$x_i = \mu_X \exp[z_i V_X] \tag{4.9}$$

4.1.6 General procedure for generating random numbers from an arbitrary distribution

In the preceding sections, we considered generation of sample values of random variables for some of the most common distributions used in structural reliability analysis. A general procedure can be formulated, which, theoretically, is applicable to any type of distribution function.

Consider a random variable X with a CDF $F_X(x)$. To generate sample values x_i for the random variable, the following steps can be taken:

1. Generate a sample value u_i for a uniformly distributed random variable between 0 and 1.
2. Calculate a sample value x_i from the following formula:

$$x_i = F_X^{-1}(u_i) \tag{4.10}$$

where F_X^{-1} is the inverse of F_X.

The above procedure is completely general. However, in some cases, it is difficult to determine a closed-form solution for the inverse CDF.

4.1.7 Accuracy of probability estimates

In Section 4.1.1, Example 4.1, we discussed how simulation results can be used to estimate a probability of failure. It is important to recognize that this estimate of probability is indeed only an estimate. However, the estimate improves as the number of simulations increases. In this section, we look at the relationship between the uncertainty in the estimated probability and the number of simulations.

The estimated probability, $\overline{P}$, is calculated as the ratio

$$\overline{P} = \frac{n}{N} \tag{4.11}$$

where N is the total number of simulations and n is the number of times (out of N simulations) that a particular criterion was achieved. For example, suppose 100 simulations are conducted to see how many times $Y < 0$ for Example 4.1. Of those 100 simulations, suppose 5 simulations result in $Y < 0$. Thus, $n = 5$ and $N = 100$; hence, our estimate of the probability $P(Y < 0)$ is 5/100 = 0.05. Now, suppose a completely new simulation is done with $N = 100$. This time, we get $n = 7$ and our estimate of $P(Y < 0)$ is 7/100 = 0.07. The point is that the estimated probability $\overline{P}$ is a "sample estimate." By that, we mean that the calculated value will vary from sample to sample. Therefore, the estimated probability itself can be treated as a random variable with its own mean, standard deviation, and/or coefficient of variation.

Let P_{true} be the theoretically correct probability that we are trying to estimate by calculating $\overline{P}$. It can be shown (e.g., Soong and Grigoriu, 1993) that the expected value, variance, and coefficient of variation of the estimated probability $\overline{P}$ are as follows:

$$E[\overline{P}] = P_{true}; \quad \sigma_{\overline{P}}^2 = \frac{1}{N}\left[P_{true}(1-P_{true})\right]; \quad V_{\overline{P}} = \sqrt{\frac{(1-P_{true})}{N(P_{true})}} \tag{4.12}$$

Observe that the "uncertainty" in the estimate of the probability decreases as the total number of simulations, N, increases. These relationships provide a way to determine how many simulations are required to estimate a probability and limit the uncertainty in the estimate.

Example 4.4

Suppose we want to be able to estimate probabilities as low as 10^{-2} and keep the coefficient of variation of our estimate at or below 10%. How many simulations are required?

By rearranging the expression for coefficient of variation in Equation 4.12, we can solve for the number of simulations N as

$$N = \frac{1-P_{true}}{V_{\overline{P}}^2(P_{true})} = \frac{1-10^{-2}}{(0.10)^2(10^{-2})} = 9900$$

To estimate relatively small probabilities using Monte Carlo simulation while controlling the uncertainty in the estimate, a large number of simulations is needed. The required sample size depends on the desired coefficient of variation and the relative magnitude of the probability to be estimated.

Example 4.5

Consider the function

$$Y = R - Q$$

where R and Q are both random variables. R is lognormally distributed with a mean value of 200 and a standard deviation of 20. Q follows an extreme Type I distribution with a mean value of 100 and a standard deviation of 12. The goal is to determine the probability of $Y < 0$ using simulation. The approach is as follows:

1. Determine the number (N) of desired simulated values. To keep the example simple, we will choose $N = 25$.
2. Generate $N = 25$ uniform random variables for simulating R, and another $N = 25$ uniform random variables for simulating Q.

For this example, we will use column 1 of Table 4.1 for simulating R and column 2 of the same table for simulating Q.

3. Generate 25 values of R and 25 values of Q. These will be referred to as r_i and q_i (i = 1, 2, ..., 25). To generate the r_i values, we can use Equation 4.6. Note that this requires us to determine $\mu_{\ln R}$ and $\sigma_{\ln R}$. Using Equations 4.7 and 4.8, $\mu_{\ln R} = 5.29$ and $\sigma_{\ln R} = 0.0998$. To simulate values of Q, we need to use Equations 2.58, 2.62, 2.63, and 4.10. Equation 4.10 requires the inverse of the Type I distribution, which can be obtained in closed form. The result is

$$q_i = F_Q^{-1}(u_i) = 94.6 - \frac{\ln\left(-\ln\left(u_i\right)\right)}{0.107}$$

4. Calculate N values of y_i using $y_i = r_i - q_i$. The simulated values of r_i, q_i, and y_i are shown in Table 4.4.

Table 4.4 Simulated values of R, Q, and Y = R – Q for Example 4.5

u_i	z_i	r_i	u_i	q_i	$y_i = r_i - q_i$
0.050203	–1.643	168.3	0.269082	92.1	76.3
0.619129	0.303	204.4	0.472640	97.3	107.1
0.872402	1.138	222.2	0.422864	96.0	126.2
0.376568	–0.315	192.2	0.467299	97.2	95.1
0.139927	–1.081	178.1	0.415784	95.8	82.2
0.318491	–0.472	189.2	0.523667	98.7	90.5
0.987671	2.247	248.2	0.243629	91.4	156.8
0.033265	–1.835	165.2	0.741569	105.9	59.3
0.234626	–0.724	184.5	0.408673	95.6	88.9
0.623157	0.314	204.7	0.021790	82.1	122.6
0.957884	1.727	235.6	0.547472	99.3	136.3
0.518906	0.047	199.3	0.749779	106.2	93.0
0.442305	–0.145	195.5	0.681600	103.6	91.9
0.445845	–0.136	195.7	0.140538	88.3	107.4
0.834284	0.971	218.5	0.888607	114.6	104.0
0.811213	0.882	216.6	0.105136	87.0	129.6
0.935728	1.520	230.8	0.067202	85.3	145.5
0.450423	–0.125	195.9	0.864772	112.6	83.3
0.579058	0.199	202.3	0.363628	94.5	107.8
0.662648	0.420	206.8	0.863948	112.6	94.3
0.039918	–1.752	166.5	0.858242	112.2	54.4
0.414075	–0.217	194.1	0.432234	96.2	97.8
0.103214	–1.263	174.8	0.412091	95.7	79.1
0.112308	–1.214	175.7	0.749199	106.2	69.5
0.821833	0.922	217.5	0.260933	91.8	125.6

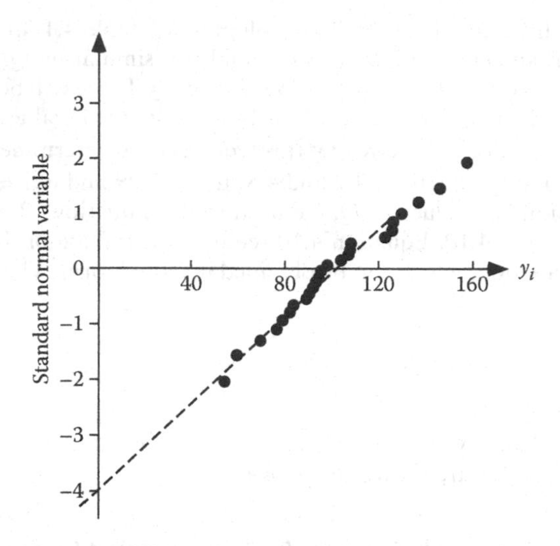

Figure 4.6 Simulated data from Example 4.5 plotted on normal probability paper.

5. Estimate the probability of $Y < 0$ using the simulation data. By observing the values in the far right column of Table 4.4, we see that there are no values of y_i below zero. Thus, Equation 4.11 would lead to $P(Y < 0) = 0$ which is not reasonable. As will be shown in Example 5.9 of Chapter 5, it turns out that the probability of $Y < 0$ for this problem is on the order of 10^{-4} to 10^{-5}. A calculation similar to that done in Example 4.4 would quickly tell us that $N = 25$ is simply too small to make a reliable estimate of such a small probability using Equation 4.11. As an alternative, we can plot the data on normal probability paper as shown in Figure 4.6. A *very rough* estimate of the failure probability can be obtained by extrapolating the curve to the left until it reaches the y axis. This extrapolation is shown by the dashed line in Figure 4.6. The intersection point corresponds to a value of the standard normal variable in the neighborhood of −4. Therefore, the estimate of the failure probability is $\Phi(-4) = 3.17 \times 10^{-5}$.

4.1.8 Simulation of correlated normal random variables

In the preceding sections, we covered simulation of random variables, and it was implicitly assumed that the variables were uncorrelated. In practice, the variables may be correlated, so our simulation procedure must be able to simulate the correlation. A transformation technique for simulating correlated normal random variables is covered in this section. Although it is strictly valid for normal random variables only, it can be used for other types of random variables as an *approximation.*

Let $X_1, X_2, \ldots, X_n$ be *correlated normal* random variables. The mean values and covariance matrix are given by

$$\{\mu_X\} = \{\mu_{X_1}, \mu_{X_2}, \ldots, \mu_{X_n}\} \tag{4.13}$$

$$[C_X] = \begin{bmatrix} \mathrm{Cov}(X_1, X_1) & \mathrm{Cov}(X_1, X_2) & \cdots & \cdots & \mathrm{Cov}(X_1, X_n) \\ \mathrm{Cov}(X_2, X_1) & \mathrm{Cov}(X_2, X_2) & & & \mathrm{Cov}(X_2, X_n) \\ \vdots & \vdots & \ddots & & \vdots \\ \vdots & \vdots & & \ddots & \vdots \\ \mathrm{Cov}(X_n, X_1) & \mathrm{Cov}(X_n, X_2) & \cdots & \cdots & \mathrm{Cov}(X_n, X_n) \end{bmatrix} \tag{4.14}$$

To generate correlated random numbers for $X_1, X_2, \ldots, X_n$, it is necessary to first generate a set of *uncorrelated* random numbers $Y_1, Y_2, \ldots, Y_n$ using the techniques discussed earlier. Then, $X_1, X_2, \ldots, X_n$ are calculated using the variable transformation

$$\{X\} = [T]\{Y\} \tag{4.15}$$

where $[T]$ is a transformation matrix. To apply this approach, we need to determine the matrix $[T]$ as well as the mean and variance values for the uncorrelated Y_i variables. To do this, we must use some concepts from linear algebra. The following discussion assumes that the reader has some familiarity with the concepts of eigenvalues and eigenvectors. For information on these concepts, see the reference by Kaplan (1991) or any other text covering linear algebra.

Let $[A]$ be an $n \times n$ matrix (i.e., a square matrix with n rows and n columns) that is symmetric. A *diagonal* matrix $[D]$ and a square matrix $[T]$ can be found such that the following relationships hold:

$$[D] = [T]^T[A][T] \tag{4.16}$$

$$[A] = [T][D][T]^T \tag{4.17}$$

The superscript T denotes transpose. The matrix $[T]$ contains the *orthonormal eigenvectors* corresponding to the eigenvalues of the matrix $[A]$. The diagonal matrix $[D]$ contains the *eigenvalues* of $[A]$.

In the present context of simulating random variables, the matrix $[A]$ is the covariance matrix $[C_X]$ of the original, correlated variables $\{X\}$. The matrix $[T]$ is made up of the orthonormal eigenvectors corresponding to the eigenvalues of the matrix $[C_X]$. Thus, the first column of $[T]$ contains the orthonormal eigenvector corresponding to the first eigenvalue, the second column contains the eigenvector corresponding to the second eigenvalue, and so on. The matrix $[T]$ is an *orthogonal* matrix, meaning that its inverse is equal to its transpose. The diagonal matrix $[D]$ corresponds to the covariance matrix $[C_Y]$ of the uncorrelated variables $\{Y\}$. With these changes in notation, Equations 4.16 and 4.17 become

$$\left[C_Y\right]=[T]^{\mathrm{T}}\left[C_X\right][T]=\begin{bmatrix} \sigma_{Y_1}^2 & 0 & \cdots & \cdots & 0 \\ 0 & \sigma_{Y_2}^2 & & & \\ \vdots & \vdots & \ddots & & \vdots \\ \vdots & \vdots & & \ddots & \vdots \\ 0 & 0 & \cdots & \cdots & \sigma_{Y_n}^2 \end{bmatrix} \tag{4.18}$$

$$[C_X] = [T][C_Y][T]^{\mathrm{T}} \tag{4.19}$$

The diagonal elements of $[C_Y]$ contain the variances of the uncorrelated Y variables needed to do the simulation. The mean values of the Y_i variables can be obtained using

$$\{\mu_Y\} = [T]^{\mathrm{T}}\{\mu_X\} \tag{4.20}$$

Once simulated values of $\{Y\}$ are obtained, Equation 4.15 can be used to obtain simulated values of $\{X\}$.

Example 4.6

Suppose we have two correlated normal random variables X_1 and X_2. The variable X_1 has a mean value of 10 and a coefficient of variation of 10%. The variable X_2 has a mean value of 15 and a coefficient of variation of 20%. The correlation coefficient between the two variables is $\rho_{12} = 0.5$. Determine the mean values and variances of the transformed coordinates Y_1 and Y_2 that must be considered to simulate X_1 and X_2.

Based on the given information, the standard deviations of X_1 and X_2 are

$$\sigma_{X_1} = V_{X_1}\mu_{X_1} = (0.10)(10) = 1; \qquad \sigma_{X_2} = V_{X_2}\mu_{X_2} = (0.20)(15) = 3$$

The vector of mean values and the covariance matrix for the original variables X_1 and X_2 are

$$\begin{Bmatrix} \mu_{X_1} \\ \mu_{X_2} \end{Bmatrix} = \begin{Bmatrix} 10 \\ 15 \end{Bmatrix}$$

$$[C_X] = \begin{bmatrix} \sigma_{X_1}^2 & \rho_{12}\sigma_{X_1}\sigma_{X_2} \\ \rho_{12}\sigma_{X_1}\sigma_{X_2} & \sigma_{X_2}^2 \end{bmatrix} = \begin{bmatrix} 1 & 1.5 \\ 1.5 & 9 \end{bmatrix}$$

By doing an eigenvalue and eigenvector analysis of $[C_X]$, we find the transformation matrix $[T]$ to be

$$[T] = \begin{bmatrix} 0.984 & 0.178 \\ -0.178 & 0.984 \end{bmatrix}$$

(Note that the eigenvectors must be suitably normalized to qualify as orthonormal eigenvectors.) With this information, the mean values and covariance matrix for the Y variables can be found using Equations 4.18 and 4.20 as follows:

$$\begin{Bmatrix} \mu_{Y_1} \\ \mu_{Y_2} \end{Bmatrix} = [T]^T \begin{Bmatrix} \mu_{X_1} \\ \mu_{X_2} \end{Bmatrix} = \begin{bmatrix} 0.984 & -0.178 \\ 0.178 & 0.984 \end{bmatrix} \begin{Bmatrix} 10 \\ 15 \end{Bmatrix} = \begin{Bmatrix} 7.16 \\ 16.5 \end{Bmatrix}$$

$$[C_Y] = [T]^T[C_X][T] = \begin{bmatrix} 0.984 & -0.178 \\ 0.178 & 0.984 \end{bmatrix} \begin{bmatrix} 1 & 1.5 \\ 1.5 & 9 \end{bmatrix} \begin{bmatrix} 0.984 & 0.178 \\ -0.178 & 0.984 \end{bmatrix}$$

$$= \begin{bmatrix} 0.728 & 0 \\ 0 & 9.27 \end{bmatrix}$$

4.2 LATIN HYPERCUBE SAMPLING

The techniques of random sampling and Monte Carlo simulation discussed in Section 4.1 are very powerful and useful techniques for performing probabilistic analyses. However, in some instances, the problem being analyzed is extremely complex, and the time needed to evaluate the problem for a single trial ($N = 1$) may be very long. As a result, the time needed to perform hundreds or thousands of simulations may be unfeasible.

The Latin hypercube method is one technique for reducing the number of simulations needed to obtain reasonable results. In this method, the range of possible values of each random input variable is partitioned into "strata," and a value from each stratum is randomly selected as a "representative" value. The representative values for each random variable are then combined so that each representative value is considered once and only once in the simulation process. In this way, all possible values of the random variables are represented in the simulation.

To be specific, let us assume that we need to simulate values of some function Y described by

$$Y = f(X_1, X_2,\ldots,X_K) \tag{4.21}$$

where $f(\ )$ is some deterministic function (but possibly not known in closed form) and $X_i (i = 1, 2, \ldots, K)$ are the random input variables. The basic steps in Latin hypercube sampling are as follows:

1. Partition the range of each X_i into N intervals. The partitioning should be done so that the probability of a value of X_i occurring in each interval is $1/N$. [This restriction simplifies the presentation but is not strictly necessary. For additional details, see Iman and Conover (1980).]
2. For each X_i variable and each of its N intervals, randomly select a representative value for the interval. In practical applications, if the number of intervals is large, the center point (i.e., the middle value) of each interval can be used instead of doing random sampling.
3. After steps 1 and 2, there will be N representative values for each of the K random variables. There are N^K possible combinations of these representative values. The objective of Latin hypercube sampling is to select N combinations such that each representative value appears once and only once in the N combinations.
4. To obtain the first combination, randomly select one of the representative values for each of the K input random variables. To obtain the second combination, randomly select one of the $N - 1$ remaining representative values of each random variable. To obtain the third combination, randomly select one of the $N - 2$ remaining representative values of each random variable. Continue this selection process until you have N combinations of values of the input random variables.
5. Evaluate Equation 4.21 for each of the N combinations of input variables generated above. This will lead to N values of the function. These values will be referred to as y_i ($i = 1, 2, \ldots, N$).

This procedure provides the simulated data. Now, we must determine how to use the data to estimate probabilities or statistical parameters for Y. The most commonly used formulas include the following:

$$\text{Estimated mean value of } Y = \bar{Y} = \frac{1}{N}\sum_{i=1}^{N} y_i \tag{4.22}$$

$$\text{Estimated } m\text{th moment of } Y = \frac{1}{N}\sum_{i=1}^{N}(y_i)^m \tag{4.23}$$

$$\text{Estimated CDF } F_Y(y) = \frac{\text{number of times } y_i \leq y}{N} \tag{4.24}$$

These formulas have the same format as those presented in Chapter 2 (Section 2.3.2) and those presented earlier in Section 4.1.1.

Example 4.7

To illustrate how Latin hypercube sampling is implemented, consider the following simple example involving two random variables.

We want to estimate the mean value of the following function of two random variables:

$$Y = 3X_1 - 7X_2$$

The variable X_1 is uniformly distributed between 2 and 4, and X_2 is normally distributed with mean $\mu = 0.6$ and standard deviation $\sigma = 0.1$. Since there are two random variables, $K = 2$.

For this simple illustration, assume that it is sufficient to define four intervals ($N = 4$) for each random variable. Each interval must have an equal probability ($1/N = 0.25$) of a value in the interval. The corresponding intervals are as follows:

X_1 (uniform):	(2, 2.5)	(2.5, 3)	(3, 3.5)	(3.5, 4)
X_2 (normal):	$(-\infty, 0.533)$	(0.533, 0.6)	(0.6, 0.667)	$(0.667, +\infty)$

Now, we must randomly select a value from each interval as a representative value of the interval. Suppose that we do this and obtain the following representative values:

X_1 (uniform):	(2.23)	(2.90)	(3.04)	(3.81)
X_2 (normal):	(0.51)	(0.55)	(0.61)	(1.5)

There are a total of $N^K = (4)^2 = 16$ possible combinations of these representative values. We must now randomly create four pairs such that each representative value appears once and only once. Suppose the following pairs are created:

$$(x_1, x_2) = (2.23, 0.55); (2.90, 1.5); (3.04, 0.61); (3.81, 0.51)$$

For each of these pairs, we must compute the corresponding value of Y:

$$y_1(2.23, 0.55) = 2.84$$

$$y_2(2.90, 1.5) = -1.8$$

$$y_3(3.04, 0.61) = 4.85$$

$$y_4(3.81, 0.51) = 7.86$$

These represent the four simulated values ($N = 4$) of Y. The sample mean is obtained using Equation 4.22:

$$\bar{Y} = \frac{1}{N}\sum_{i=1}^{4} y_i = \frac{1}{4}(2.84 - 1.8 + 4.85 + 7.86) = 3.44$$

For comparison, since the function is linear, Equation 3.2 can be used to calculate the exact mean value of the function:

$$\mu_Y = 3(\mu_{X_1}) - 7(\mu_{X_2}) = 4.8$$

The agreement between the exact and the estimated mean value is not very good because we used only four intervals for each random variable. By choosing more intervals, we could improve our estimate.

For additional details on the Latin hypercube sampling method and its advantages and disadvantages, the interested reader is referred to Dunn and Clark (1974), McKay et al. (1979), Iman and Conover (1980), and Press et al. (1992). For applications of this method to structural reliability analysis, see Han and Wen (1994) and Zhou and Nowak (1988).

4.3 ROSENBLUETH'S 2*K* + 1 POINT ESTIMATE METHOD

Complex problems may also be solved by using a point estimate method. Although there are many such methods, the $2K + 1$ method proposed by Rosenblueth (1975) is one of the easiest to implement. Loosely speaking, this method can be thought of as a simulation technique in which the number of simulations is $N = 2K + 1$ where K is the number of input random variables. The basic idea is to evaluate a function of random variables at $2K + 1$ key points and then to use this information to estimate the mean and variance (or coefficient of variation) of the function. However, the CDF of the function cannot be obtained by this method.

Consider a function Y described by

$$Y = f(X_1, X_2, \ldots, X_K) \tag{4.25}$$

where $f()$ is some deterministic function (but possibly not known in closed form) and the $X_i (i = 1, 2, \ldots, K)$ are the random input variables. The steps in Rosenblueth's $2K + 1$ method are as follows:

1. Determine the mean value $\left(\mu_{X_i}\right)$ and standard deviation $\left(\sigma_{X_i}\right)$ for each of the K input random variables.
2. Define y_0 as the value of Equation 4.25 when all input variables are equal to their mean values; i.e.,

$$y_0 = f\left(\mu_{X_1}, \mu_{X_2}, \ldots, \mu_{X_K}\right) \tag{4.26}$$

3. Evaluate the function Y at $2K$ additional points as follows. For each random variable X_i, evaluate the function at two values of X_i, which are shifted from the mean value μ_{X_i} by $\pm\sigma_{X_i}$ while all other variables are assumed to be equal to their mean values. These values of the function will be referred to as y_i^+ and y_i^-. The subscript denotes the variable that is shifted, and the superscript indicates the direction of the shift. In mathematical notation,

$$y_i^+ = f\left(\mu_{X_1}, \mu_{X_2}, \ldots, \mu_{X_i} + \sigma_{X_i}, \ldots \mu_{X_K}\right) \tag{4.27a}$$

$$y_i^- = f\left(\mu_{X_1}, \mu_{X_2}, \ldots, \mu_{X_i} - \sigma_{X_i}, \ldots \mu_{X_K}\right) \tag{4.27b}$$

4. For each random variable, calculate the following two quantities based on y_i^+ and y_i^-:

$$\bar{y}_i = \frac{y_i^+ + y_i^-}{2} \tag{4.28a}$$

$$V_{y_i} = \frac{y_i^+ - y_i^-}{y_i^+ + y_i^-} \tag{4.28b}$$

5. Calculate the estimated mean and coefficient of variation of Y as follows:

$$\bar{Y} = y_0 \prod_{i=1}^{K} \left(\frac{\bar{y}_i}{y_0}\right) \tag{4.29a}$$

$$V_Y = \sqrt{\left\{\prod_{i=1}^{K}\left(1+V_{y_i}^2\right)\right\}-1} \tag{4.29b}$$

There are two distinct advantages to this method. First, it is not necessary to know the distributions of the input random variables; only the first two moments are needed. Second, the number of function evaluations (i.e., "simulations") is relatively small compared to Latin hypercube sampling or general Monte Carlo simulation. For a discussion of the assumptions and theoretical basis for this method, see the original paper by Rosenblueth (1975). For example applications of this method in structural reliability calculations, see Tantawi (1986).

Example 4.8

For the function considered in Example 4.7, estimate the mean and coefficient of variation using Rosenblueth's $2K + 1$ method.

The function is

$$Y = 3X_1 - 7X_2$$

where X_1 is uniformly distributed between 2 and 4 and X_2 is normally distributed with mean $\mu = 0.6$ and standard deviation $\sigma = 0.1$. Since there are two random variables, $K = 2$. For X_1, based on the given distribution information, the mean value is 3 and the variance is 1/3.

The steps in the procedure are outlined below.

1. Determine mean values and standard deviations. This has been done.
2. Calculate y_0. Using Equation 4.26,

$$y_0 = 3\mu_{X_1} - 7\mu_{X_2} = 3(3) - 7(0.6) = 4.8$$

3. Evaluate the function at $2K = 2(2) = 4$ additional points using Equations 4.27a and 4.27b.

$$y_1^+ = f(\mu_{X_1} + \sigma_{X_1}, \mu_{X_2}) = 3\left(3+\sqrt{1/3}\right) - 7(0.6) = 6.53$$

$$y_1^- = f(\mu_{X_1} - \sigma_{X_1}, \mu_{X_2}) = 3\left(3-\sqrt{1/3}\right) - 7(0.6) = 3.07$$

$$y_2^+ = f(\mu_{X_1}, \mu_{X_2} + \sigma_{X_2}) = 3(3) - 7(0.6+0.1) = 4.10$$

$$y_2^- = f(\mu_{X_1}, \mu_{X_2} - \sigma_{X_2}) = 3(3) - 7(0.6-0.1) = 5.50$$

4. Use Equations 4.28a and 4.28b to calculate some intermediate quantities.

$$\bar{y}_1 = \frac{y_1^+ + y_1^-}{2} = \frac{6.53 + 3.07}{2} = 4.80$$

$$\bar{y}_2 = \frac{y_2^+ + y_2^-}{2} = \frac{4.10 + 5.50}{2} = 4.80$$

(By coincidence, these two values are the same as y_0. In general, they are different.)

$$V_{y_1} = \frac{y_1^+ - y_1^-}{y_1^+ + y_1^-} = \frac{6.53 - 3.07}{6.53 + 3.07} = 0.360$$

$$V_{y_2} = \frac{y_2^+ - y_2^-}{y_2^+ + y_2^-} = \frac{4.10 - 5.50}{4.10 + 5.50} = -0.146$$

5. Use Equations 4.29a and 4.29b to estimate the mean and coefficient of variation of Y.

$$\bar{Y} = y_0 \prod_{i=1}^{2} \left(\frac{\bar{y}_i}{y_0} \right) = 4.80 \left(\frac{4.80}{4.80} \right) \left(\frac{4.80}{4.80} \right) = 4.80$$

$$V_Y = \sqrt{\left\{ \prod_{i=1}^{2} \left(1 + V_{y_i}^2\right) \right\} - 1} = \sqrt{\left(1 + (0.360)^2\right)\left(1 + (-0.146)^2\right) - 1} = 0.392$$

For comparison purposes, consider the exact mean and coefficient of variation. In Example 4.7, the exact mean value of this function was found to be 4.80. From Equation 3.4, the exact variance (assuming no correlation) is

$$\sigma_Y^2 = \sum_{i=1}^{2} a_i^2 \sigma_{X_i}^2 = (3)^2 \left(\sqrt{1/3}\right)^2 + (-7)^2 (0.1)^2 = 3.49$$

Therefore, the exact coefficient of variation is

$$V_Y = \frac{\sigma_Y}{\mu_Y} = \frac{\sqrt{3.49}}{4.80} = 0.389$$

For this simple problem, the estimated and exact mean value are identical. The estimated and exact values of the coefficient of variation are very close.

PROBLEMS

4.1 Consider the following limit state function:

$$g(R, D, L) = R - D - L$$

where R is the load-carrying capacity, D is the deal load effect, and L is the live load effect. The load-carrying capacity is lognormally distributed with a nominal value of 245 k-ft, a bias factor of 1.12, and a coefficient of variation of 13%. (The bias factor is defined as the ratio of the mean value to the nominal value.) Both the dead load and the live load are normally distributed variables. The statistics of the dead load are as follows: nominal value = 60 k-ft, bias factor = 1.03, and COV = 8%. The statistics of the live load are as follows: nominal value = 110 k-ft, bias factor = 0.85, and COV = 13.5%. All three variables are uncorrelated. Calculate the probability of failure using the Monte Carlo simulation method (to simplify the computations, limit the number of simulations to 25) (1 k-ft = 1.356 kN m).

4.2 Consider the following limit state equation for a reinforced concrete beam:

$$g = A_S F_y \left(d - 0.59 \frac{A_S F_y}{f_c' b} \right) - [D + L]$$

The parameters of the random variables in this equation are listed in Table P4.1.

The other parameters can be treated as deterministic variables. Their values are as follows: b = 15 in, d = 24 in, A_S = 8 in^2. Use Monte Carlo simulation to predict the probability of failure (1 ft = 0.3048 m, 1 in = 25.40 mm, 1 kip = 4.448 kN, 1000 psi = 1 ksi, 1 ksi = 6.895 MPa, 1 k-ft = 1.356 kN m).

Table P4.1 Parameters of the random variables in the equation in Problem 4.2

Parameter	*Mean*	*COV*	*Distribution*
f_c'	3750 psi	12.5%	Normal
F_y	42.5 ksi	11.5%	Lognormal
D	100 k-ft	10.5%	Normal
L	200 k-ft	17.5%	Type I extreme

4.3 Consider Example 4.3. In that example, we used different sets of five uniformly distributed random numbers to generate five values of P and five values of w. Suppose we combined the two sets of five uniformly distributed random numbers to make a set of 10 numbers, and use this one set to generate 10 values of P and 10 values of w. In other words, each uniformly distributed number would be used twice. Do you see any problems with doing this? Explain your answer.

4.4 Consider Example 4.2. Using the computer program of your choice,
   A. Generate a set of 100 values of dead load.
   B. Calculate the sample mean and standard deviation of your simulated data and compare them with the values given in the problem.
   C. Plot the relative frequency histogram for the data. Does it resemble the PDF of a normal distribution? Should it?
   D. Plot the cumulative frequency histogram for the data. Does it resemble the CDF of a normal distribution? Should it?
   E. Plot the data on normal probability paper.
   F. Choose a new seed for your random number generator, and repeat steps A–E. Compare the results of the two simulations.
   G. Repeat steps A and B for 1000 simulated data points.

4.5 Using the computer program of your choice,
   A. Simulate 100 values of uniformly distributed random integers between –3 and 4 inclusive.
   B. Plot a relative frequency histogram of the simulated data. Comment on what the ideal histogram should look like in this case, and comment on how your histogram compares with the ideal histogram.
   C. Repeat steps A and B for 1000 simulated values.

4.6 Use Monte Carlo simulation to solve Problem 3.6 at the end of Chapter 3.

4.7 Using the software program of your choice, solve Example 4.3 using 200 simulated data points.

4.8 Use Rosenblueth's method to solve Problem 3.4 (parts A and B) at the end of Chapter 3.

4.9 Use Rosenblueth's method to find the mean and coefficient of variation of the performance function discussed in Example 3.2.

4.10 Use Rosenblueth's method to find the mean and coefficient of variation of the function Y discussed in Example 3.3.

4.11 Using the software program of your choice,
   A. Solve Example 4.5 by Monte Carlo simulation. Estimate the required number of simulations needed to calculate the failure probability (estimated to be on the order of 10^{-4}) with a coefficient of variation of 20%. Use this number of simulations in your solution.

B. Change the seed for your random number generator, and repeat step A. Comment on any similarities or differences.

4.12 Using the random number generator of your choice, generate 1000 uniformly distributed random numbers between 0 and 1. Plot a relative frequency histogram using an interval size of 0.1. Does the histogram look as expected? Repeat the calculation for a larger number of simulated values. Also try changing the seed value of the generator.

Chapter 5

Structural safety analysis

Before we begin a structural reliability analysis, we must first determine how we define such terms as "safety" and "reliability." The first section of this chapter introduces the notion of a "limit state," which is useful in defining safety and reliability. Then, in later sections, some techniques for structural reliability analysis are described and examples are presented.

5.1 LIMIT STATES

5.1.1 Definition of failure

The term "failure" means different things to different people. We could say that a structure fails if it cannot perform its intended function. However, this is a vague definition because we have not specified the function of the structure.

To illustrate this point, consider a simply supported steel hot-rolled beam such as the one shown in Figure 5.1. We could state that the beam fails when the maximum deflection exceeds $\delta_{critical}$. However, a steel beam may "fail" by developing a plastic hinge (see Figure 5.2), losing overall stability, or by local buckling of the compression flange or web (see Figure 5.3). It is obvious that the term "failure" can have different meanings. Before attempting a structural reliability analysis, failure must be clearly defined.

The concept of a "limit state" is used to help define failure in the context of structural reliability analyses. A *limit state* is a boundary between desired and undesired performance of a structure. This boundary is often represented mathematically by a *limit state function* or *performance function* (to be discussed in Section 5.1.2). For example, in bridge structures, failure could be defined as the inability to carry traffic. This undesired performance can occur by many modes of failure: cracking, corrosion, excessive deformations, exceeding load-carrying capacity for shear or bending moment, or local or overall buckling. Some members may fail in a brittle manner, whereas others may fail in a ductile fashion. In the traditional

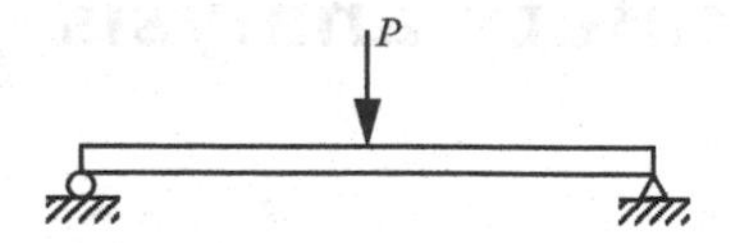

Figure 5.1 A simply supported beam.

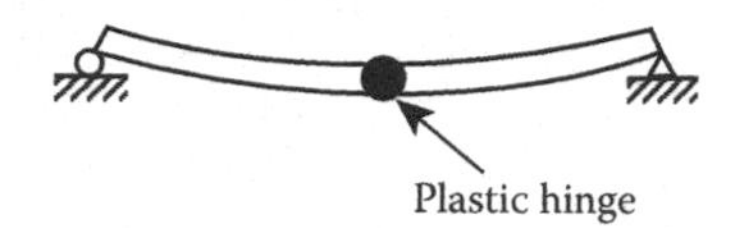

Figure 5.2 Development of a plastic hinge in beam.

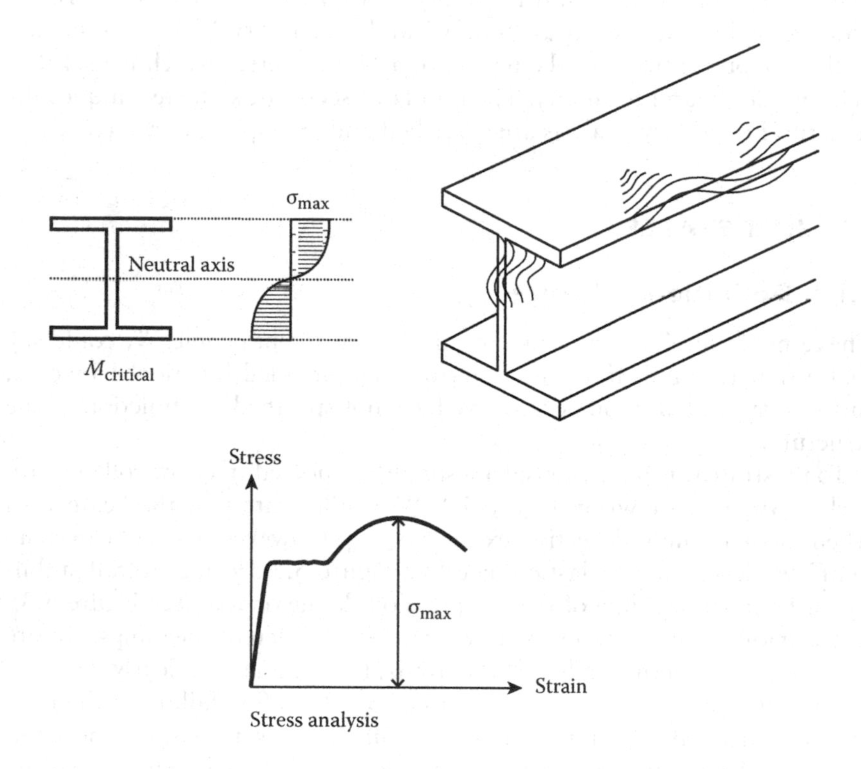

Figure 5.3 Local buckling in a steel beam.

approach, each mode of failure is considered separately, and each mode can be defined using the concept of a limit state.

In structural reliability analyses, three types of limit states are considered—ultimate limit states, serviceability limit states, and fatigue limit states. Each of these is described below.

1. *Ultimate limit states* (ULSs) are related to the loss of load-carrying capacity. Examples of modes of failure in this category include
   Exceeding the moment carrying capacity
   Formation of a plastic hinge
   Crushing of concrete in compression
   Shear failure of the web in a steel beam
   Loss of the overall stability
   Buckling of flange
   Buckling of web
   Weld rupture
2. *Serviceability limit states* (SLSs) are related to gradual deterioration, user's comfort, or maintenance costs. They may or may not be directly related to structural integrity. Examples of modes of failure would include the following:
   Excess deflection: Deflection is a rather controversial limit state. The acceptable limits are subjective, and they may depend on human perception. A building with visible deflections (horizontal or vertical) is not acceptable by the public, even though it may be structurally safe. Excessive deflections may interfere with the operation of precise instruments sensitive to movement. For example, for bridge girders, the current practice is to limit deflections to a fraction of the span length; for example, $L/800$, where L = span length. The deflection limit often governs the design.
   Excess vibration: Vibration is another SLS that is difficult to quantify. The acceptability criteria are also highly subjective and often depend on human perception. In a building, the occupants may not tolerate excessive vibration; a vibrating bridge, however, may be acceptable if pedestrians are not involved. The design for vibration may require a complicated dynamic analysis. In many current design codes, vibration is not considered in a direct form. Indirectly, the codes impose a limit on static deflection, and this is also intended to serve as a limit for vibration.
   Permanent deformations: Each time the load exceeds the elastic limit, a permanent deformation may result. Accumulation of these permanent deformations can lead to serviceability problems. Therefore, in some design codes, a limit is imposed on permanent deformations. For example, consider a multispan bridge with continuous girders as shown in Figure 5.4. Each time the strain exceeds the yield strain, there is some permanent strain left in the

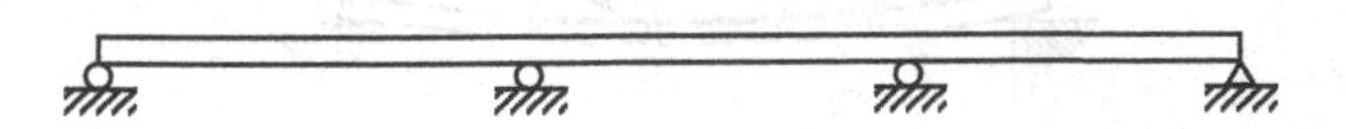

Figure 5.4 Continuous bridge girder.

section (see Figure 5.5). This strain accumulates and eventually causes the formation of a "kink," as shown in Figure 5.6.

Cracking: Cracks, such as those shown in Figure 5.7, by themselves do not necessarily affect the structural performance of concrete structures. However, they lead to steel corrosion, spalled concrete, salt (deicing agent) penetration, and irreversible loss of concrete tensile strength. To define acceptable cracking standards, many questions must be answered. What is acceptable with regard to cracking? Are acceptable cracks limited by size? Width? Length? How frequently can the cracks open?

3. *Fatigue limit states* (FLSs) are related to loss of strength under repeated loads. FLSs are related to the accumulation of damage and eventual failure under repeated loads. It has been observed that a structural component can fail under repeated loads at a level lower than the ultimate load. The failure mechanism involves the formation and propagation of cracks until their rupture. This may result in structural collapse. FLSs occur in steel components and reinforcement bars

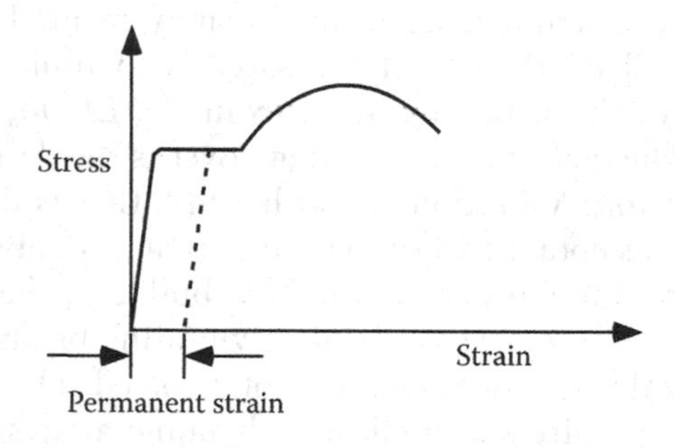

Figure 5.5 Permanent strain.

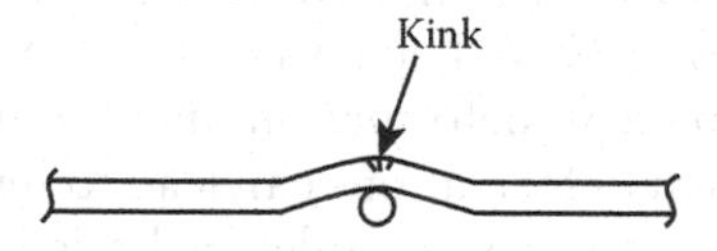

Figure 5.6 Formation of a kink in a continuous steel beam.

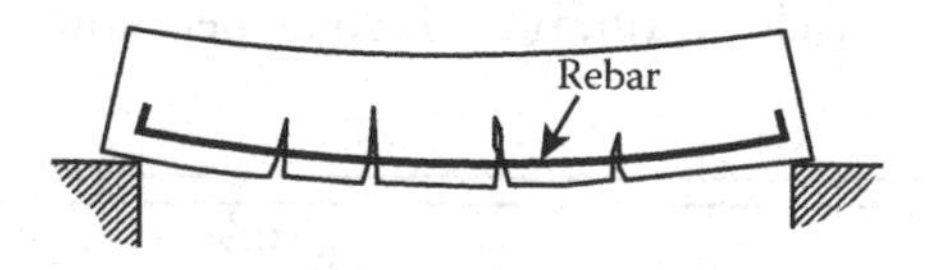

Figure 5.7 Cracks in a reinforced concrete beam.

in concrete, particularly those in tension. Welding affects the fatigue resistance of components and connections. Fatigue failures have also been reported in the prestressing strands of post-tensioned concrete bridges. In any fatigue analysis, the critical factors are both the magnitude and frequency of load.

5.1.2 Limit state functions (performance functions)

A traditional notion of the "safety margin" or "margin of safety" is associated with the ULSs. For example, a mode of beam failure could be when the moment due to loads exceeds the moment-carrying capacity. Let R represent the resistance (moment carrying capacity) and Q represent the load effect (total applied moment). It is sometimes helpful to think of R as the "capacity" and Q as the "demand." A *performance function*, or *limit state function*, can be defined for this mode of failure as

$$g(R,Q) = R - Q \tag{5.1}$$

The *limit state*, corresponding to the boundary between desired and undesired performance, would be when $g = 0$. If $g \geq 0$, the structure is safe (desired performance); if $g < 0$, the structure is not safe (undesired performance). The probability of failure, P_f, is equal to the probability that the undesired performance will occur. Mathematically, this can be expressed in terms of the performance function as

$$P_f = P(R - Q < 0) = P(g < 0) \tag{5.2}$$

If both R and Q are continuous random variables, then each has a probability density function (PDF) such as shown in Figure 5.8. Furthermore, the quantity $R - Q$ is also a random variable with its own PDF. This is also

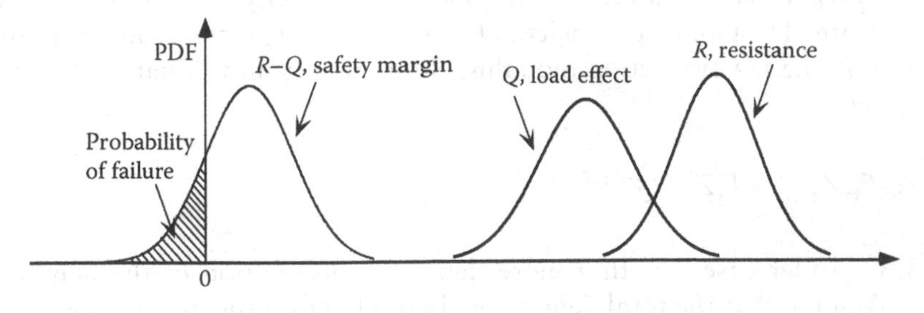

Figure 5.8 PDFs of load, resistance, and safety margin.

shown in Figure 5.8. The probability of failure corresponds to the shaded area in Figure 5.8.

Now, let us generalize the concepts just introduced. All realizations of a structure can be put into one of two categories:

(I) safe (load effect ≤ resistance)
(II) failure (load effect > resistance)

The state of the structure can be described using various parameters X_1, X_2, ..., X_n, which are load and resistance parameters such as dead load, live load, length, depth, compressive strength, yield strength, moment of inertia, etc. A *limit state function*, or *performance function*, is a function $g(X_1, X_2, ..., X_n)$ of these parameters such that

$g(X_1, X_2, ..., X_n) > 0$ for a safe structure
$g(X_1, X_2, ..., X_n) = 0$ border or boundary between safe and unsafe
$g(X_1, X_2, ..., X_n) < 0$ for failure

Each limit state function is associated with a particular limit state. Different limit states may have different limit state functions. Here are some examples of limit state functions:

1. Let Q = total load effect (total demand) and R = resistance (or capacity). Then, the limit state function can be defined as

$$g(R,Q) = R - Q \tag{5.3}$$

or

$$g(R,Q) = R/Q - 1 \tag{5.4}$$

2. Consider case 1 above for the moment capacity of a compact steel beam. The moment capacity is $R = F_yZ$ where F_y is the yield stress and Z is the plastic section modulus. Substituting into Equation 5.3, we get

$$g(F_y,Z,Q) = F_yZ - Q \tag{5.5}$$

3. Consider case 2 with a more definitive description of the demand. Assume that the total demand or load effect on the beam is made up

of contributions from dead load (D), live load (L), wind load (W), and earthquake load (E). If $Q = D + L + W + E$, then Equation 5.5 is

$$g(F_y,Z,D,L,W,E) = F_yZ - D - L - W - E. \tag{5.6}$$

In general, the performance function (limit state function) can be a function of many variables: load components, influence factors, resistance parameters, material properties, dimensions, analysis factors, and so on. A direct calculation of P_f using Equation 5.2 is often very difficult, if not impossible. Therefore, it is convenient to measure structural safety in terms of a *reliability index*. This will be discussed in Section 5.3.

5.2 FUNDAMENTAL CASE

5.2.1 Probability of failure

We now examine how to determine the probability of failure for the relatively simple performance function given earlier by

$$g(R,Q) = R - Q \tag{5.1}$$

In general, if the random variables R and Q are *correlated*, the probability of failure must be calculated by integrating the *joint* PDF over all values of R and Q that make the performance function negative. Mathematically, this means

$$P_f = \iint_{g(R,Q)<0} f_{RQ}(r,q)\,dq\,dr \tag{5.7}$$

If R and Q are *statistically independent*, this double integral can be simplified as follows. First, as discussed in Section 2.8, the joint PDF of *statistically independent* R and Q can be written as $f_{RQ}(r,q) = f_Q(q)f_R(r)$. The integration over the region $g(R,Q) < 0$ can be accomplished in one of two ways. One way is to integrate with respect to r over all values that satisfy the failure condition $r < q$ and then integrate with respect to q over all possible values. This gives

$$P_f = \int_{-\infty}^{+\infty} f_Q(q)\left[\int_{-\infty}^{q} f_R(r)\,dr\right]dq = \int_{-\infty}^{+\infty} f_Q(q)F_R(q)\,dq \tag{5.8a}$$

Alternately, integration with respect to q can be done first for all q values that satisfy the failure condition $q > r$, and then integration with respect to r over all possible values can be done. This gives

$$P_f = \int_{-\infty}^{+\infty} f_R(r) \left[\int_{r}^{+\infty} f_Q(q)\,\mathrm{d}q \right] \mathrm{d}r$$

$$= \int_{-\infty}^{+\infty} f_R(r) \left[1 - \int_{-\infty}^{r} f_Q(q)\,\mathrm{d}q \right] \mathrm{d}r$$

$$= \int_{-\infty}^{+\infty} f_R(r) \left[1 - F_Q(r) \right] \mathrm{d}r$$

$$= \int_{-\infty}^{+\infty} f_R(r)\,\mathrm{d}r - \int_{-\infty}^{+\infty} f_R(r) F_Q(r)\,\mathrm{d}r$$

$$= 1 - \int_{-\infty}^{+\infty} f_R(r) F_Q(r)\mathrm{d}r \tag{5.8b}$$

[The development of Equations 5.8a and 5.8b used properties of the PDF and cumulative distribution function (CDF) functions discussed in Sections 2.1 and 2.2.] The integrals in Equations 5.7 and 5.8 are often very difficult to evaluate. The integration requires special numerical techniques, and the accuracy of these techniques may not be adequate. Therefore, in practice, the probability of failure is calculated indirectly using other procedures that are discussed in Section 5.3.

Before leaving this section, we provide an interpretation of Equation 5.8b in the context of Figure 5.9 showing the PDFs of the random variables R and Q. In Figure 5.9, the variable X can represent either R or Q as appropriate. Recall from calculus that a definite integral represents an infinite summation. In Equation 5.8b, the product $f_R(r)\mathrm{d}r$ represents the probability that R lies between r and $r + \mathrm{d}r$, and $F_Q(r)$ is equivalent to $P(Q \le R|R = r)$ when Q and R are statistically independent. [In Figure 5.9, the quantity

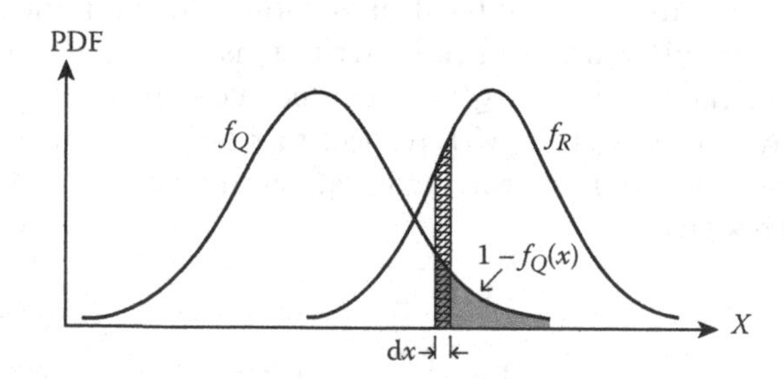

Figure 5.9 PDFs of load (Q) and resistance (R).

$f_R(r)dr$ is related to the area of the vertical strip, and the quantity $1 - F_Q(r)$ is the area of the shaded region to the right of the vertical strip.] If we interpret $f_R(r)dr$ as the probability $P(R = r_i)$ where r_i is a discrete value of R in the interval $(r, r + dr)$, Equation 5.8b can be interpreted as a summation over all possible values of r_i as follows:

$$P_f = 1 - \sum_i P(Q \le R | R = r_i) P(R = r_i) \tag{5.9}$$

By virtue of the total probability theorem (Equation 2.101), the summation represents $P(Q \le R)$. Substituting this into Equation 5.9 and using the basic probability properties discussed in Sections 2.1 and 2.2, we obtain the definition of failure probability that we started with

$$\begin{aligned} P_f &= 1 - P(Q \le R) \\ &= P(Q > R) \end{aligned} \tag{5.10}$$

5.2.2 Space of state variables

To begin our analysis, we need to define the state variables of the problem. The *state variables* are the basic load and resistance parameters used to formulate the performance function. For n state variables, the limit state function is a function of n parameters.

If all loads (or load effects) are represented by the variable Q and total resistance (or capacity) by R, then the space of state variables is a two-dimensional space as shown in Figure 5.10. Within this space, we can

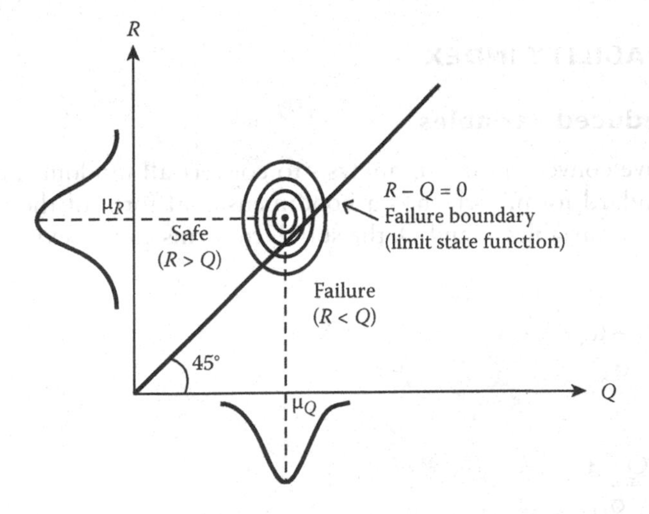

Figure 5.10 Safe domain and failure domain in a two-dimensional state space.

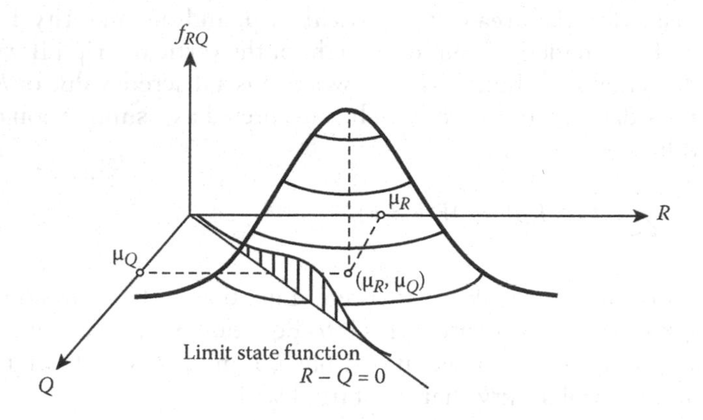

Figure 5.11 Three-dimensional representation of a possible joint density function f_{RQ}.

separate the "safe domain" from the "failure domain"; the boundary between the two domains is described by the limit state function $g(R,Q) = 0$.

Since both R and Q are random variables, we can define a joint density function $f_{RQ}(r,q)$. A general joint density function is plotted in Figure 5.11. Again, the limit state function separates the safe and failure domains. The probability of failure is calculated by integration of the joint density function over the failure domain [i.e., the region in which $g(R,Q) < 0$] as shown in Equation 5.7.

As noted earlier, this probability is often very difficult to evaluate; thus, the concept of a reliability index is used to quantify structural reliability.

5.3 RELIABILITY INDEX

5.3.1 Reduced variables

It will prove convenient in our analysis to convert all random variables to their "standard form," which is a nondimensional form of the variables. For the basic variables R and Q, the standard forms can be expressed as

$$Z_R = \frac{R - \mu_R}{\sigma_R} \tag{5.11a}$$

$$Z_Q = \frac{Q - \mu_Q}{\sigma_Q} \tag{5.11b}$$

The variables Z_R and Z_Q are sometimes called *reduced variables*. By rearranging Equations 5.11, the resistance R and the load Q can be expressed in terms of the reduced variables as follows:

$$R = \mu_R + Z_R\sigma_R \tag{5.12a}$$

$$Q = \mu_Q + Z_Q\sigma_Q \tag{5.12b}$$

The limit state function $g(R,Q) = R - Q$ can be expressed in terms of the reduced variables by using Equations 5.12. The result is

$$g(Z_R, Z_Q) = \mu_R + Z_R\sigma_R - \mu_Q - Z_Q\sigma_Q = (\mu_R - \mu_Q) + Z_R\sigma_R - Z_Q\sigma_Q \tag{5.13}$$

For any specific value of $g(Z_R, Z_Q)$, Equation 5.13 represents a straight line in the space of reduced variables Z_R and Z_Q. The line of interest to us in reliability analysis is the line corresponding to $g(Z_R,Z_Q) = 0$ because this line separates the safe and failure domains in the space of reduced variables.

5.3.2 General definition of the reliability index

In Chapter 3 (Example 3.1), a version of the reliability index was defined as the reciprocal of the coefficient of variation (COV). In the context of the present discussion, we will define the reliability index as the *shortest* distance from the origin of reduced variables to the line $g(Z_R,Z_Q) = 0$. This definition, which was introduced by Hasofer and Lind (1974), is illustrated in Figure 5.12.

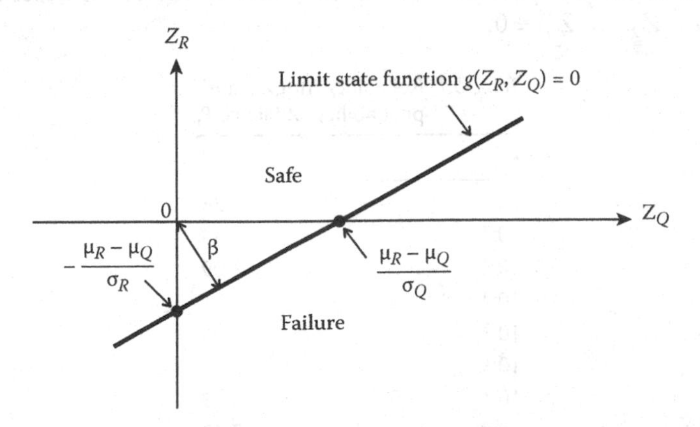

Figure 5.12 Reliability index defined as the shortest distance in the space of reduced variables.

Using geometry, we can calculate the reliability index (shortest distance) from the following formula:

$$\beta = \frac{\mu_R - \mu_Q}{\sqrt{\sigma_R^2 + \sigma_Q^2}} \tag{5.14}$$

Note that β as shown in Equation 5.14 is the reciprocal of the COV of the function $g(R,Q) = R - Q$ when R and Q are uncorrelated. For *normally distributed* random variables R and Q, it can be shown (following the procedure presented in Example Problems 3.1 and 3.2) that the reliability index is related to the probability of failure by

$$\beta = -\Phi^{-1}(P_f) \text{ or } P_f = \Phi(-\beta) \tag{5.15}$$

Table 5.1 provides an indication of how β varies with P_f and vice versa based on Equation 5.15.

The above definition for a two-variable case can be generalized for n variables as follows: Consider a limit state function $g(X_1, X_2, \ldots, X_n)$ where the X_i variables are all uncorrelated. The Hasofer–Lind reliability index is defined as follows:

1. Define the set of reduced variables $\{Z_1, Z_2, \ldots, Z_n\}$ using

$$Z_i = \frac{X_i - \mu_{X_i}}{\sigma_{X_i}} \tag{5.16}$$

2. Redefine the limit state function by expressing it in terms of the reduced variables $(Z_1, Z_2, \ldots, Z_n)$.
3. The reliability index is the shortest distance from the origin in the n-dimensional space of reduced variables to the curve described by $g(Z_1, Z_2, \ldots, Z_n) = 0$.

Table 5.1 Reliability index β and probability of failure P_f

P_f	β
10^{-1}	1.28
10^{-2}	2.33
10^{-3}	3.09
10^{-4}	3.71
10^{-5}	4.26
10^{-6}	4.75
10^{-7}	5.19
10^{-8}	5.62
10^{-9}	5.99

5.3.3 First-order, second-moment reliability index

5.3.3.1 Linear limit state functions

Consider a *linear* limit state function of the form

$$g(X_1, X_2, \ldots, X_n) = a_0 + a_1X_1 + a_2X_2 + \ldots + a_nX_n = a_0 + \sum_{i=1}^{n} a_iX_i \quad (5.17)$$

where the a_i terms (i = 0, 1, 2, ... n) are constants and the X_i terms are *uncorrelated* random variables. If we apply the three-step procedure outlined above for determining the Hasofer–Lind reliability index, we would obtain the following expression for β:

$$\beta = \frac{a_0 + \sum_{i=1}^{n} a_i\mu_{X_i}}{\sqrt{\sum_{i=1}^{n} (a_i\sigma_{X_i})^2}} \quad (5.18)$$

Observe that the reliability index, β, in Equation 5.18 depends only on the means and standard deviations of the random variables. Therefore, this β is called a *second-moment* measure of structural safety because only the first two moments (mean and variance) are required to calculate β. There is no explicit relationship between β and the type of probability distributions of the random variables. If the random variables are all normally distributed and uncorrelated, then this formula is *exact* in the sense that β and P_f are related by Equation 5.15. Otherwise, Equation 5.15 provides only an approximate means of relating β to a probability of failure. When Equation 5.15 is truly an approximate relationship, the probability of failure is sometimes referred to as a "notional" probability of failure. As noted by Ellingwood et al. (1980), the notional probability "...should be interpreted, at best, in a comparative sense as opposed to a classical or relative frequency sense."

Example 5.1

Consider the simply supported beam shown in Figure 5.13. The beam is subjected to a concentrated live load P and a uniformly distributed dead load w. The loads are random variables.

In this problem, we will assume that P, w, and the yield stress, F_y, are random quantities; the length L and the plastic section modulus Z will be assumed to be precisely known (deterministic).

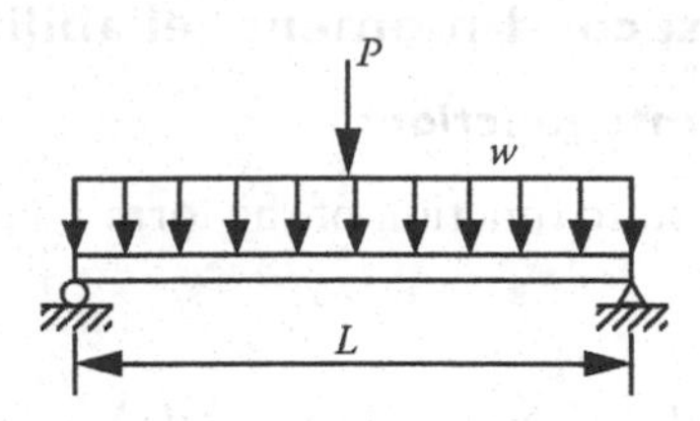

Figure 5.13 Beam considered in Example 5.1.

The distribution parameters for P, w, and F_y are given below. A quantity known as the "bias factor" (denoted by λ) is specified for each of the random variables. It is defined as the ratio of the mean value of a variable to its nominal value (i.e., the value specified in a standard or code). The length L is 18 ft, and the plastic section modulus is 80 in^3 (1 ft = 0.3048 m; 1 in = 25.40 mm; 1 kip = 4.448 kN; 1 ksi = 6.895 MPa; 1 k-ft = 1.356 kN m).

nominal (design) value of $w = w_n = 3.0$ k/ft = 0.25 k/in
bias factor for $w = \lambda_W = 1.0$
$\mu_W = \lambda_W w_n = 3.0$ k/ft = 0.25 k/in
$V_W = 10\% \Rightarrow \sigma_W = V_W \mu_W = 0.3$ k/ft = 0.025 k/in

nominal (design) value of $P = p_n = 12.0$ k
bias factor for $P = \lambda_P = 0.85$
$\mu_P = \lambda_P p_n = 10.2$ k
$V_P = 11\% \Rightarrow \sigma_P = V_P \mu_P = 1.12$ k

nominal (design) value of $F_y = f_y = 36$ ksi
bias factor for $F_y = \lambda_F = 1.12$
$\mu_F = \lambda_F f_y = 40.3$ ksi
$V_F = 11.5\% \Rightarrow \sigma_F = V_F \mu_F = 4.64$ ksi

Calculate the reliability index.

The limit state function for beam bending can be expressed as

$$g(P,w,F_y) = F_y Z - \frac{PL}{4} - \frac{wL^2}{8}.$$

Substituting for L and Z (and converting all units to inches), the limit state function can be rewritten as

$$g(P,w,F_y) = 80F_y - 54P - 5832w \text{ (units: k, in)}$$

Since the limit state function is linear, Equation 5.18 can be used to determine the reliability index β:

$$\beta = \frac{a_0 + \sum_{i=1}^{n} a_i \mu_{X_i}}{\sqrt{\sum_{i=1}^{n} (a_i \sigma_{X_i})^2}} = \frac{80(40.3) - 54(10.2) - 5832(0.25)}{\sqrt{((80)(4.64))^2 + ((-54)(1.12))^2 + ((-5832)(0.025))^2}}$$

$$= \frac{1215.2}{403.37}$$

$$= 3.01$$

5.3.3.2 Nonlinear limit state functions

Now, consider the case of a *nonlinear* limit state function. When the function is nonlinear, we can obtain an approximate answer by linearizing the nonlinear function using a Taylor series expansion. The result is

$$g(X_1, X_2, \ldots, X_n) \approx g(x_1^*, x_2^*, \ldots, x_n^*) + \sum_{i=1}^{n} (X_i - x_i^*) \left. \frac{\partial g}{\partial X_i} \right|_{\text{evaluated at } (x_1^*, x_2^*, \ldots, x_n^*)} \tag{5.19}$$

where $(x_1^*, x_2^*, \ldots, x_n^*)$ is the point about which the expansion is performed. One choice for this linearization point is the point corresponding to the mean values of the random variables. Thus, Equation 5.19 becomes

$$g(X_1, X_2, \ldots, X_n) \approx g(\mu_{X_1}, \mu_{X_2}, \ldots, \mu_{X_n}) + \sum_{i=1}^{n} (X_i - \mu_{X_i}) \left. \frac{\partial g}{\partial X_i} \right|_{\text{evaluated at mean values}} \tag{5.20}$$

Since Equation 5.20 is a linear function of the X_i variables, it can be rewritten to look exactly like Equation 5.17. Thus, Equation 5.18 can be used as an approximate solution for the reliability index β. After some algebraic manipulations, the following expression for β results:

$$\beta = \frac{g(\mu_{X_1}, \mu_{X_2}, \ldots, \mu_{X_n})}{\sqrt{\sum_{i=1}^{n} (a_i \sigma_{X_i})^2}} \quad \text{where} \quad a_i = \left. \frac{\partial g}{\partial X_i} \right|_{\text{evaluated at mean values}} \tag{5.21}$$

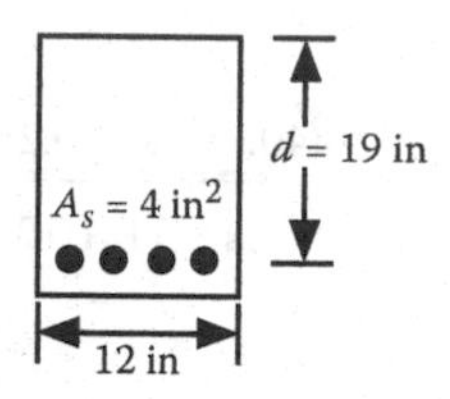

Figure 5.14 Cross section of reinforced concrete beam considered in Example 5.2.

The reliability index defined in Equation 5.21 is called a *first-order, second-moment, mean value reliability index*. It is a long name, but the underlying meaning of each part of the name is very important:

first order	we use first-order terms in the Taylor series expansion
second moment	only means and variances are needed
mean value	the Taylor series expansion is about the mean values

Example 5.2

Consider the reinforced concrete beam shown in Figure 5.14.

The moment-carrying capacity of the section is calculated using

$$M = A_s f_y \left(d - 0.59 \frac{A_s f_y}{f_c' b} \right) = A_s f_y d - 0.59 \frac{(A_s f_y)^2}{f_c' b},$$

where A_s is the area of steel, f_y is the yield strength of the steel, f_c' is the compressive strength of the concrete, b is the width of the section, and d is the depth of the section. We want to examine the limit state of exceeding the beam capacity in bending. The limit state function would be

$$g(A_s, f_y, f_c', Q) = A_s f_y d - 0.59 \frac{(A_s f_y)^2}{f_c' b} - Q,$$

where Q is the moment (load effect) due to the applied load. The random variables in the problem are Q, f_y, f_c', and A_s. The distribution parameters and design parameters are given in Table 5.2, where λ is

Table 5.2 Parameters of variables in Example 5.2

	Mean	*Nominal*	λ	σ	*V*
f_y	44 ksi	40 ksi	1.10	4.62 ksi	0.105
A_S	4.08 in^2	4 in^2	1.02	0.08 in^2	0.02
f_c'	3.12 ksi	3 ksi	1.04	0.44 ksi	0.14
Q	2052 k-in	2160 k-in	0.95	246 k-in	0.12

(1 ft = 0.3048 m; 1 in = 25.40 mm; 1 kip = 4.448 kN; 1 ksi = 6.895 MPa; 1 k-ft = 1.356 kN m; 1 k-in = 0.1130 kN m)

the bias factor (ratio of mean value to nominal value). The values of d and b are assumed to be deterministic constants. Calculate the reliability index, β.

For this problem, the limit state function is nonlinear, so we need to apply Equation 5.19 or 5.20. The Taylor expansion about the mean values yields the following linear function:

$$g(A_s, f_y, f_c', Q) \approx \left[\mu_{A_s}\mu_{f_y} d - 0.59 \frac{\left(\mu_{A_s}\mu_{f_y}\right)^2}{\mu_{f_c'} b} - \mu_Q \right]$$

$$+\left(A_s - \mu_{A_s}\right)\frac{\partial g}{\partial A_s}\bigg|_{\text{evaluated at mean values}}$$

$$+\left(f_y - \mu_{f_y}\right)\frac{\partial g}{\partial f_y}\bigg|_{\text{evaluated at mean values}}$$

$$+\left(f_c' - \mu_{f_c'}\right)\frac{\partial g}{\partial f_c'}\bigg|_{\text{evaluated at mean values}}$$

$$+\left(Q - \mu_Q\right)\frac{\partial g}{\partial Q}\bigg|_{\text{evaluated at mean values}}$$

To calculate β, the partial derivatives must be determined and the limit state function must be evaluated at the mean values of the random variables:

$$g(\mu_{A_s}, \mu_{f_y}, \mu_{f_c'}, \mu_Q) = \mu_{A_s}\mu_{f_y} d - 0.59 \frac{\left(\mu_{A_s}\mu_{f_y}\right)^2}{\mu_{f_c'} b} - \mu_Q = 851.0 \text{ k-in}$$

$$a_1 = \frac{\partial g}{\partial A_s}\bigg|_{\text{mean values}} = \left[f_y d - 0.59 \frac{(2A_s f_y^2)}{f_c' b} \right]_{\text{mean values}} = 587.1 \text{ k/in}$$

$$a_2 = \frac{\partial g}{\partial f_y}\bigg|_{\text{mean values}} = \left[A_s d - 0.59 \frac{(2 f_y A_s^2)}{f_c' b} \right]_{\text{mean values}} = 54.44 \text{ in}^3$$

$$a_3 = \frac{\partial g}{\partial f_c'}\bigg|_{\text{mean values}} = \left[0.59 \frac{(A_s f_y)^2}{(f_c')^2 b} \right]_{\text{mean values}} = 162.8 \text{ in}^3$$

$$a_4 = \frac{\partial g}{\partial Q}\bigg|_{\text{mean values}} = -1\bigg|_{\text{mean values}} = -1$$

Substituting these results into Equation 5.21, we get

$$\beta = \frac{g(\mu_{A_s}, \mu_{f_y}, \mu_{f_c'}, \mu_Q)}{\sqrt{\left((587.1)(\sigma_{A_s})\right)^2 + \left((54.44)(\sigma_{f_y})\right)^2 + \left((162.8)(\sigma_{f_c'})\right)^2 + \left((-1)(\sigma_Q)\right)^2}}$$

$$= \frac{851.0}{\sqrt{\left((587.1)(0.08)\right)^2 + \left((54.44)(4.62)\right)^2 + \left((162.8)(0.44)\right)^2 + \left((-1)(246)\right)^2}}$$

$$= \frac{851.0}{362.1} = 2.35$$

5.3.4 Comments on the first-order, second-moment mean value index

The first-order, second-moment mean value method is based on approximating nonnormal CDFs of the state variables by normal variables, as shown in Figure 5.15 for the simple case in which $g(R,Q) = R - Q$. The method has both advantages and disadvantages in structural reliability analysis.

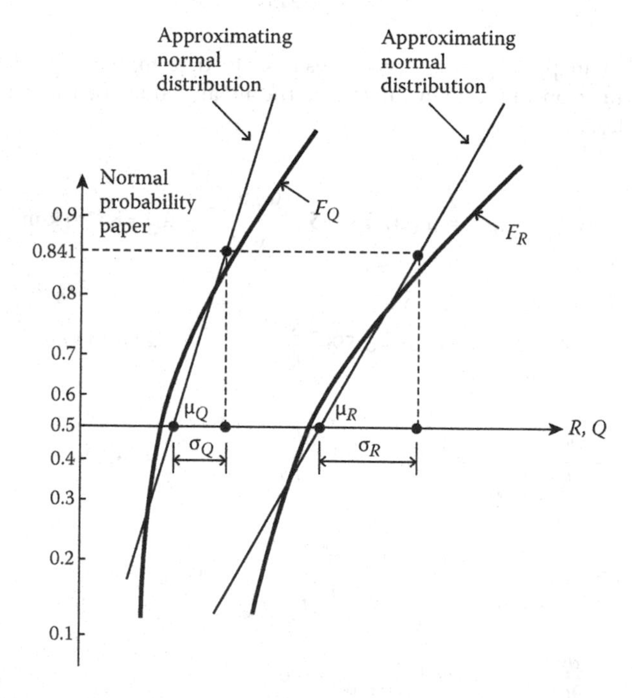

Figure 5.15 Mean value second-moment formulation.

Advantages

1. It is easy to use.
2. It does not require knowledge of the distributions of the random variables.

Disadvantages

1. Results are inaccurate if the tails of the distribution functions cannot be approximated by a normal distribution.
2. There is an invariance problem: the value of the reliability index depends on the specific form of the limit state function.

The invariance problem is best clarified by an example.

Example 5.3

Consider the steel beam shown in Figure 5.16.

The steel beam is assumed to be compact with parameters Z (plastic modulus) and yield stress F_y. There are four random variables to consider: P, L, Z, and F_y. It is assumed that the four variables are uncorrelated. The means and covariance matrix are given as

$$\{\mu_X\} = \begin{Bmatrix} \mu_P \\ \mu_L \\ \mu_Z \\ \mu_{F_y} \end{Bmatrix} = \begin{Bmatrix} 10\,\text{kN} \\ 8\,\text{m} \\ 100\times10^{-6}\,\text{m}^3 \\ 600\times10^{3}\,\text{kN/m}^2 \end{Bmatrix}$$

$$[C_X] = \begin{bmatrix} 4\,\text{kN}^2 & 0 & 0 & 0 \\ 0 & 10\times10^{-3}\,\text{m}^2 & 0 & 0 \\ 0 & 0 & 400\times10^{-12}\,\text{m}^6 & 0 \\ 0 & 0 & 0 & 10\times10^{9}\,(\text{kN/m}^2)^2 \end{bmatrix}$$

To begin, consider a limit state function in terms of moments. We can write

$$g_1(Z, F_y, P, L) = ZF_y - \frac{PL}{4}$$

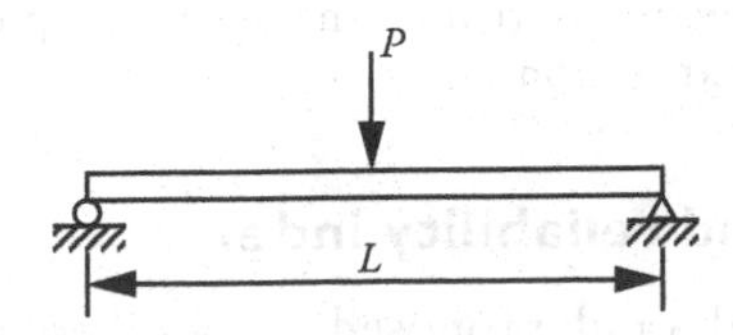

Figure 5.16 Steel beam considered in Example 5.3.

Now recall that the purpose of the limit state function is to define the boundary between the safe and unsafe domains, and the boundary corresponds to $g = 0$. Hence, if we divide g_1 by a positive quantity (for example, Z), then we are not changing the boundary or the regions in which the limit state function is positive or negative. Thus, an alternate limit state function (with units of stress) would be

$$g_2(Z,F_y,P,L) = F_y - \frac{PL}{4Z} = \frac{g_1(Z,F_y,P,L)}{Z}.$$

Since both functions satisfy the requirements for a limit state function, they are both valid. Now, we calculate the reliability index for both functions.

For the function g_1, since it is nonlinear, the calculation of the reliability index for g_1 is given by Equation 5.21. The limit state function is linearized about the means. The results are

$$g_1 \approx \left[\mu_Z\mu_{F_y} - \frac{\mu_P\mu_L}{4}\right] + \mu_{F_y}(Z-\mu_Z) + \mu_Z(F_y - \mu_{F_y}) - \frac{\mu_L}{4}(P-\mu_P)$$

$$-\frac{\mu_P}{4}(L-\mu_L)$$

$$\beta = 2.48$$

For g_2, which is also nonlinear, we use Equation 5.21 and again linearize about the mean values. The results are

$$g_2 \approx \left[\mu_{F_y} - \frac{\mu_P\mu_L}{4\mu_Z}\right] + \frac{\mu_P\mu_L}{4(\mu_Z)^2}(Z-\mu_Z) + (1)(F_y - \mu_{F_y}) - \frac{\mu_L}{4\mu_Z}(P-\mu_P)$$

$$-\frac{\mu_P}{4\mu_Z}(L-\mu_L)$$

$$\beta = 3.48.$$

This example clearly demonstrates the "invariance" in the mean value second-moment reliability index. In this example, the same fundamental limit state forms the basis for both limit state functions. Therefore, the probability of failure (as reflected by the reliability index) should be the same. It is possible to remove the invariance problem, and this is discussed in the next section.

5.3.5 Hasofer–Lind reliability index

In 1974, Hasofer and Lind proposed a modified reliability index that did not exhibit the invariance problem illustrated in Example 5.3. The "correction" is to evaluate the limit state function at a point known as the "design point" instead of the mean values. The "design point" is a point on the failure surface $g = 0$. Since this design point is generally not known in advance, an iteration technique must be used (in general) to solve for the reliability index.

Consider a limit state function $g(X_1, X_2, ..., X_n)$ where the random variables X_i are all *uncorrelated*. (If the variables are correlated, then a transformation can be used to obtain uncorrelated variables. See Example Problem 5.15.) The limit state function is rewritten in terms of the standard form of the variables (reduced variables) using

$$Z_i = \frac{X_i - \mu_{X_i}}{\sigma_{X_i}} \tag{5.22}$$

As before, the Hasofer–Lind reliability index is defined as the shortest distance from the origin of the reduced variable space to the limit state function $g = 0$.

Thus far, nothing has changed from the previous presentation of the reliability index. In fact, if the limit state function is linear, then the reliability index is still calculated as in Equation 5.18:

$$\beta = \frac{a_0 + \sum_{i=1}^{n} a_i \mu_{X_i}}{\sqrt{\sum_{i=1}^{n} (a_i \sigma_{X_i})^2}} \tag{5.18}$$

If the limit state function is *nonlinear*, however, iteration is required to find the design point $\{z_1^*, z_2^*, ..., z_n^*\}$ in reduced variable space such that β still corresponds to the shortest distance. This concept is illustrated in Figures 5.17 through 5.19 for the case of two random variables.

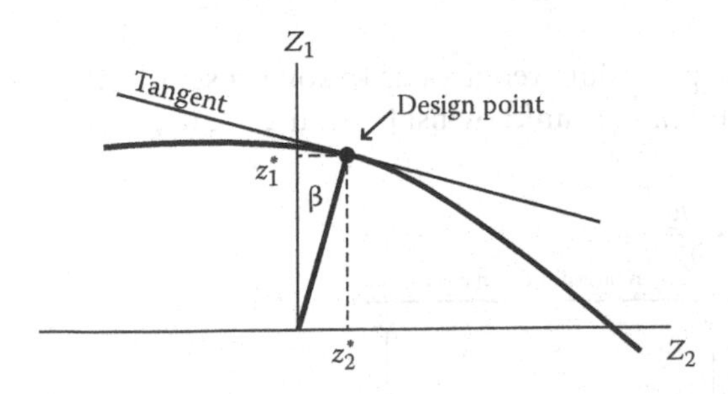

Figure 5.17 Hasofer–Lind reliability index.

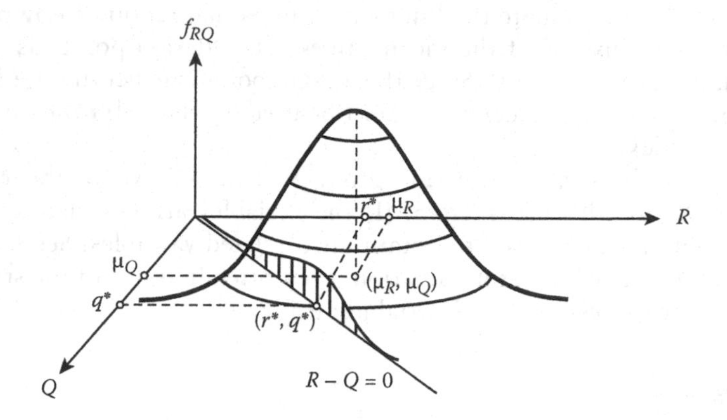

Figure 5.18 Design point on the failure boundary for the linear limit state function $g = R - Q$.

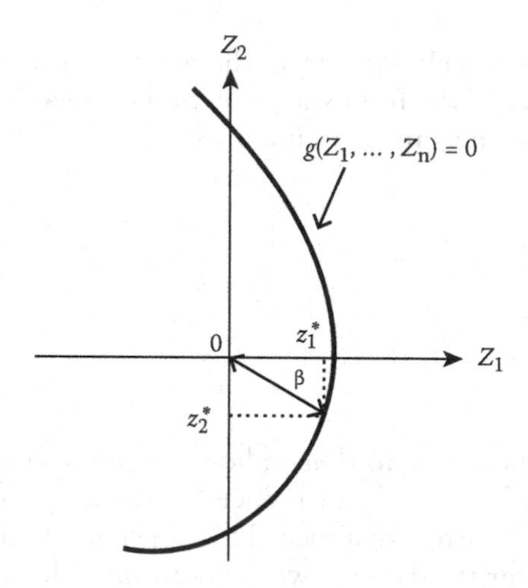

Figure 5.19 Design point and reliability index for a highly nonlinear limit state function.

The iterative procedure requires us to solve a set of $(2n + 1)$ simultaneous equations with $(2n + 1)$ unknowns: β, α_1, α_2, …, α_n, $z_1^*, z_2^*, \ldots, z_n^*$, where

$$\alpha_i = \frac{-\left.\dfrac{\partial g}{\partial Z_i}\right|_{\text{evaluated at design point}}}{\sqrt{\displaystyle\sum_{k=1}^{n}\left(\left.\dfrac{\partial g}{\partial Z_k}\right|_{\text{evaluated at design point}}\right)^2}} \tag{5.23a}$$

$$\frac{\partial g}{\partial Z_i} = \frac{\partial g}{\partial X_i}\frac{\partial X_i}{\partial Z_i} = \frac{\partial g}{\partial X_i}\sigma_{X_i} \tag{5.23b}$$

$$\sum_{i=1}^{n}(\alpha_i)^2 = 1 \tag{5.23c}$$

$$z_i^* = \beta\alpha_i \tag{5.24}$$

$$g(z_1^*, z_2^*, \ldots, z_n^*) = 0 \tag{5.25}$$

Equation 5.23b is just an application of the chain rule of differentiation. Equation 5.23c is a requirement on the values of the α_i variables, which can be confirmed by looking at Equation 5.23a. Equation 5.25 is a mathematical statement of the requirement that the design point must be on the failure boundary.

There are two alternative procedures presented below for performing the iterative analysis: the simultaneous equation procedure and the matrix procedure. The steps in each procedure are summarized below.

Simultaneous Equation Procedure

1. Formulate the limit state function and appropriate parameters for all random variables involved.
2. Express the limit state function in terms of reduced variates Z_i.
3. Use Equation 5.24 to express the limit state function in terms of β and α_i.
4. Calculate the n α_i values. Use Equation 5.24 here also to express each α_i as a function of all α_i and β.
5. Conduct the initial cycle: Assume numerical values of β and all α_i, noting that the α_i values must satisfy Equation 5.23c.
6. Use the numerical values of β and α_i on the right-hand sides of the equations formed in steps 3 and 4 above.
7. Solve the $n + 1$ simultaneous equations in step 6 for β and α_i.
8. Go back to step 6 and repeat. Iterate until the β and α_i values converge.

Matrix Procedure

1. Formulate the limit state function and appropriate parameters for all random variables X_i ($i = 1, 2, \ldots, n$) involved.
2. Obtain an initial design point $\{x_i^*\}$ by assuming values for $n - 1$ of the random variables X_i. (Mean values are often a reasonable initial choice.) Solve the limit state equation $g = 0$ for the remaining

random variable. This ensures that the design point is on the failure boundary.

3. Determine the reduced variates $\{z_i^*\}$ corresponding to the design point $\{x_i^*\}$ using

$$z_i^* = \frac{x_i^* - \mu_{X_i}}{\sigma_{X_i}} \tag{5.26}$$

4. Determine the partial derivatives of the limit state function with respect to the reduced variates using Equation 5.23b. For convenience, define a column vector $\{G\}$ as the vector whose elements are these partial derivatives multiplied by -1; i.e.,

$$\{G\} = \begin{Bmatrix} G_1 \\ G_2 \\ \vdots \\ G_n \end{Bmatrix}, \quad \text{where} \quad G_i = -\left.\frac{\partial g}{\partial Z_i}\right|_{\text{evaluated at design point}} \tag{5.27}$$

5. Calculate an estimate of β using the following formula:

$$\beta = \frac{\{G\}^T\{z^*\}}{\sqrt{\{G\}^T\{G\}}}, \quad \text{where} \quad \{z^*\} = \begin{Bmatrix} z_1^* \\ z_2^* \\ \vdots \\ z_n^* \end{Bmatrix} \tag{5.28}$$

The superscript T denotes transpose. If the limit state equation is *linear*, then Equation 5.28 reduces to Equation 5.18.

6. Calculate a column vector containing the sensitivity factors using

$$\{\alpha\} = \frac{\{G\}}{\sqrt{\{G\}^T\{G\}}} \tag{5.29}$$

7. Determine a new design point in reduced variates for $n - 1$ of the variables using

$$z_i^* = \alpha_i \beta \tag{5.30}$$

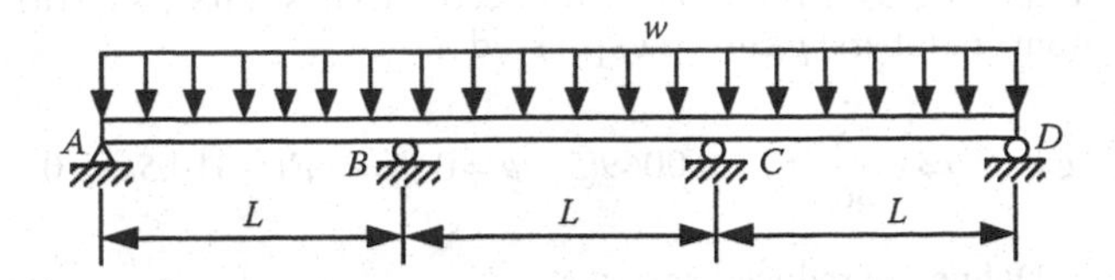

Figure 5.20 Beam considered in Example 5.4.

8. Determine the corresponding design point values in original coordinates for the $n - 1$ values in step 7 using

$$x_i^* = \mu_{X_i} + z_i^* \sigma_{X_i} \tag{5.31}$$

9. Determine the value of the remaining random variable (i.e., the one not found in steps 7 and 8) by solving the limit state function $g = 0$.
10. Repeat steps 3–9 until β and the design point $\{x_i^*\}$ converge.

Example 5.4

Calculate the Hasofer–Lind reliability index for the three-span continuous beam shown in Figure 5.20.

To solve the problem, we will follow the steps in the simultaneous equation procedure.

1. The random variables in the problem are distributed load (w), span length (L), modulus of elasticity (E), and moment of inertia (I). The limit state to be considered is deflection, and the allowable deflection is specified as $L/360$. The maximum deflection is $0.0069wL^4/EI$, and it occurs at $0.446L$ from either end (AISC, 2006). The limit state function is

$$g(w, L, E, I) = \frac{L}{360} - 0.0069\frac{wL^4}{EI}.$$

The means and standard deviations of the random variables are listed in Table 5.3.

Table 5.3 Parameters of random variables in Example 5.4

Variable	*Mean*	*Standard deviation*
w	10 kN/m	0.4 kN/m
L	5 m	~0
E	2×10^7 kN/m^2	0.5×10^7 kN/m^2
I	8×10^{-4} m^4	1.5×10^{-4} m^4

2. Express g as a function of reduced variates. First, substituting some numbers, g can be expressed as

$$g=0 \;\Rightarrow\; \frac{5}{360}EI-0.0069(5^4)w=0 \;\Rightarrow\; EI-310.5w=0$$

Define the reduced variates:

$$Z_1=\frac{I-\mu_I}{\sigma_I};\quad Z_2=\frac{E-\mu_E}{\sigma_E};\quad Z_3=\frac{w-\mu_w}{\sigma_w}$$

$$I=\mu_I+Z_1\sigma_I;\quad E=\mu_E+Z_2\sigma_E;\quad w=\mu_w+Z_3\sigma_w$$

Substitute into g.

$$(\mu_E+Z_2\sigma_E)(\mu_I+Z_1\sigma_I)-310.5(\mu_w+Z_3\sigma_w)=0$$

$$(2\times10^7+Z_2(0.5\times10^7))(8\times10^{-4}+Z_1(1.5\times10^{-4}))$$
$$-310.5(10+Z_3(0.4))=0$$

$$(3000)Z_1+(4000)Z_2+(750)Z_1Z_2-(124.2)Z_3+12{,}895=0$$

3. Formulate g in terms of β and α_i.

$$z_i^*=\beta\alpha_i$$
$$\Downarrow$$
$$3000\beta\alpha_1+4000\beta\alpha_2+750\beta^2\alpha_1\alpha_2-124.2\beta\alpha_3+12{,}895=0$$

$$\beta=\frac{-12{,}895}{3000\alpha_1+4000\alpha_2+750\beta\alpha_1\alpha_2-124.2\alpha_3}$$

4. Calculate α_i values.

$$\alpha_1=\frac{-(3000+750\beta\alpha_2)}{\sqrt{(3000+750\beta\alpha_2)^2+(4000+750\beta\alpha_1)^2+(-124.2)^2}}$$

$$\alpha_2=\frac{-(4000+750\beta\alpha_1)}{\sqrt{(3000+750\beta\alpha_2)^2+(4000+750\beta\alpha_1)^2+(-124.2)^2}}$$

$$\alpha_3=\frac{-(-124.2)}{\sqrt{(3000+750\beta\alpha_2)^2+(4000+750\beta\alpha_1)^2+(-124.2)^2}}$$

5. The iterations start with a guess for β, α_1, α_2, and α_3. For example, let us start with

$$\alpha_1 = \alpha_2 = -\sqrt{0.333} = -0.58; \quad \alpha_3 = \sqrt{0.333} = 0.58$$

and let $\beta = 3$.

6–8. The iterations are summarized in Table 5.4. Notice that between iterations 5 and 6, the values change very little; thus, the solution has converged. Faster convergence occurs when the correct signs are used for each α_i (+ for load, – for resistance). Thus, the calculated reliability index is approximately 3.17.

Example 5.5

Repeat Example 5.4 using the matrix procedure.

To illustrate the matrix procedure, we will go through each step in detail for the first iteration and then summarize the results of additional iterations in a table.

1. The limit state equation and random variable parameters are the same as in Example 5.4. For convenience, let $X_1 = I$, $X_2 = E$, and $X_3 = w$. The limit state equation in terms of X_1, X_2, and X_3 is

$$g(X_1, X_2, X_3) = \frac{L}{360} - 0.0069\frac{X_3 L^4}{X_2 X_1}.$$

2. For the first iteration, we will assume x_1^* and x_2^* are the mean values of X_1 and X_2. The value of x_3^* will be obtained by solving the limit state equation $g = 0$. Thus,

$$x_1^* = 8\times10^{-4}; \quad x_2^* = 2\times10^7 \quad \Rightarrow \quad x_3^* = \frac{L}{360}\left[\frac{x_2^* x_1^*}{(0.0069)L^4}\right]$$

$$= 0.4026\frac{x_2^* x_1^*}{L^3} = 51.53$$

Table 5.4 Iterations for Example 5.4

	Initial guess	*Iteration #* 1	2	3	4	5	6
β	3	3.664	3.429	3.213	3.175	3.173	3.173
α_1	−0.58	−0.532	−0.257	−0.153	−0.168	−0.179	−0.182
α_2	−0.58	−0.846	−0.965	−0.988	−0.985	−0.983	−0.983
α_3	+0.58	0.039	0.047	0.037	0.034	0.034	0.034

3. Determine reduced variates z_i^* for $i = 1$ to 3.

$$z_1^* = \frac{x_1^* - \mu_{X_1}}{\sigma_{X_1}} = 0; \quad z_2^* = \frac{x_2^* - \mu_{X_2}}{\sigma_{X_2}} = 0;$$

$$z_3^* = \frac{x_3^* - \mu_{X_3}}{\sigma_{X_3}} = 103.8$$

4. Determine the $\{G\}$ vector that involves the partial derivatives of g with respect to the reduced variables Z_i.

$$G_1 = -\left.\frac{\partial g}{\partial Z_1}\right|_{\{z_i^*\}} = -\left.\frac{\partial g}{\partial X_1}\right|_{\{x_i^*\}} \sigma_{X_1} = -0.0069 \frac{x_3^* L^4}{x_2^* (x_1^*)^2} \sigma_{X_1}$$

$$G_2 = -\left.\frac{\partial g}{\partial Z_2}\right|_{\{z_i^*\}} = -\left.\frac{\partial g}{\partial X_2}\right|_{\{x_i^*\}} \sigma_{X_2} = -0.0069 \frac{x_3^* L^4}{x_1^* (x_2^*)^2} \sigma_{X_2}$$

$$G_3 = -\left.\frac{\partial g}{\partial Z_3}\right|_{\{z_i^*\}} = -\left.\frac{\partial g}{\partial X_3}\right|_{\{x_i^*\}} \sigma_{X_3} = 0.0069 \frac{L^4}{x_2^* x_1^*} \sigma_{X_3}$$

Plugging in the information for the first iteration, we find the vector $\{G\}$ to be

$$\{G\} = \begin{Bmatrix} -2.604 \times 10^{-3} \\ -3.472 \times 10^{-3} \\ 1.078 \times 10^{-4} \end{Bmatrix}.$$

5. Calculate β:

$$\beta = \frac{\{G\}^T \{z^*\}}{\sqrt{\{G\}^T \{G\}}} = 2.578.$$

6. Calculate $\{\alpha\}$:

$$\{\alpha\} = \frac{\{G\}}{\sqrt{\{G\}^T \{G\}}} = \begin{Bmatrix} -0.600 \\ -0.800 \\ 0.025 \end{Bmatrix}.$$

7. Determine a new design point in reduced coordinates for $n - 1$ of the random variables. To be consistent with what we did in step 2, we will find z_1^* and z_2^*:

$$z_1^* = \alpha_1\beta = (-0.600)(2.578) = -1.546$$

$$z_2^* = \alpha_2\beta = (-0.800)(2.578) = -2.062$$

8. For the reduced variates from step 7, determine the corresponding values of the design point in original coordinates:

$$x_1^* = \mu_{X_1} + z_1^*\sigma_{X_1} = (8\times10^{-4}) + (-1.546)(1.5\times10^{-4}) = 5.68\times10^{-4}$$

$$x_2^* = \mu_{X_2} + z_2^*\sigma_{X_2} = (2\times10^{7}) + (-2.062)(0.5\times10^{7}) = 9.69\times10^{6}$$

9. Determine the updated value of x_3^* from the limit state equation $g = 0$. We can use the formula presented in step 2 for this. The result is

$$x_3^* = 17.7.$$

10. Repeat steps 3–9 until convergence is achieved. The results of the first iteration and subsequent iterations are summarized in Table 5.5. Note that the value of β (within numerical precision) obtained using this procedure is the same value obtained using the simultaneous equation procedure.

Example 5.6

Determine the required mean value of the moment of inertia for the beam in Figure 5.21 so that the reliability index is 3.0. The limit state

Table 5.5 Summary of iterations for Example Problem 5.5

	Iteration #						
	1	*2*	*3*	*4*	*5*	*6*	*7*
x_1^*	8×10^{-4}	5.68×10^{-4}	5.75×10^{-4}	6.70×10^{-4}	7.05×10^{-4}	7.12×10^{-4}	7.14×10^{-4}
x_2^*	2×10^{7}	9.69×10^{6}	5.37×10^{6}	4.54×10^{6}	4.42×10^{6}	4.38×10^{6}	4.37×10^{6}
x_3^*	51.5	17.7	9.95	9.80	10.0	10.0	10.0
↓							
β	2.58	3.29	3.21	3.18	3.18	3.18	3.18
↓							
x_1^*	5.68×10^{-4}	5.75×10^{-4}	6.70×10^{-4}	7.05×10^{-4}	7.12×10^{-4}	7.14×10^{-4}	7.14×10^{-4}
x_2^*	9.69×10^{6}	5.37×10^{6}	4.54×10^{6}	4.42×10^{6}	4.38×10^{6}	4.37×10^{6}	4.37×10^{6}
x_3^*	17.7	9.95	9.80	10.0	10.0	10.0	10.0

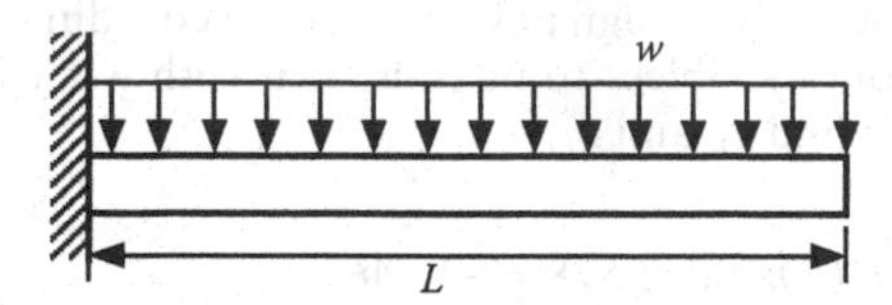

Figure 5.21 Beam considered in Example 5.6.

to be considered is the deflection limit state, and the maximum allowable deflection is $L/180$. For the given loading, the maximum deflection will occur at the free end, and its value is

$$\Delta_{max} = \frac{wL^4}{8EI}.$$

We will use the simultaneous equation procedure to solve this problem. Even though we are not trying to find β in this problem, the basic steps in the iteration are the same.

1. The variables are w, E, and I (uncorrelated). The parameters are given in Table 5.6. The length L is assumed to be deterministic (zero standard deviation).
   The limit state function is

$$g = \frac{L}{180} - \frac{wL^4}{8EI}$$

$$g = 0 \;\Rightarrow\; \frac{10}{180} - \frac{w(10^4)}{8EI} = 0 \;\Rightarrow\; EI - 22{,}500w = 0.$$

2. Express g in terms of reduced variates.

$$Z_1 = \frac{w - \mu_w}{\sigma_w}; \quad Z_2 = \frac{E - \mu_E}{\sigma_E}; \quad Z_3 = \frac{I - \mu_I}{\sigma_I} = \frac{I - \mu_I}{V_I \mu_I}$$

Table 5.6 Parameters of variables in Example 5.6

Variable	*Mean*	*Standard deviation*
w	12 kN/m	1 kN/m
L	10 m	~0
I	To be determined	10% of mean
E	200×10^6 kN/m^2	20×10^6 kN/m^2

$$w = \mu_w + Z_1\sigma_w; \quad E = \mu_E + Z_2\sigma_E; \quad I = \mu_I + Z_3\sigma_I$$

$$g = 0 \Rightarrow (\mu_E + Z_2\sigma_E)(\mu_I + Z_3\sigma_I) - 22{,}500(\mu_w + Z_1\sigma_w) = 0$$

$$(200\times10^6 + Z_2(20\times10^6))(\mu_I + Z_3(0.1\mu_I)) - 22{,}500(12 + Z_1(1)) = 0$$

$\Downarrow$ normalize

$$100\mu_I + 10\mu_I Z_2 + 10\mu_I Z_3 + \mu_I Z_3 Z_2 - 0.135 - (1.125\times10^{-2})Z_1 = 0$$

3. Formulate g in terms of β and α_i. In this problem, β is given in advance. Instead of having an equation for β, we will have an equation for μ_I.

$$z_i^* = \beta\alpha_i = 3\alpha_i$$

$$100\mu_I + 10\mu_I(3\alpha_2) + 10\mu_I(3\alpha_3) + \mu_I(3\alpha_3)(3\alpha_2) - 0.135 - (1.125\times10^{-2})(3\alpha_1) = 0$$

Solving for μ_I,

$$\mu_I = \frac{0.135 + (3.375\times10^{-2})\alpha_1}{100 + 30\alpha_2 + 30\alpha_3 + 9\alpha_3\alpha_2}$$

4. Calculate α_i values.

$$\alpha_1 = \frac{-(-1.125\times10^{-2})}{\sqrt{(-1.125\times10^{-2})^2 + (10\mu_I + \mu_I(3\alpha_3))^2 + (10\mu_I + \mu_I(3\alpha_2))^2}}$$

$$\alpha_2 = \frac{-(10\mu_I + \mu_I(3\alpha_3))}{\sqrt{(-1.125\times10^{-2})^2 + (10\mu_I + \mu_I(3\alpha_3))^2 + (10\mu_I + \mu_I(3\alpha_2))^2}}$$

$$\alpha_3 = \frac{-(10\mu_I + \mu_I(3\alpha_2))}{\sqrt{(-1.125\times10^{-2})^2 + (10\mu_I + \mu_I(3\alpha_3))^2 + (10\mu_I + \mu_I(3\alpha_2))^2}}$$

5. For this problem, the four unknowns are μ_I and α_1, α_2, α_3. The iterations start with a guess for μ_I, α_1, α_2, α_3.

6–8. The iterations are summarized in Table 5.7. The solution is $\beta = 3$ when $\mu_I = 2.29 \times 10^{-3}$ m^4.

Table 5.7 Summary of iterations for Example Problem 5.6

	Initial guess	Iteration # 1	2	3	4	5
μ_I	5×10^{-4}	2.266×10^{-3}	2.027×10^{-3}	2.288×10^{-3}	2.286×10^{-3}	2.289×10^{-3}
α_1	0.58	0.888	0.363	0.439	0.395	0.397
α_2	−0.58	−0.326	−0.659	−0.635	−0.650	−0.649
α_3	−0.58	−0.326	−0.659	−0.635	−0.650	−0.649

Example 5.7

Repeat Example 5.6 using the matrix procedure.

The matrix procedure is set up to calculate β using given information on all the parameters of the problem. As we will find out, the procedure can be modified to solve this type of problem by modifying step 5. As before, we will go through each step in detail for the first iteration and then summarize the results of additional iterations in a table.

1. The limit state equation and random variable parameters are the same as in Example 5.6. For convenience, let $X_1 = I$, $X_2 = E$, and $X_3 = w$. The limit state equation in terms of X_1, X_2, and X_3 is

$$g(X_1, X_2, X_3) = \frac{L}{180} - \frac{1}{8}\frac{X_3 L^4}{X_2 X_1}.$$

2. For the first iteration, we will assume x_2^* and x_3^* are the mean values of X_2 and X_3. The value of x_1^* will be obtained by solving the limit state equation $g = 0$. Thus,

$$x_3^* = 12; \quad x_2^* = 200 \times 10^6 \quad \Rightarrow \quad x_1^* = \frac{180}{8}\frac{x_3^* L^3}{x_2^*} = 1.35 \times 10^{-3}.$$

3. Determine reduced variates z_i^* for $i = 1$ to 3.

$$z_1^* = \frac{x_1^* - \mu_{X_1}}{\sigma_{X_1}} = \frac{1.35 \times 10^{-3} - \mu_I}{V_I \mu_I}; \quad z_2^* = \frac{x_2^* - \mu_{X_2}}{\sigma_{X_2}} = 0;$$

$$z_3^* = \frac{x_3^* - \mu_{X_3}}{\sigma_{X_3}} = 0$$

4. Determine the $\{G\}$ vector, which involves the partial derivatives of g with respect to the reduced variables Z_i.

$$G_1 = -\left.\frac{\partial g}{\partial Z_1}\right|_{\{z_i^*\}} = -\left.\frac{\partial g}{\partial X_1}\right|_{\{x_i^*\}} \sigma_{X_1} = -\frac{1}{8}\frac{x_3^* L^4}{x_2^*\left(x_1^*\right)^2} V_I \mu_I$$

$$G_2 = -\left.\frac{\partial g}{\partial Z_2}\right|_{\{z_i^*\}} = -\left.\frac{\partial g}{\partial X_2}\right|_{\{x_i^*\}} \sigma_{X_2} = -\frac{1}{8}\frac{x_3^* L^4}{x_1^*\left(x_2^*\right)^2} \sigma_{X_2}$$

$$G_3 = -\left.\frac{\partial g}{\partial Z_3}\right|_{\{z_i^*\}} = -\left.\frac{\partial g}{\partial X_3}\right|_{\{x_i^*\}} \sigma_{X_3} = \frac{1}{8}\frac{L^4}{x_2^* x_1^*} \sigma_{X_3}$$

Plugging in the information for the first iteration, we find the vector {G} to be

$$\{G\} = \begin{Bmatrix} -4.115\mu_I \\ -5.556\times10^{-3} \\ 4.630\times10^{-3} \end{Bmatrix}$$

5. Calculate β. Here is where the modification comes in. Instead of calculating β, we already know the target β. Hence, we need to use this equation to predict a value of μ_I *for each iteration.*

$$\beta = 3 = \frac{\{G\}^T\{z^*\}}{\sqrt{\{G\}^T\{G\}}} \quad \Rightarrow \quad \text{find } \mu_I$$

This is easier said than done. However, if you carry out the matrix multiplication and do some algebra, the expression is a quadratic equation in μ_I. Skipping the intermediate algebra, the result for the first iteration is

$\mu_I = 2.193 \times 10^{-3}$.

6. Calculate $\{\alpha\}$: Here, we must use the estimated μ_I from step 5.

$$\{\alpha\} = \frac{\{G\}}{\sqrt{\{G\}^T\{G\}}} = \begin{Bmatrix} -0.780 \\ -0.480 \\ 0.400 \end{Bmatrix}$$

7. Determine a new design point in reduced coordinates for $n - 1$ of the random variables. To be consistent with what we did in step 2, we will find z_2^* and z_3^*.

$$z_2^* = \alpha_2\beta = (-0.480)(3) = -1.441$$

$$z_3^* = \alpha_3\beta = (0.400)(3) = 1.200$$

8. For the reduced variates from step 7, determine the corresponding values of the design point in original coordinates.

$$x_2^* = \mu_{X_2} + z_2^*\sigma_{X_2} = (200\times10^6) + (-1.441)(20\times10^6) = 1.712\times10^8$$

$$x_3^* = \mu_{X_3} + z_3^*\sigma_{X_3} = (12) + (1.200)(1) = 13.20$$

9. Determine the updated value of x_1^* from the limit state equation $g = 0$. We can use the formula presented in step 2 for this. The result is

$$x_1^* = 1.735\times10^{-3}.$$

10. Repeat until convergence is achieved. The results of the first iteration outlined above and subsequent iterations are summarized in Table 5.8.

For $\beta = 3$, the mean value of the moment of inertia must be about 2.28×10^{-3} m^4. This is the same result (within numerical roundoff error) that we obtained using the simultaneous equation procedure.

Table 5.8 Summary of iterations for Example Problem 5.7

	Iteration #				
	1	*2*	*3*	*4*	*5*
x_1^*	1.35×10^{-3}	1.74×10^{-3}	1.82×10^{-3}	1.84×10^{-3}	1.84×10^{-3}
x_2^*	2.0×10^{8}	1.71×10^{8}	1.63×10^{8}	1.62×10^{8}	1.61×10^{8}
x_3^*	12	13.2	13.2	13.2	13.2
↓					
μ_I	2.19×10^{-3}	2.29×10^{-3}	2.29×10^{-3}	2.30×10^{-3}	2.28×10^{-3}
↓					
x_1^*	1.74×10^{-3}	1.82×10^{-3}	1.84×10^{-3}	1.84×10^{-3}	1.84×10^{-3}
x_2^*	1.71×10^{8}	1.63×10^{8}	1.62×10^{8}	1.61×10^{8}	1.61×10^{8}
x_3^*	13.2	13.2	13.2	13.2	13.2

5.4 RACKWITZ–FIESSLER PROCEDURE

5.4.1 Modified matrix procedure

We have now covered methods for calculating reliability indices using information on the means and standard deviations of the random variables. Detailed information on the type of distribution for each random variable was not needed. If we know the distributions of the random variables, then we can improve our procedures to calculate reliability indices. In this section, we introduce the Rackwitz–Fiessler procedure for calculating reliability indices. The method requires the knowledge of the probability distributions for all the variables involved.

The basic idea behind the procedure begins with the calculation of "equivalent normal" values of the mean and standard deviation for each nonnormal random variable. We use these equivalent normal parameters in our analysis. Suppose that a particular random variable X with mean μ_X and standard deviation σ_X is described by a CDF $F_X(x)$ and a PDF $f_X(x)$. To obtain the equivalent normal mean μ_X^e and standard deviation σ_X^e, we require that the CDF and PDF of the actual function be equal to the normal CDF and normal PDF at the value of the variable x^* on the failure boundary described by $g = 0$. Mathematically, these requirements are expressed as

$$F_X(x^*) = \Phi\left(\frac{x^* - \mu_X^e}{\sigma_X^e}\right) \tag{5.32}$$

$$f_X(x^*) = \frac{1}{\sigma_X^e}\phi\left(\frac{x^* - \mu_X^e}{\sigma_X^e}\right) \tag{5.33}$$

where Φ is the CDF for the standard normal distribution and Φ is the PDF for the standard normal distribution. Equation 5.32 simply requires the cumulative probabilities to be equal at x^*. Equation 5.33 is obtained by differentiating both sides of Equation 5.32 with respect to x^*. By manipulating these equations, we can obtain expressions for μ_X^e and σ_X^e as follows:

$$\mu_X^e = x^* - \sigma_X^e\left[\Phi^{-1}\left(F_X(x^*)\right)\right] \tag{5.34}$$

$$\sigma_X^e = \frac{1}{f_X(x^*)}\phi\left(\frac{x^* - \mu_X^e}{\sigma_X^e}\right) = \frac{1}{f_X(x^*)}\phi\left[\Phi^{-1}\left(F_X(x^*)\right)\right] \tag{5.35}$$

The basic steps in the iteration procedure are the same as those in the matrix procedure given in Section 5.3.5 except that we must add an additional step to calculate the equivalent normal parameters. The same procedure applies to both linear and nonlinear limit state functions. The steps in the matrix procedure implementing the Rackwitz–Fiessler modification are as follows:

1. Formulate the limit state function. Determine the probability distributions and appropriate parameters for all random variables $X_i (i = 1, 2, \ldots, n)$ involved.
2. Obtain an initial design point $\{x_i^*\}$ by assuming values for $n - 1$ of the random variables X_i. (Mean values are often a reasonable choice.) Solve the limit state equation $g = 0$ for the remaining random variable. This ensures that the design point is on the failure boundary.
3. For each of the design point values x_i^* corresponding to a nonnormal distribution, determine the equivalent normal mean $\mu_{X_i}^e$ and standard deviation $\sigma_{X_i}^e$ using Equations 5.34 and 5.35. If one or more x_i^* values correspond to a normal distribution, then the equivalent normal parameters are simply the actual parameters.
4. Determine the reduced variates $\{z_i^*\}$ corresponding to the design point $\{x_i^*\}$ using

$$z_i^* = \frac{x_i^* - \mu_{X_i}^e}{\sigma_{X_i}^e} \tag{5.36}$$

5. Determine the partial derivatives of the limit state function with respect to the reduced variates using Equation 5.23b. For convenience, define a column vector $\{G\}$ as the vector whose elements are these partial derivatives multiplied by –1; i.e.,

$$\{G\} = \begin{Bmatrix} G_1 \\ G_2 \\ \vdots \\ G_n \end{Bmatrix}, \quad \text{where} \quad G_i = -\left.\frac{\partial g}{\partial Z_i}\right|_{\text{evaluated at design point}} \tag{5.37}$$

6. Calculate an estimate of β using the following formula:

$$\beta = \frac{\{G\}^T \{z^*\}}{\sqrt{\{G\}^T \{G\}}}, \quad \text{where} \quad \{z^*\} = \begin{Bmatrix} z_1^* \\ z_2^* \\ \vdots \\ z_n^* \end{Bmatrix} \tag{5.38a}$$

The superscript T denotes transpose. For *linear* performance functions, Equation 5.38a simplifies to

$$\beta = \frac{a_0 + \sum_{i=1}^{n} a_i \mu_{X_i}^{e}}{\sqrt{\sum_{i=1}^{n} (a_i \sigma_{X_i}^{e})^2}} \tag{5.38b}$$

7. Calculate a column vector containing the sensitivity factors using

$$\{\alpha\} = \frac{\{G\}}{\sqrt{\{G\}^{T}\{G\}}} \tag{5.39}$$

8. Determine a new design point in reduced variates for $n - 1$ of the variables using

$$z_i^* = \alpha_i \beta \tag{5.40}$$

9. Determine the corresponding design point values in original coordinates for the $n - 1$ values in step 7 using

$$x_i^* = \mu_{X_i}^{e} + z_i^* \sigma_{X_i}^{e} \tag{5.41}$$

10. Determine the value of the remaining random variable (i.e., the one not found in steps 8 and 9) by solving the limit state function $g = 0$.
11. Repeat steps 3–10 until β and the design point $\{x_i^*\}$ converge.

The procedure is best illustrated with examples.

Example 5.8

Show that the equivalent normal parameters for a lognormal random variable X can be calculated using

$$\sigma_X^{e} = x^* \sigma_{\ln X} \qquad \mu_X^{e} = x^* \left[1 - \ln(x^*) + \mu_{\ln X}\right],$$

where x^* is the design point value and $\sigma_{\ln X}$ and $\mu_{\ln X}$ are the distribution parameters for the lognormal distribution.

From Chapter 2, we know that the PDF and CDF for a lognormal random variable X are

$$f_X(x^*) = \frac{1}{x^* \sigma_{\ln X}} \phi\left(\frac{\ln(x^*) - \mu_{\ln X}}{\sigma_{\ln X}}\right)$$

$$F_X(x^*) = \Phi\left(\frac{\ln(x^*) - \mu_{\ln X}}{\sigma_{\ln X}}\right),$$

where ϕ and Φ are the PDF and CDF, respectively, for a standard normal random variable. Using these relationships with Equations 5.34 and 5.35, the equivalent mean and standard deviation are

$$\begin{aligned}
\sigma_X^e &= \frac{1}{f_X(x^*)} \phi\left[\Phi^{-1}\left(F_X(x^*)\right)\right] \\
&= \frac{1}{f_X(x^*)} \phi\left[\Phi^{-1}\left(\Phi\left(\frac{\ln(x^*) - \mu_{\ln X}}{\sigma_{\ln X}}\right)\right)\right] \\
&= \frac{1}{f_X(x^*)} \phi\left(\frac{\ln(x^*) - \mu_{\ln X}}{\sigma_{\ln X}}\right) \\
&= \frac{1}{f_X(x^*)}\left[x^* \sigma_{\ln X} f_X(x^*)\right] \\
&= x^* \sigma_{\ln X}
\end{aligned}$$

$$\begin{aligned}
\mu_X^e &= x^* - \sigma_X^e\left[\Phi^{-1}\left(F_X(x^*)\right)\right] \\
&= x^* - \sigma_X^e\left[\Phi^{-1}\left(\Phi\left(\frac{\ln(x^*) - \mu_{\ln X}}{\sigma_{\ln X}}\right)\right)\right] \\
&= x^* - \sigma_X^e\left(\frac{\ln(x^*) - \mu_{\ln X}}{\sigma_{\ln X}}\right) \\
&= x^* - x^* \sigma_{\ln X}\left(\frac{\ln(x^*) - \mu_{\ln X}}{\sigma_{\ln X}}\right) \\
&= x^*\left[1 - \ln(x^*) + \mu_{\ln X}\right]
\end{aligned}$$

Example 5.9

The modified matrix procedure is demonstrated on a simple case of two uncorrelated variables. Let R be the resistance and Q be the load effect. The limit state function is

$$g(R,Q) = R - Q.$$

The variable R is lognormally distributed with mean $\mu_R = 200$ and standard deviation $\sigma_R = 20$. The variable Q follows an extreme Type I distribution with $\mu_Q = 100$ and $\sigma_Q = 12$. Our objective is to calculate β.

The first cycle of the iteration proceeds as follows. All subsequent iterations are shown in Table 5.9.

1. Formulate limit state function and probability distributions. This has been done.
2. Guess an initial design point. Try $r^* = 150$ as an arbitrary guess. Then, from the limit state equation $g = 0$, we get $q^* = 150$.
3. Determine equivalent normal parameters. For R, since it is lognormal, we know that

$$\sigma_{\ln R}^2 = \ln\left(1 + \frac{\sigma_R^2}{\mu_R^2}\right) = 9.95 \times 10^{-3} \quad \Rightarrow \quad \sigma_{\ln R} = 0.0998$$

$$\mu_{\ln R} = \ln(\mu_R) - 0.5\sigma_{\ln R}^2 = 5.29.$$

Therefore, using the results of Example 5.8,

$$\sigma_R^e = r^* \sigma_{\ln R} = (150)(0.0998) = 15.0$$

$$\begin{aligned} \mu_R^e &= r^*\left[1 - \ln(r^*) + \mu_{\ln R}\right] \\ &= (150)\left[1 - \ln(150) + 5.29\right] \\ &= 192. \end{aligned}$$

Since Q follows an extreme Type I distribution, we know (from Chapter 2) that

$$F_Q(q) = \exp\left[-\exp\left(-a(q-u)\right)\right]$$

$$f_Q(q) = a\left\{\exp\left(-a(q-u)\right)\right\}\exp\left[-\exp\left(-a(q-u)\right)\right],$$

where a and u are distribution parameters that are related to the mean and standard deviation of Q by

$$u = \mu_Q - \frac{0.5772}{a}; \quad a = \sqrt{\frac{\pi^2}{6\sigma_Q^2}}.$$

Plugging in the values of μ_Q and σ_Q, we find $a = 0.107$ and $u = 94.6$. The values of the CDF and PDF at q^* are

$$F_Q(q^*) = 0.997; \quad f_Q(q^*) = 2.86 \times 10^{-4}.$$

Thus, the equivalent normal parameters for Q are

$$\sigma_Q^e = \frac{1}{f_Q(q^*)} \phi\left[\Phi^{-1}\left(F_Q(q^*)\right)\right] = \frac{1}{2.86 \times 10^{-4}} \phi\left[\Phi^{-1}(0.997)\right] = 28.9$$

$$\mu_Q^e = q^* - \sigma_Q^e\left[\Phi^{-1}\left(F_Q(q^*)\right)\right] = 69.5.$$

4. Determine the values of the reduced variates. Let z_1^* be the reduced variate for r^* and let z_2^* be the reduced variate for q^*.

$$z_1^* = \frac{r^* - \mu_R^e}{\sigma_R^e} = -2.83; \quad z_2^* = \frac{q^* - \mu_Q^e}{\sigma_Q^e} = 2.78$$

5. Determine the $\{G\}$ vector.

$$G_1 = -\left.\frac{\partial g}{\partial Z_1}\right|_{\{z_i^*\}} = -\left.\frac{\partial g}{\partial R}\right|_{\{x_i^*\}} \sigma_R^e = -1\sigma_R^e$$

$$G_2 = -\left.\frac{\partial g}{\partial Z_2}\right|_{\{z_i^*\}} = -\left.\frac{\partial g}{\partial Q}\right|_{\{x_i^*\}} \sigma_Q^e = +1\sigma_Q^e$$

6. Calculate an estimate of β.

$$\beta = \frac{\{G\}^T\{z^*\}}{\sqrt{\{G\}^T\{G\}}} = 3.78$$

7. Calculate $\{\alpha\}$:

$$\{\alpha\} = \frac{\{G\}}{\sqrt{\{G\}^T\{G\}}} = \begin{Bmatrix} -0.460 \\ 0.888 \end{Bmatrix}.$$

Table 5.9 Iterations for Example Problem 5.9

	Iteration #		
	1	2	3
r^*	150	166	168
q^*	150	166	168
↓			
β	3.78	3.76	3.76
↓			
r^*	166	168	168
q^*	166	168	168

8. Determine new values of z_i^* for $n - 1$ of the variables. We will find z_1^* using

$$z_1^* = \alpha_1\beta = (-0.460)(3.78) = -1.74$$

9. Determine the value of r^* using the updated z_1^*:

$$r^* = \mu_R^e + z_1^*\sigma_R^e = 166.$$

10. Determine the value of q^* using the limit state equation $g = 0$. For this case, $q^* = r^* = 166$.
11. Iterate until the value of β and the design point converge. The results of subsequent iterations are shown in Table 5.9. Observe that iteration is necessary even when the limit state function is linear. This is because we are using the equivalent normal parameters in the iteration, and they change during the iteration process.

Example 5.10

The limit state function $g(R,Q) = R - Q$ is frequently used to simplify complex reliability problems. The capacity R is often modeled using a lognormal random variable, and the total load effect Q is modeled as a normal random variable. There is a convenient form of Equation 5.38b that can be used to simplify the iteration process in this special case, which is derived below.

Since Q is normal, $\mu_Q^e = \mu_Q$ and $\sigma_Q^e = \sigma_Q$. From Example 5.8, the equivalent normal parameters for the lognormal random variable R are

$$\sigma_R^e = r^*\sigma_{\ln R} \qquad \mu_R^e = r^*\left[1 - \ln(r^*) + \mu_{\ln R}\right].$$

If the COV of R, V_R, is less than 0.20, we can simplify these expressions by using the approximations presented in Equations 2.50 and 2.51 in Chapter 2.

$$\sigma_R^e \approx r^* V_R \qquad \mu_R^e \approx r^*\left[1-\ln(r^*)+\ln(\mu_R)\right]$$

$$\approx r^*\left[1-\ln\left(\frac{r^*}{\mu_R}\right)\right]$$

Substituting these expressions into Equation 5.38b, we get

$$\beta = \frac{a_0 + \sum_{i=1}^{n} a_i \mu_{X_i}^e}{\sqrt{\sum_{i=1}^{n} (a_i \sigma_{X_i}^e)^2}}$$

$$= \frac{0+(1)\mu_R^e + (-1)\mu_Q^e}{\sqrt{\left((1)\sigma_R^e\right)^2 + \left((-1)\sigma_Q^e\right)^2}}$$

$$= \frac{r^*\left[1-\ln\left(r^*/\mu_R\right)\right] - \mu_Q}{\sqrt{(r^* V_R)^2 + (\sigma_Q)^2}}.$$

In selecting the value of r^*, it is convenient to select a value that is some number of standard deviations away from the true mean value. Mathematically, this means that

$$r^* = \mu_R - k\sigma_R = \mu_R[1 - kV_R]$$

or, alternatively,

$$r^* = \lambda_R R_n[1 - kV_R],$$

where k measures the shift away from the mean value in standard deviation units, λ_R is the bias factor, and R_n is the nominal design value of R. When expressed in this form, the ratio r^*/μ_R is equal to $(1 - kV_R)$. Substituting these results into the expression for β above gives

$$\beta = \frac{r^*\left[1-\ln\left(r^*/\mu_R\right)\right]-\mu_Q}{\sqrt{\left(r^*V_R\right)^2+\left(\sigma_Q\right)^2}}$$

$$= \frac{\lambda_R R_n\left[1-kV_R\right]\left[1-\ln(1-kV_R)\right]-\mu_Q}{\sqrt{\left(\lambda_R R_n\left[1-kV_R\right]V_R\right)^2+\left(\sigma_Q\right)^2}}.$$

This expression shows how the parameters of R and Q and the parameter k influence the value of β.

Example 5.11

Calculate the reliability index for a steel compact beam with the following limit state function

$$g(Z,F_y,M) = ZF_y - M$$

The distributions and distribution parameters are listed in Table 5.10.

We will follow the step-by-step matrix procedure to solve the problem. The steps shown below are for the first cycle of iteration. The results of subsequent steps are summarized in Table 5.11.

1. Formulate the limit state function and probability distributions. This has been done. In this analysis, we will refer to Z as variable X_1, F_y as variable X_2, and M as variable X_3.
2. Guess an initial design point. We will assume values for x_1^* and x_2^* and use $g = 0$ to find x_3^*. Try $x_1^* = 100$ and $x_2^* = 40$. Then, using $g = 0$, we find x_3^* must be 4000.
3. Determine equivalent normal parameters. Since $X_1(Z)$ is normal, we do not need to find equivalent normal parameters. For $X_2(F_y)$, since it is lognormal, we know that

$$\sigma_{\ln F_y}^2 = \ln\left(1+\frac{\sigma_{F_y}^2}{\mu_{F_y}^2}\right) = 9.95\times10^{-3} \quad\Rightarrow\quad \sigma_{\ln F_y} = 0.0998$$

$$\mu_{\ln F_y} = \ln(\mu_{F_y}) - 0.5\sigma_{\ln F_y}^2 = 3.68.$$

Table 5.10 Parameters of variables in Example 5.11

Variable	*Distribution*	*Mean*	*COV*
Z	Normal	100	4%
F_y	Lognormal	40	10%
M	Extreme Type I	2000	10%

Therefore,

$$F_{F_y}(x_2^*) = \Phi\left(\frac{\ln x_2^* - \mu_{\ln F_y}}{\sigma_{\ln F_y}}\right) = \Phi(0.0499) = 0.520$$

$$f_{F_y}(x_2^*) = \frac{1}{\sqrt{2\pi}}\frac{1}{\sigma_{\ln F_y} x_2^*}\exp\left[-\frac{1}{2}\left(\frac{\ln x_2^* - \mu_{\ln F_y}}{\sigma_{\ln F_y}}\right)^2\right] = 0.0999.$$

The equivalent normal parameters for $X_2(F_y)$ are therefore

$$\sigma_{F_y}^e = \frac{1}{f_{F_y}(x_2^*)}\phi\left[\Phi^{-1}\left(F_{F_y}(x_2^*)\right)\right] = \frac{1}{0.0999}\phi(0.0499) = 3.99$$

$$\mu_{F_y}^e = x_2^* - \sigma_{F_y}^e\left[\Phi^{-1}\left(F_{F_y}(x_2^*)\right)\right] = 39.8.$$

(We could have used the expressions derived in Example 5.8 to find these parameters.) Now we need to go through similar calculations for M. Since M follows an extreme Type I distribution, we know from Equations 2.58 and 2.59 that

$$F_M(m) = \exp\left[-\exp\left(-a(m-u)\right)\right]$$

$$f_M(m) = a\left\{\exp\left(-a(m-u)\right)\right\}\exp\left[-\exp\left(-a(m-u)\right)\right],$$

where a and u are distribution parameters that are related to the mean and standard deviation of M by

$$u = \mu_M - \frac{0.5772}{a}; \quad a = \sqrt{\frac{\pi^2}{6\sigma_M^2}}.$$

Plugging in the values of μ_M and σ_M, we find $a = 0.00641$ and $u = 1910$. The values of the CDF and PDF at x_3^* are

$$F_M(x_3^*) = 9.99\times10^{-1}; \quad f_M(x_3^*) = 9.69\times10^{-9}.$$

Thus, the equivalent normal parameters for M are

$$\sigma_M^e = \frac{1}{f_M(x_3^*)}\phi\left[\Phi^{-1}\left(F_M(x_3^*)\right)\right] = \frac{1}{9.69\times10^{-9}}\phi\left[\Phi^{-1}(0.999)\right] = 759$$

$$\mu_M^e = x_3^* - \sigma_M^e\left[\Phi^{-1}\left(F_M(x_3^*)\right)\right] = 456.$$

4. Determine the values of the reduced variates

$$z_1^* = \frac{x_1^* - \mu_Z}{\sigma_Z} = 0; \quad z_2^* = \frac{x_2^* - \mu_{F_y}^e}{\sigma_{F_y}^e} = 0.0501; \quad z_2^* = \frac{x_3^* - \mu_M^e}{\sigma_M^e} = 4.69.$$

5. Determine the $\{G\}$ vector.

$$G_1 = -\left.\frac{\partial g}{\partial Z_1}\right|_{\{z_i^*\}} = -\left.\frac{\partial g}{\partial X_1}\right|_{\{x_i^*\}} \sigma_Z = -x_2^* \sigma_Z$$

$$G_2 = -\left.\frac{\partial g}{\partial Z_2}\right|_{\{z_i^*\}} = -\left.\frac{\partial g}{\partial X_2}\right|_{\{x_i^*\}} \sigma_{F_y}^e = -x_1^* \sigma_{F_y}^e$$

$$G_3 = -\left.\frac{\partial g}{\partial Z_3}\right|_{\{z_i^*\}} = -\left.\frac{\partial g}{\partial X_3}\right|_{\{x_i^*\}} \sigma_M^e = (1)\sigma_M^e$$

6. Calculate an estimate of β.

$$\beta = \frac{\{G\}^T\{z^*\}}{\sqrt{\{G\}^T\{G\}}} = 4.04$$

7. Calculate $\{\alpha\}$:

$$\{\alpha\} = \frac{\{G\}}{\sqrt{\{G\}^T\{G\}}} = \begin{Bmatrix} -0.183 \\ -0.457 \\ 0.870 \end{Bmatrix}.$$

8. Determine new values of z_i^* for $n - 1$ of the variables. We will find z_1^* and z_2^* using

$$z_1^* = \alpha_1\beta = (-0.183)(4.04) = -0.739$$

$$z_2^* = \alpha_2\beta = (-0.457)(4.04) = -1.85.$$

Table 5.11 Iterations for Example Problem 5.11

	1	2	3	4	5
x_1^*	100	97.0	96.8	96.8	96.8
x_2^*	40	32.4	32.9	32.9	32.9
x_3^*	4000	3143	3186	3189	3189
↓					
β	4.04 ↗	4.02 ↗	4.02 ↗	4.03 ↗	4.03
↓					
x_1^*	97.0	96.8	96.8	96.8	96.8
x_2^*	32.4	32.9	32.9	32.9	32.9
x_3^*	3143	3186	3189	3189	3189

9. Determine the value of x_1^* and x_2^* using the updated z_1^* and z_2^*:

$$x_1^* = \mu_Z + z_1^* \sigma_Z = 97.0$$

$$x_2^* = \mu_{F_y}^e + z_2^* \sigma_{F_y}^e = 32.4.$$

10. Determine the value of x_3^* using the limit state equation $g = 0$. For this case, $x_3^* = 3143$.
11. Iterate until the value of β and the design point converge. The results of subsequent iterations are shown in Table 5.11. The final value of β is 4.03.

5.4.2 Graphical procedure

A graphical version of the Rackwitz–Fiessler procedure (i.e., using equivalent normal parameters) can be applied when the CDFs of the basic variables are available as plots on normal probability paper. Each nonnormal variable is approximated by a normal distribution, which is represented by a straight line. The value of the CDF of the approximating normal variable is the same at the design point as that of the original distribution. On normal probability paper, this means that the straight line intersects with the original CDF at the design point. Furthermore, at the design point, the PDFs of the original variable and approximating normal are the same. Since the PDF is a tangent (first derivative) of the CDF, the straight line (approximating normal) is tangent to the original CDF at the design point. The parameters of the approximating normal distribution (mean and standard deviation) can be read directly from the graph. This graphical approach is illustrated in Figure 5.22.

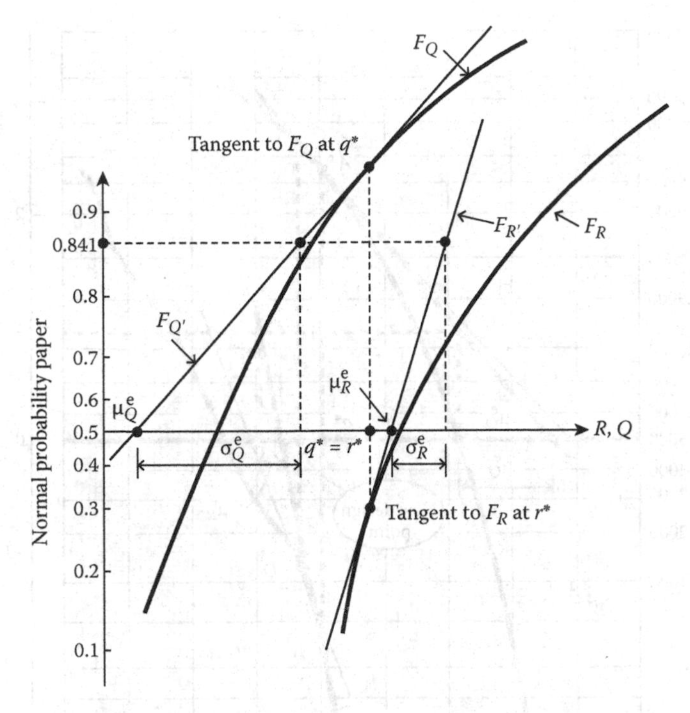

Figure 5.22 **Graphical illustration of Rackwitz–Fiessler procedure.**

Example 5.12

We will demonstrate the calculation of the reliability index using the graphical procedure for the limit state function

$$g(R,Q) = R - Q,$$

where R is the variable representing resistance and Q is the variable representing the load effect. The CDFs for R and Q are plotted on normal probability paper as shown in Figure 5.23.

The first step, as in the analytical procedure, is to guess the initial design point (r^*, q^*). For the considered limit state function, $r^* = q^*$, the initial value of r^* and q^* can be taken somewhere between the means of R and Q. The final value of the design point can be expected to be closer to the variable with a lower degree of variation (steeper tail on the normal probability paper).

The points corresponding to the design point are marked on the original CDFs of R and Q. A ruler is used to plot straight lines representing the approximating normal distributions. These lines must be tangent to the CDFs and they must pass through the marked points. The accuracy of plotting the tangent lines depends on the quality of the drafting equipment and the physical conditions of the individual.

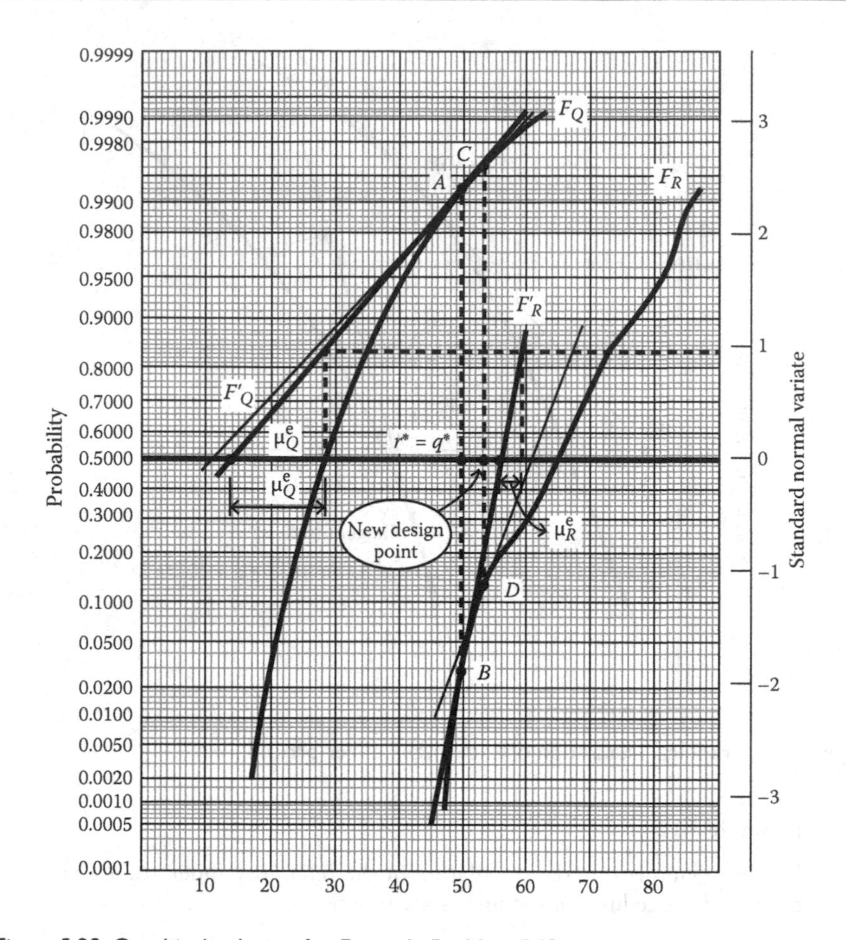

Figure 5.23 Graphical solution for Example Problem 5.12.

The basic steps in the procedure are as follows:

1. Guess the initial value of the design point. For this example, assume $r^* = q^* = 50$ ksi (1 ksi = 6.895 MPa). Mark points A and B on the plots of F_R and F_Q, respectively.
2. Plot tangents to F_R and F_Q at A and B.
3. Read $\mu_R^e, \sigma_R^e, \mu_Q^e, \sigma_Q^e$ directly from the graph (probability paper):

$$\mu_R^e = 56 \text{ ksi} \quad \sigma_R^e = 3.5 \text{ ksi}$$

$$\mu_Q^e = 14 \text{ ksi} \quad \sigma_Q^e = 14.5 \text{ ksi.}$$

4. Calculate β. Using Equation 5.38b,

$$\beta = \frac{\mu_R^e - \mu_Q^e}{\sqrt{\left(\sigma_R^e\right)^2 + \left(\sigma_Q^e\right)^2}} = \frac{56-14}{\sqrt{(3.5)^2 + (14.5)^2}} = 2.82.$$

5. Calculate a new design point. This is done using Equation 5.31 with $z_i^* = \alpha_i \beta$. In this case, calculating α_R using Equations 5.23a and 5.23b, we get

$$r^* = \mu_R^e - \frac{\left(\sigma_R^e\right)^2 \beta}{\sqrt{(\sigma_R^e)^2 + (\sigma_Q^e)^2}} = 56 - \frac{(3.5)^2(2.82)}{\sqrt{(3.5)^2 + (14.5)^2}} = 53.7 \text{ ksi}$$

   The value of q^* is found from the requirement $g = 0$. Therefore, $q^* = r^*$.

6. Plot tangents to F_R and F_Q at C and D using the updated design point.
7. Read $\mu_R^e, \sigma_R^e, \mu_Q^e, \sigma_Q^e$ directly from the graph (probability paper):

$$\mu_R^e = 61 \text{ ksi} \qquad \sigma_R^e = 6.5 \text{ ksi}$$

$$\mu_Q^e = 11.5 \text{ ksi} \qquad \sigma_Q^e = 15.5 \text{ ksi}.$$

8. Calculate new β and design point,

$$\beta = 2.94$$

$$r^* = q^* = 53.6.$$

9. The process would continue until β converges to a value.

5.4.3 Correlated random variables

Thus far, we have considered limit state equations in which the random variables are all uncorrelated. However, in many practical applications, some of the random variables may be correlated, and this correlation can have a significant impact on the calculated reliability index.

To deal with correlated random variables, we can take two approaches:

1. Use a coordinate transformation such as discussed in Chapter 4, Section 4.1.8. This approach can become messy when dealing with the Rackwitz–Fiessler procedure involving equivalent normal parameters.
2. Modify the procedure presented in Section 5.4.1 by introducing a correlation matrix $[\rho]$ into Equations 5.38a and 5.39. The correlation

matrix $[\rho]$ is the matrix of correlation coefficients for the random variables involved in the limit state equation. The modified equations are as follows:

$$\beta = \frac{\{G\}^T\{z^*\}}{\sqrt{\{G\}^T\{G\}}} \quad \text{changes to} \quad \beta = \frac{\{G\}^T\{z^*\}}{\sqrt{\{G\}^T[\rho]\{G\}}} \tag{5.42}$$

$$\{\alpha\} = \frac{\{G\}}{\sqrt{\{G\}^T\{G\}}} \quad \text{changes to} \quad \{\alpha\} = \frac{[\rho]\{G\}}{\sqrt{\{G\}^T[\rho]\{G\}}} \tag{5.43}$$

Example 5.13

You are given the following limit state equation:

$$g(X_1,X_2,X_3) = 25X_1 - 30X_2 - 20X_3$$

The distributions, distribution parameters, and correlations are given in Table 5.12. Determine the reliability index β.

We will follow the step-by-step procedure to solve the problem. The steps shown below are for the first cycle of iteration. The results of subsequent steps will be summarized in Table 5.13.

Table 5.12 Parameters for variables in Example 5.13

Variable	*Distribution*	*Mean*	*Standard deviation*
X_1	Normal	14	1.3
X_2	Lognormal	5	0.75
X_3	Normal	6	0.3

Correlation coefficient for X_2 and X_3 is 0.7.

Correlation coefficient for X_1 and X_3 is 0.2.

Correlation coefficient for X_2 and X_1 is 0.5.

1. Formulate limit state function and probability distributions. This has been done.
2. Guess an initial design point. We will assume values for x_1^* and x_2^* and then use the limit state condition $g = 0$ to find x_3^*. Try $x_1^* = 14$ and $x_2^* = 5$. Then, using $g = 0$, we find x_3^* must be 10.
3. Determine equivalent normal parameters. Since X_1 and X_3 are normal, we do not have to find equivalent normal parameters. For X_2, since it is lognormal, we can use the formulas derived in Example 5.8. Skipping the math, the equivalent normal parameters for the first iteration are

$$\sigma_{X_2}^e = 0.7458 \quad \mu_{X_2}^e = 4.994.$$

4. Determine the values of the reduced variates using Equation 5.36:

$$z_1^* = 0; \quad z_2^* = 0.07458; \quad z_3^* = 13.33.$$

5. Determine the {G} vector.

$$G_1 = -\left.\frac{\partial g}{\partial Z_1}\right|_{\{z_i^*\}} = -\left.\frac{\partial g}{\partial X_1}\right|_{\{x_i^*\}} \sigma_{X_1} = -25\sigma_{X_1}$$

$$G_2 = -\left.\frac{\partial g}{\partial Z_2}\right|_{\{z_i^*\}} = -\left.\frac{\partial g}{\partial X_2}\right|_{\{x_i^*\}} \sigma_{X_2}^e = 30\sigma_{X_2}^e$$

$$G_3 = -\left.\frac{\partial g}{\partial Z_3}\right|_{\{z_i^*\}} = -\left.\frac{\partial g}{\partial X_3}\right|_{\{x_i^*\}} \sigma_{X_3} = 20\sigma_{X_3}$$

6. Calculate an estimate of β.

$$\beta = \frac{\{G\}^T\{z^*\}}{\sqrt{\{G\}^T[\rho]\{G\}}} = 2.615 \quad \text{where} \quad [\rho] = \begin{bmatrix} 1 & 0.5 & 0.2 \\ 0.5 & 1 & 0.7 \\ 0.2 & 0.7 & 1 \end{bmatrix}$$

7. Calculate {α}:

$$\{\alpha\} = \frac{[\rho]\{G\}}{\sqrt{\{G\}^T[\rho]\{G\}}} = \begin{Bmatrix} -0.6439 \\ 0.3306 \\ 0.4854 \end{Bmatrix}.$$

8. Determine new values of z_i^* for $n - 1$ of the variables. We will find z_1^* and z_2^*.

$$z_1^* = \alpha_1\beta = (-0.6439)(2.615) = -1.684$$

$$z_2^* = \alpha_2\beta = (0.3306)(2.615) = 0.8643$$

9. Determine the values of x_1^* and x_2^* using the updated z_1^* and z_2^*:

$$x_1^* = \mu_{X_1} + z_1^*\sigma_{X_1} = 11.81$$

$$x_2^* = \mu_{X_2}^e + z_2^*\sigma_{X_2}^e = 5.589.$$

Table 5.13 Iterations for Example Problem 5.13

	Iteration #							
	1	*2*	*3*	*4*	*5*	*6*	*7*	*8*
x_1^*	14	11.81	12.05	12.11	12.13	12.13	12.13	12.13
x_2^*	5	5.589	5.768	5.814	5.827	5.828	5.831	5.830
x_3^*	10	6.381	6.406	6.420	6.415	6.426	6.414	6.421
↓								
β	2.615 ↗	2.575 ↗	2.567 ↗	2.572 ↗	2.565 ↗	2.573 ↗	2.569 ↗	2.571
↓								
x_1^*	11.81	12.05	12.11	12.13	12.13	12.13	12.13	12.13
x_2^*	5.589	5.768	5.814	5.827	5.828	5.831	5.830	5.831
x_3^*	6.381	6.406	6.420	6.415	6.426	6.414	6.421	6.416

10. Determine the value of x_3^* using the limit state equation $g = 0$. For this case, $x_3^* = 6.381$.
11. Iterate until the value of β and the design point converge. The results of subsequent iterations are shown in Table 5.13.

Although there is still a bit of variation after eight iterations, it appears that the solution is converging to $\beta \cong 2.57$.

Example 5.14

Calculate the reliability index, β, for the limit state function

$$g(X_1,X_2) = 3X_1 - 2X_2$$

where

$$\mu_{X_1} = 16.6; \quad \sigma_{X_1} = 2.45$$

$$\mu_{X_2} = 18.8; \quad \sigma_{X_2} = 2.83$$

and $Cov(X_1,X_2) = 2.0$. We do not have any information on the distributions of X_1 and X_2; thus, we will assume both are normally distributed.

We will follow the step-by-step procedure to solve the problem The steps shown below are for the first cycle of iteration. The results of subsequent steps are summarized in Table 5.14.

1. Formulate limit state function and probability distributions. This has been done.
2. Guess an initial design point. We will assume a value for x_1^* of 17. From $g = 0$, the estimate of x_2^* is 25.5.
3. Determining equivalent normal parameters is not necessary since we are assuming both variables are normally distributed.

Table 5.14 Iterations for Example 5.14

	Iteration #	
	1	2
x_1^*	17	13.8
x_2^*	25.5	20.7
↓		↗
β	1.55	1.55
↓		
x_1^*	13.8	13.8
x_2^*	20.7	20.7

4. Determine the values of the reduced variates

$$z_1^* = 0.163; \quad z_2^* = 2.37.$$

5. Determine the $\{G\}$ vector.

$$G_1 = -\left.\frac{\partial g}{\partial Z_1}\right|_{\{z_i^*\}} = -\left.\frac{\partial g}{\partial X_1}\right|_{\{x_i^*\}} \sigma_{X_1} = (-3)\sigma_{X_1}$$

$$G_2 = -\left.\frac{\partial g}{\partial Z_2}\right|_{\{z_i^*\}} = -\left.\frac{\partial g}{\partial X_2}\right|_{\{x_i^*\}} \sigma_{X_2} = 2\sigma_{X_2}$$

6. Calculate an estimate of β.

$$[\rho] = \begin{bmatrix} 1 & \dfrac{\mathrm{Cov}(X_1, X_2)}{\sigma_{X_1}\sigma_{X_2}} \\ \dfrac{\mathrm{Cov}(X_1, X_2)}{\sigma_{X_1}\sigma_{X_2}} & 1 \end{bmatrix}$$

$$= \begin{bmatrix} 1 & \dfrac{2}{(2.45)(2.83)} \\ \dfrac{2}{(2.45)(2.83)} & 1 \end{bmatrix} = \begin{bmatrix} 1 & 0.288 \\ 0.288 & 1 \end{bmatrix}$$

$$\beta = \frac{\{G\}^{\mathrm{T}}\{z^*\}}{\sqrt{\{G\}^{\mathrm{T}}[\rho]\{G\}}} = 1.55$$

7. Calculate $\{\alpha\}$:

$$\{\alpha\} = \frac{[\rho]\{G\}}{\sqrt{\{G\}^T[\rho]\{G\}}} = \begin{Bmatrix} -0.726 \\ 0.449 \end{Bmatrix}$$

8. Determine new values of z_i^* for $n - 1$ of the variables. We will find z_1^*:

$$z_1^* = \alpha_1\beta = (-0.726)(1.55) = -1.13$$

9. Determine the value of x_1^* using the value of z_1^*:

$$x_1^* = \mu_{X_1} + z_1^*\sigma_{X_1} = 13.8.$$

10. Determine the value of x_2^* using the limit state equation $g = 0$. For this case, $x_2^* = 20.7$
11. Iterate until the value of β and the design point converge. The results of subsequent iterations is shown in Table 5.14.

Observe that we have the correct answer after one iteration. This occurs because we have a linear performance function and because we did not have to deal with equivalent normal parameters (since all random variables are normal). To check this result, we can use the results from Chapter 3 to calculate β. Since g is a linear function of random variables, we know that the mean and variance of g are

$$\mu_g = a_0 + \sum_{i=1}^{n} a_i\mu_{X_i} = 3\mu_{X_1} - 2\mu_{X_2} = 12.2$$

$$\sigma_g^2 = \sum_{i=1}^{n}\sum_{j=1}^{n} a_i a_j \mathrm{Cov}(X_i, X_j) = \sum_{i=1}^{n}\sum_{j=1}^{n} a_i a_j \rho_{X_i X_j}\sigma_{X_i}\sigma_{X_j}$$

$$= (3)^2(1)(2.45)^2 + 2[(3)(-2)(0.288)(2.45)(2.83)] + (-2)^2(1)(2.83)^2$$

$$= 62.1$$

Thus, we can calculate β as

$$\beta = \frac{\mu_g}{\sigma_g} = \frac{12.2}{\sqrt{62.1}} = 1.55$$

In fact, it can be shown (e.g., Ang and Tang, 1984) that for *linear* performance functions of correlated *normal* random variables, the reliability index can be calculated directly as

$$\beta = \frac{a_0 + \sum_{i=1}^{n} a_i \mu_{X_i}}{\sum_{i=1}^{n}\sum_{j=1}^{n} a_i a_j \rho_{X_i X_j} \sigma_{X_i} \sigma_{X_j}}$$

Example 5.15

This example provides another illustration of how the coordinate transformation technique (discussed in Chapter 4) can be applied to convert a set of correlated normal random variables to a set of uncorrelated normal random variables. We will solve Example Problem 5.14 using the coordinate transformation method.

First, we must form the mean vector and covariance matrix of the original (correlated) variables. These are

$$\{\mu_X\} = \begin{Bmatrix} 16.6 \\ 18.8 \end{Bmatrix}; \quad [C_X] = \begin{bmatrix} 6 & 2 \\ 2 & 8 \end{bmatrix}.$$

Then, we do an eigenvalue and eigenvector analysis of the matrix $[C_X]$ and use the orthonormal eigenvectors to find the transformation matrix $[T]$. The result is

$$[T] = \begin{bmatrix} 0.851 & 0.526 \\ -0.526 & 0.851 \end{bmatrix}.$$

With $[T]$, we can convert the correlated variables X_1 and X_2 to uncorrelated variables Y_1 and Y_2 as follows:

$$\{Y\} = [T]^T\{X\} \quad \Rightarrow \begin{Bmatrix} Y_1 \\ Y_2 \end{Bmatrix} = \begin{bmatrix} 0.851 & -0.526 \\ 0.526 & 0.851 \end{bmatrix} \begin{Bmatrix} X_1 \\ X_2 \end{Bmatrix}$$

$$\{X\} = [T]\{Y\} \quad \Rightarrow \begin{Bmatrix} X_1 \\ X_2 \end{Bmatrix} = \begin{bmatrix} 0.851 & 0.526 \\ -0.526 & 0.851 \end{bmatrix} \begin{Bmatrix} Y_1 \\ Y_2 \end{Bmatrix}.$$

By substituting the foregoing expressions for X_1 and X_2 into the limit state equation, we obtain the following limit state equation in terms of uncorrelated variables Y_1 and Y_2:

$$g(Y_1, Y_2) = 3.605Y_1 - 0.124Y_2.$$

Since this is a linear function of uncorrelated normal random variables (since we assumed the X_i variables to be normally distributed), we can apply the following formula to calculate β:

$$\beta = \frac{a_0 + \sum_{i=1}^{n} a_i \mu_{Y_i}}{\sqrt{\sum_{i=1}^{n} (a_i \sigma_{Y_i})^2}}.$$

The required means and variances of Y_1 and Y_2 can be found as follows:

$$\{\mu_Y\} = [T]^T \{\mu_X\} = \begin{Bmatrix} 4.24 \\ 24.7 \end{Bmatrix}$$

$$[C_Y] = [T]^T [C_X][T] = \begin{bmatrix} 4.76 & 0 \\ 0 & 9.24 \end{bmatrix}.$$

Hence, substituting into the expression for β we obtain

$$\beta = \frac{a_0 + \sum_{i=1}^{n} a_i \mu_{Y_i}}{\sqrt{\sum_{i=1}^{n} (a_i \sigma_{Y_i})^2}} = \frac{0 + (3.605)(4.24) + (-0.124)(24.7)}{\sqrt{(3.605)^2 (4.76) + (-0.124)^2 (9.24)}}$$

$$= \frac{12.2}{\sqrt{62.0}} = 1.55,$$

which, as expected, agrees with our previous result.

5.5 RELIABILITY ANALYSIS USING SIMULATION

The preceding sections have presented techniques for determining the reliability index β for some common forms of limit state functions encountered in practice. However, more complicated limit state functions are sometimes encountered, and application of the simultaneous equation approach or the matrix approach to find β is either extremely difficult or impossible. In those instances, Monte Carlo simulation provides the only feasible way to determine β or the probability of failure.

The steps in the Monte Carlo simulation are the same as those described in Chapter 4. First, simulated values of the random variables in the limit state equation are generated. Then, these values are used to simulate values of the limit state function itself. Finally, Equation 4.11 can be used to estimate the probability of failure. Alternatively, the simulated values of the limit state function can be plotted on normal probability paper. The probability of failure, corresponding to $P[g() < 0]$, can be found by reading the probability value at the location where the plotted data curve intersects a vertical line passing through the origin. Equation 5.15 tells us that $-\beta$ corresponds to the value of the standard normal variable at the intersection point.

If there are too few data points and the plotted curve does not intersect the vertical axis, the plotted curve can be extrapolated as illustrated in Example Problem 4.5. Extrapolation is not recommended in general; increasing the number of simulations is preferred. However, using extrapolation is sometimes the only feasible way to obtain an estimate of β.

Example 5.16

Consider the following limit state equation involving resistance (R), dead load effect (D), and live load effect (L):

$$g(R,D,L) = R - (D + L) = R - D - L$$

The distributions and parameters of the random variables are as follows:

R is lognormal	$V_R = 13\%$	$\mu_R = 2300$ kN
D is normal	$V_D = 10\%$	$\mu_D = 900$ kN
L is extreme Type I	$V_L = 25\%$	$\mu_L = 675$ kN

Determine the probability of failure using Monte Carlo simulation.

The steps are as follows:

1. Determine the necessary distribution information. For R, we need $\mu_{\ln R}$ and $\sigma_{\ln R}$. Using Equations 2.48 and 2.49, we get

   $$\mu_{\ln R} = 7.732 \quad \sigma_{\ln R} = 0.1295.$$

   For L, we need the distribution parameters u and α. Using Equations 2.62 and 2.63, we get

   $$\alpha = 7.597 \times 10^{-3} \quad u = 5.991 \times 10^{2}.$$

2. Determine the number of simulated values of the limit state function to be generated. To begin, we will arbitrarily choose $N = 1000$. Since there are three variables that must be simulated each time we evaluate the function, we need a total of $3N = 3000$ uniformly distributed random numbers, u_i.

3. Simulate values of r_i, d_i, and $l_i (i = 1, 2, \ldots 1000)$. The values of r_i can be obtained using Equation 4.6. The values of d_i can be obtained using Equation 4.5. To simulate the values of l_i, we need to take the inverse of the Type I distribution as was shown in Example 4.5. The resulting expression is

$$l_i = 5.991 \times 10^2 - \frac{\ln(-\ln(u_i))}{7.597 \times 10^{-3}}.$$

   Histograms of the simulated values of R, D, and L are shown in Figure 5.24. The sample mean and sample standard deviation for each simulated variable are also shown for comparison with the distribution parameters given above.
4. Calculate a value of the limit state function, g_i, for each set of values $\{r_i, d_i, l_i\}$.
5. The failure probability corresponds to

   $P[g(R,D,L) < 0]$.

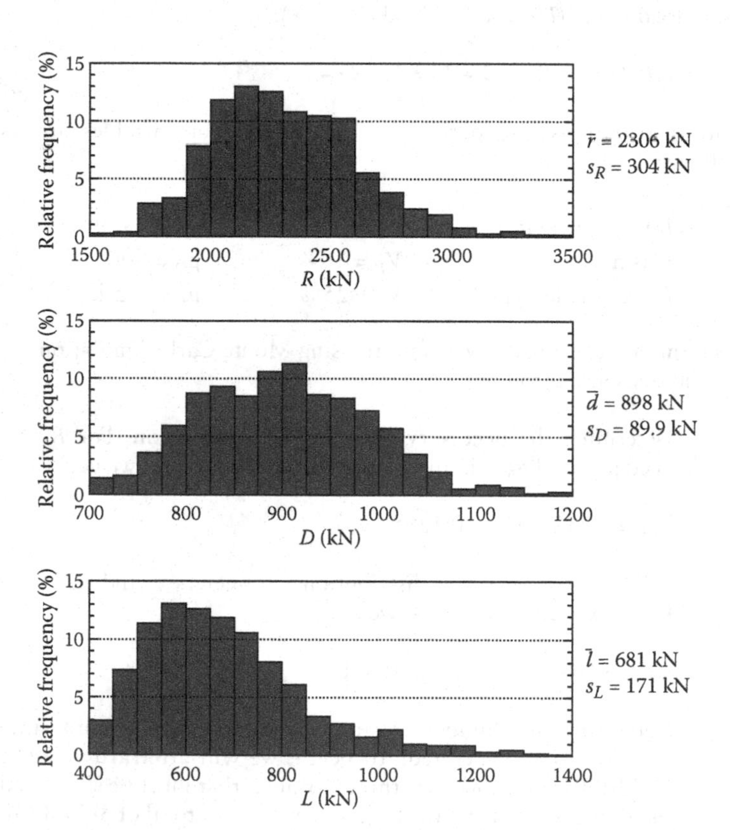

Figure 5.24 Relative frequency histograms of the simulated variables in Example 5.16.

To estimate this probability, the simulated values of g_i are plotted on normal probability paper as shown in Figure 5.25. The failure probability corresponds to the point where the plotted curve intersects the vertical axis passing through the origin. The intersection point occurs at a value of the standard normal variate ≈ −2 on the vertical axis. Thus, the estimate of β is 2.

We can also make an estimate of the failure probability using Equation 4.11. From the simulation results, 24 of the 1000 simulated values of g_i are less than zero. Therefore, the estimated failure probability is 24/1000 = 0.024. Using Equation 5.15, this probability corresponds to β = 1.98. An estimate of the COV for this calculated probability is, based on Equation 4.12,

$$V_{\bar{P}} = \sqrt{\frac{1-(0.024)}{1000(0.024)}} = 0.20$$

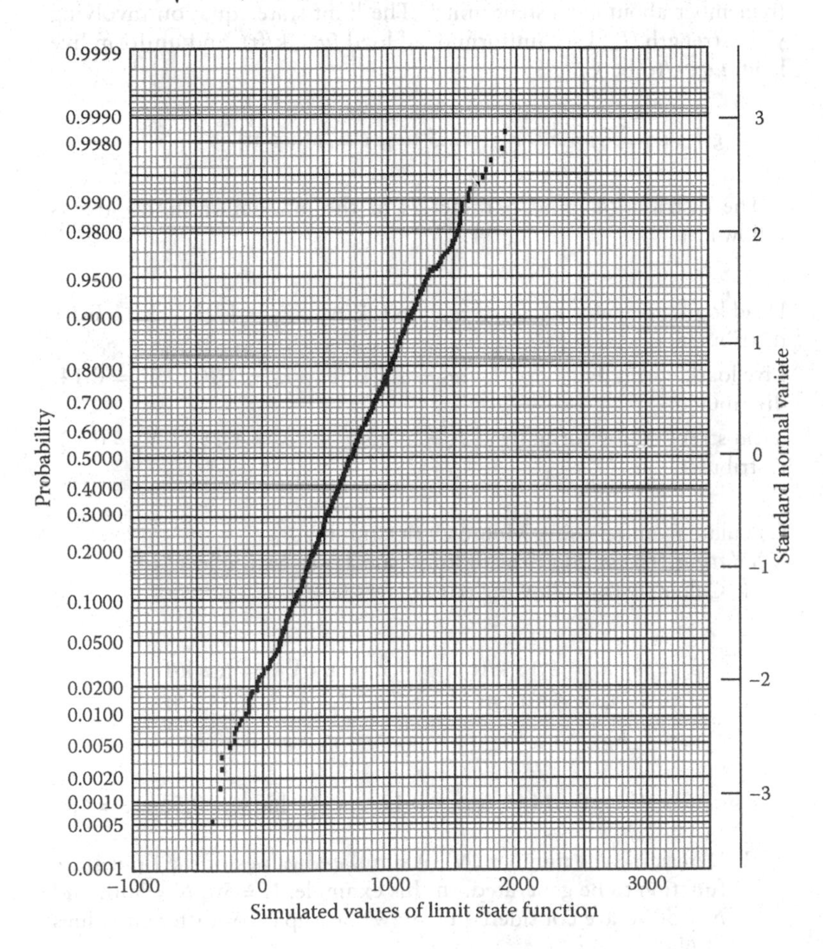

Figure 5.25 Simulated values of the limit state function considered in Example 5.16.

Example 5.17

Consider the 10 ft long steel beam with cross section W 18 × 106 (plastic modulus $Z = 60$ in^3) under dead load and live load. General limit state function is as follows:

$$g(R,D,L) = R - (D + L).$$

For a uniformly loaded, simply supported beam, this limit state function becomes:

$$g = ZF_y - (w_D + w_L)\ L^2/8.$$

We can substitute deterministic values for Z and L into the equation (remember about consistent units). The limit state equation involving yield strength (F_y, ksi), uniform dead load (w_D, k/ft), and uniform live load (w_L, k/ft) becomes:

$$g(F_y, w_D, w_L) = 5.0\ F_y - 12.5\ (w_D + w_L)\ \text{[kip-ft]}.$$

The distributions and parameters of the random variables are as follows:

Dead load, normally distributed	$w_D = 6.0$ k/ft	$\lambda_D = 1.03$	$V_D = 0.10$
Live load, normally distributed	$w_L = 8.5$ k/ft	$\lambda_L = 0.80$	$V_L = 0.14$
Yield strength, lognormally distributed	$F_y = 50$ ksi	$\lambda_F = 1.11$	$V_F = 0.12$

Calculate P_f and β using Monte Carlo simulation.

Solution. The steps are as follows (similar to Example 5.16):

1. Calculate the following parameters:

$$\mu_D = w_D\ \lambda_D = 6.18 \text{ k/ft} \qquad \sigma_D = \mu_D\ V_D = 0.62 \text{ k/ft}$$
$$\mu_L = w_L\ \lambda_L = 6.80 \text{ k/ft} \qquad \sigma_L = \mu_L\ V_L = 0.95 \text{ k/ft}$$
$$\mu_F = F_y\ \lambda_F = 55.5 \text{ ksi} \qquad \sigma_F = \mu_F\ V_F = 6.66 \text{ ksi}$$

$$\sigma_{\ln(F)}{}^2 = V_F{}^2 = 0.0144 \quad \sigma_{\ln(f)} = 0.12 \quad \mu_{\ln(F)} = \ln(\mu_F) - 0.5\sigma_{\ln(F)}{}^2 = 4.0$$

2. Determine the number (N) of simulated values of the limit state function to be generated. In this example, $N = 50$, $N = 500$, and $N = 5000$ are considered to show the impact of different values of N.

Table 5.15 Monte Carlo simulation results (15 out of 5000 runs) in Example 5.17

i	*Random*	z_i	$12.5*w_{Di}$	*Random*	z_i	$12.5*w_{Li}$	*Random*	z_i	$5.0*F_i$
1	0.8769	1.1598	86.21	0.8466	1.0220	106.88	0.0245	−1.9681	224.10
2	0.9949	2.5700	97.10	0.3427	−0.4051	88.20	0.7693	0.7366	309.68
3	0.1282	−1.1351	68.48	0.4173	−0.2089	90.77	0.9099	1.3403	332.85
4	0.8098	0.8773	84.03	0.5074	0.0187	93.74	0.0902	−1.3395	241.60
5	0.6370	0.3505	79.96	0.0980	−1.2928	76.58	0.8883	1.2174	328.00
6	0.3511	−0.3823	74.30	0.0666	−1.5018	73.84	0.9274	1.4567	337.52
7	0.8746	1.1483	86.12	0.2400	−0.7064	84.25	0.8398	0.9935	319.33
8	0.1782	−0.9221	70.13	0.5798	0.2014	96.14	0.4835	−0.0413	282.17
9	0.2392	−0.7090	71.77	0.1829	−0.9045	81.66	0.7054	0.5401	302.48
10	0.1624	−0.9847	69.64	0.6405	0.3597	98.21	0.8215	0.9212	316.58
11	0.1342	−1.1068	68.70	0.1993	−0.8440	82.45	0.8958	1.2578	329.59
12	0.6202	0.3059	79.61	0.7920	0.8132	104.14	0.9338	1.5049	339.47
13	0.5792	0.1998	78.79	0.0164	−2.1340	65.57	0.0023	−2.8403	201.91
14	0.7433	0.6535	82.30	0.5862	0.2178	96.35	0.8556	1.0609	321.92
15	0.4683	−0.0795	76.64	0.7887	0.8021	104.00	0.4816	−0.0461	282.01

3. Simulate values of D_i, L_i, R_i ($i = 1, 2, \ldots, N$).

   Normally distributed: $D_i = \mu_D + z_i \cdot \sigma_D$

   Normally distributed: $L_i = \mu_L + z_i \cdot \sigma_L$

   Lognormally distributed: $F_i = \exp(\mu_{\ln F} + z_i \cdot \sigma_{\ln F})$
   Sample results are shown in Table 5.15.
4. Calculate a value of the limit state function, g_i, for each set of values $\{D_i, L_i, R_i\}$. Sample results are shown in Table 5.16.
5. To estimate the failure probability, the simulated values of g_i are plotted on normal probability paper as shown in Figures 5.26 through 5.28 for the three values of N considered. The failure probability corresponds to the point where the plotted curve intersects the vertical axis passing through the origin. The figures show how the estimation of the failure probability improves as the number of simulations is increased.

Example 5.18

In this example, we will use simulation to generate values of three random variables DL, WL, and IL. The simulated values will then be plotted on normal probability paper. The distributions and parameters of the random variables are as follows:

Dead load effect

Normally distributed $DL = 53$ k-ft $\lambda_{DL} = 1.05$ $V_{DL} = 0.04$

Table 5.16 Sample values of limit state function (15 out of 500 runs) in Example 5.17

i	$g_i = 5.0{*}F_i - 12.5(w_{Di} + w_{Li})$	Sort(g_i)	$i/(N+1)$	z_i
1	31.02	−6.33	0.0002	−3.5401
2	124.37	−1.95	0.0004	−3.3529
3	173.61	3.96	0.0006	−3.2389
4	63.83	6.23	0.0008	−3.1560
5	171.46	9.42	0.0010	−3.0903
6	189.38	9.76	0.0012	−3.0357
7	148.96	11.02	0.0014	−2.9889
8	115.91	11.38	0.0016	−2.9479
9	149.05	11.56	0.0018	−2.9113
10	148.73	11.89	0.0020	−2.8782
11	178.43	11.93	0.0022	−2.8480
12	155.71	14.68	0.0024	−2.8202
13	57.55	14.85	0.0026	−2.7944
14	143.27	15.58	0.0028	−2.7704
15	101.37	15.60	0.0030	−2.7478

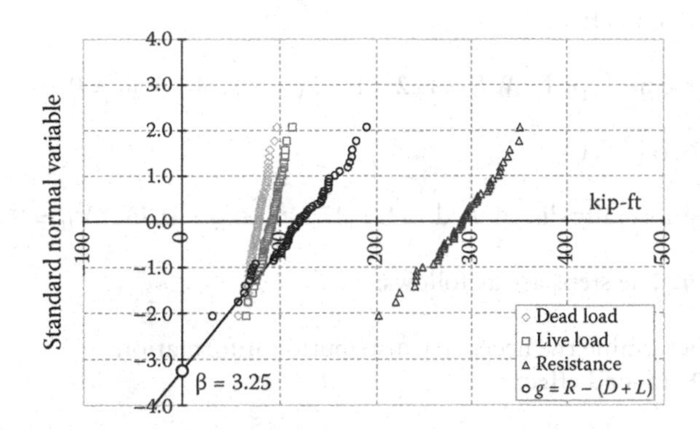

Figure 5.26 Results of simulation based on $N = 50$ runs: CDFs of D, L, R, and $g = R - D - L$ [reliability index $\beta = 3.25$; probability of failure $P_f = \Phi(-\beta) = 0.00058$].

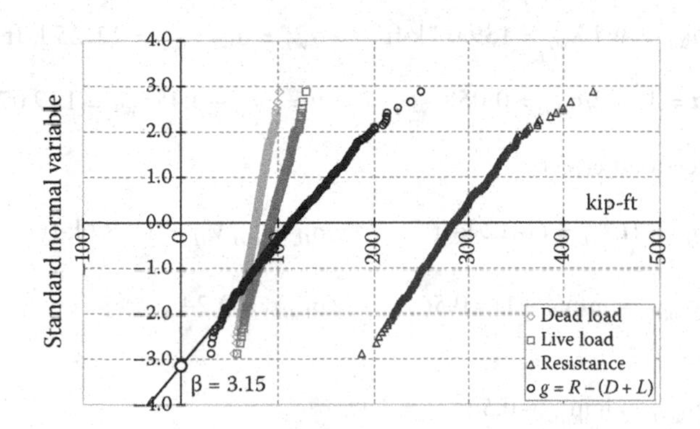

Figure 5.27 Results of simulation based on $N = 500$ runs: CDFs of D, L, R, and $g = R - D - L$ [reliability index $\beta = 3.15$; probability of failure $P_f = \Phi(-\beta) = 0.00082$].

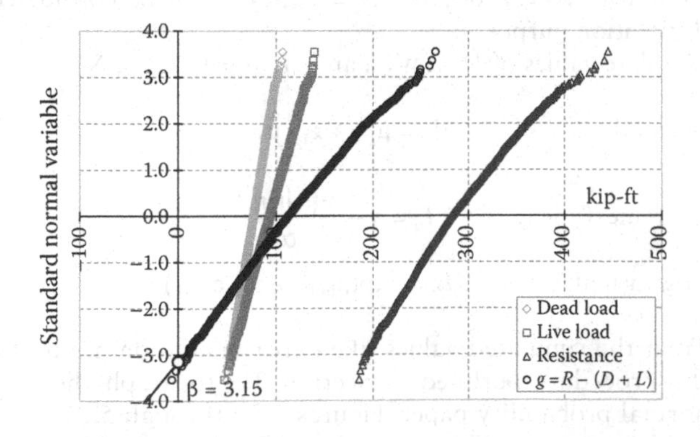

Figure 5.28 Results of simulation based on $N = 5000$ runs: CDFs of D, L, R, and $g = R - D - L$ [reliability index $\beta = 3.15$; probability of failure $P_f = \Phi(-\beta) = 0.00082$].

Wind load effect

Extreme Type I $WL = 122$ k-ft $\lambda_{WL} = 1.14$ $V_{WL} = 0.16$

Ice load effect

Lognormally distributed $IL = 87$ k-ft $\lambda_{IL} = 0.76$ $V_{IL} = 0.24$

Solution. The steps are as follows:

1. Determine the necessary distribution information:
Dead load effect:

$\mu_{DL} = DL\lambda_{DL} = 55.65$ k-ft $\qquad \sigma_{DL} = \mu_{DL}V_{DL} = 2.23$ k-ft

Wind load effect :

$\mu_{WL} = WL\lambda_{WL} = 139.08$ k-ft $\qquad \sigma_{WL} = \mu_{WL}V_{WL} = 22.25$ k-ft

$\alpha = 1.282/\sigma_{WL} = 0.058$ $\qquad u = \mu_{WL} - 0.45\sigma_{WL} = 129.07$ k-ft

Ice load effect:

$\mu_{IL} = IL\lambda_{IL} = 66.12$ k-ft $\qquad \sigma_{IL} = \mu_{IL}V_{IL} = 15.87$ k-ft

$\sigma^2_{\ln(IL)} = \ln\left(V^2_{IL} + 1\right) = 0.56$ $\qquad \sigma_{\ln(IL)} = 0.24$

$\mu_{\ln(IL)} = \ln(\mu_{IL}) - 0.5\sigma^2_{\ln(IL)} = 4.16$

2. Determine the number (N) of simulated values of the limit state function to be generated. In this example, three different values (N = 100, N = 1000, and N = 10,000) will be considered for illustration purposes.
3. Simulate values of DL_i, WL_i, and IL_i ($i = 1, 2, \ldots, N$).

Normal: $\qquad DL_i = \mu_{DL} + z_i \cdot \sigma_{DL}$

Extreme Type I: $\qquad WL_i = u - \dfrac{\ln(-\ln u_i)}{\alpha}$

Lognormal: $\qquad IL_i = \exp(\mu_{\ln IL} + z_i \cdot \sigma_{\ln IL})$

4. Treat the simulated values like experimental data, and follow the procedure outlined in Section 2.5 to graph the data on normal probability paper. Figures 5.29 through 5.31 show the results corresponding to the three N values considered in this example.

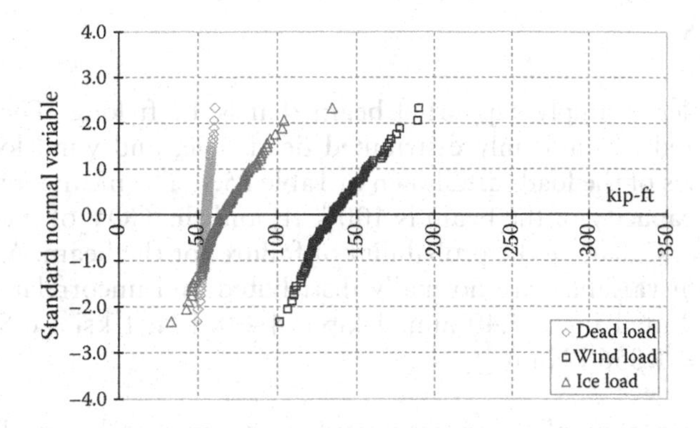

Figure 5.29 Results of simulation in Example Problem 5.18 for N = 100 runs: CDFs of *DL*, *WL*, and *IL*.

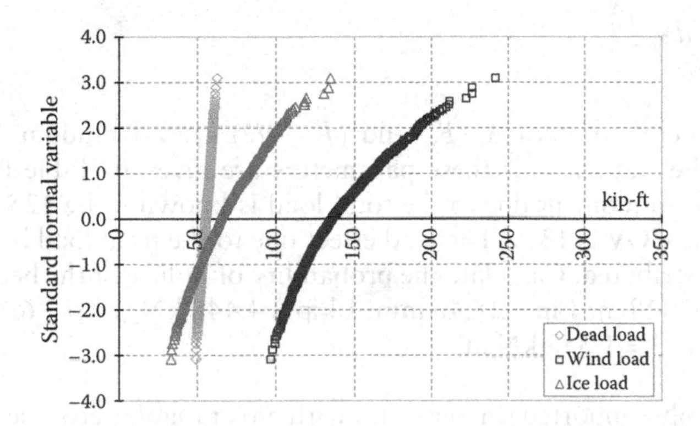

Figure 5.30 Results of simulation in Example Problem 5.18 for N = 1000 runs: CDFs of *DL*, *WL*, and *IL*.

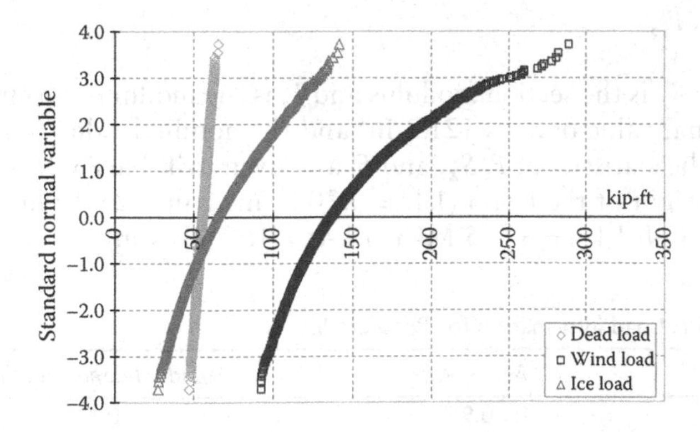

Figure 5.31 Results of simulation in Example Problem 5.18 for N = 10,000 runs: CDFs of *DL*, *WL*, and *IL*.

PROBLEMS

5.1 Consider a simply supported beam that is 12 ft long. The beam is subjected to uniformly distributed dead, live, and wind loads. The statistics of the loads are shown in Table P5.1. The mean moment carrying capacity of the beam is 100 k-ft, and the COV of the capacity is 13%. Calculate the probability of failure for the beam. Assume all random variables are normally distributed and uncorrelated (1 ft = 0.3048 m; 1 in = 25.40 mm; 1 kip = 4.448 kN; 1 ksi = 6.895 MPa; 1 k-ft = 1.356 kN m).

5.2 Consider a simply supported reinforced concrete beam. The load-carrying capacity is

$$A_s F_y \left(d - \frac{a}{2} \right),$$

where variables A_s, F_y, and $(d - a/2)$ are all random variables. The statistics for these parameters are given in Table P5.2. The mean moment due to the total load is known to be 125 k-ft, and the COV is 13%. The load effect due to the total load is normally distributed. Calculate the probability of failure of the beam (1 ft = 0.3048 m; 1 in = 25.40 mm; 1 kip = 4.448 kN; 1 ksi = 6.895 MPa; 1 k-ft = 1.356 kN m).

5.3 A simply supported timber beam with a rectangular cross section 6 in by 12 in and a span of 13 ft supports a concentrated load $P = 10$ kips at midspan. The moment-carrying capacity of the beam is

$$M_n = S_X F_r,$$

where S_X is the section modulus and F_r is the modulus of rupture. The nominal value of S_X is 121.2 in^3 and the nominal value of F_r is 2500 psi. The statistics of P, S_X, and F_r are given in Table P5.3. Neglect the self-weight of the beam (1 ft = 0.3048 m; 1 in = 25.40 mm; 1 kip = 4.448 kN; 1 ksi = 6.895 MPa; 1 k-ft = 1.356 kN m).

Table P5.1 Load information for Problem 5.1

Load	*Mean (k/ft)*	*Standard deviation (k/ft)*
Dead	0.95	0.1
Live	1.5	0.2
Wind	0.6	0.12

Table P5.2 Information on distribution parameters for Problem 5.2

Parameter	*Mean*	*COV*	*Distribution*
A_S	3 in^2	2.5%	Lognormal
F_y	44 ksi	10.5%	Lognormal
$d - a/2$	14 in	11.5%	Lognormal

A. Calculate the Hasofer–Lind reliability index by the iterative method.
B. Calculate the Hasofer–Lind reliability index using the graphical method.

5.4 Consider the limit state function

$$g(X,Y) = 3Y - X,$$

where X follows a Type I extreme value distribution and Y is lognormal. The mean value of X is 24 and the COV is 0.15. The mean value of Y is 12.5 and the COV is 0.125. Calculate the reliability index using the Rackwitz–Fiessler method (matrix approach).

5.5 Consider a timber beam in a highway bridge. The nominal beam dimensions are 6 in × 16 in; the actual dimensions are 5.5 in × 15.5 in. The total load effect considered is the combined moment due to dead load, live load, and impact. The resistance, R, is calculated using $R = S_X F_r$ where S_X is the section modulus and F_r is the modulus of rupture. The load and resistance parameters are given in Table P5.4. The design formula is

$$D + L < 0.95SF$$

(1 ft = 0.3048 m; 1 in = 25.40 mm; 1 kip = 4.448 kN; 1 ksi = 6.895 MPa; 1 k-ft = 1.356 kN m).

A. Calculate the maximum nominal values of D and L allowed by the design formula.
B. Calculate the reliability index corresponding to the maximum values of D and L determined in part A.

Table P5.3 Parameters for Problem 5.3

Variable	*Bias factor*	*COV*
F_r	1.75	0.30
S_X	1.0	0.00
P	0.875	0.13

5.6 Consider a limit state function

$$g(X,Y,Z) = \frac{X}{Y} + Z,$$

where X, Y, and Z are random variables. The statistical parameters of X, Y, and Z are given in Table P5.5. Calculate the Hasofer–Lind reliability index.

5.7 Consider the limit state equation

$$g = R/Q - 1,$$

where R and Q are both lognormal random variables. The two variables are statistically independent. Show that the reliability index can be calculated using

$$\beta = \frac{\mu_{\ln R} - \mu_{\ln Q}}{\sqrt{\sigma_{\ln R}^2 + \sigma_{\ln Q}^2}} \approx \frac{\ln\left(\frac{\mu_R}{\mu_Q}\right)}{\sqrt{V_R^2 + V_Q^2}}.$$

Specifically comment on the conditions that must be satisfied to permit the approximate equality shown above.

5.8 Verify the value of the reliability index β found in Example 5.1 using simulation. Assume all variables are normally distributed.

5.9 In Example 5.2, a first-order, second-moment mean value reliability index was calculated. Calculate the Hasofer–Lind reliability index

Table P5.4 Information for Problem 5.5

Parameter	*CDF*	*Mean*	*Bias factor*	*COV*
D (dead load)	Normal	0.10 m_Q	1.05	0.10
L (live load)	Lognormal	0.80 m_Q	1.5	0.125
I (impact)	Normal	0.10 m_Q	Undefined	0.8
S_X	Normal	216 in^3	1.00	0.05
F_r	Lognormal	5200 psi	2.75	0.29

m_Q = mean value of moment due to total load.

Table P5.5 Statistical parameters for Problem 5.6

Variable	*Mean*	*COV*
X	50.0	0.12
Y	12.0	0.11
Z	−5.0	0.14

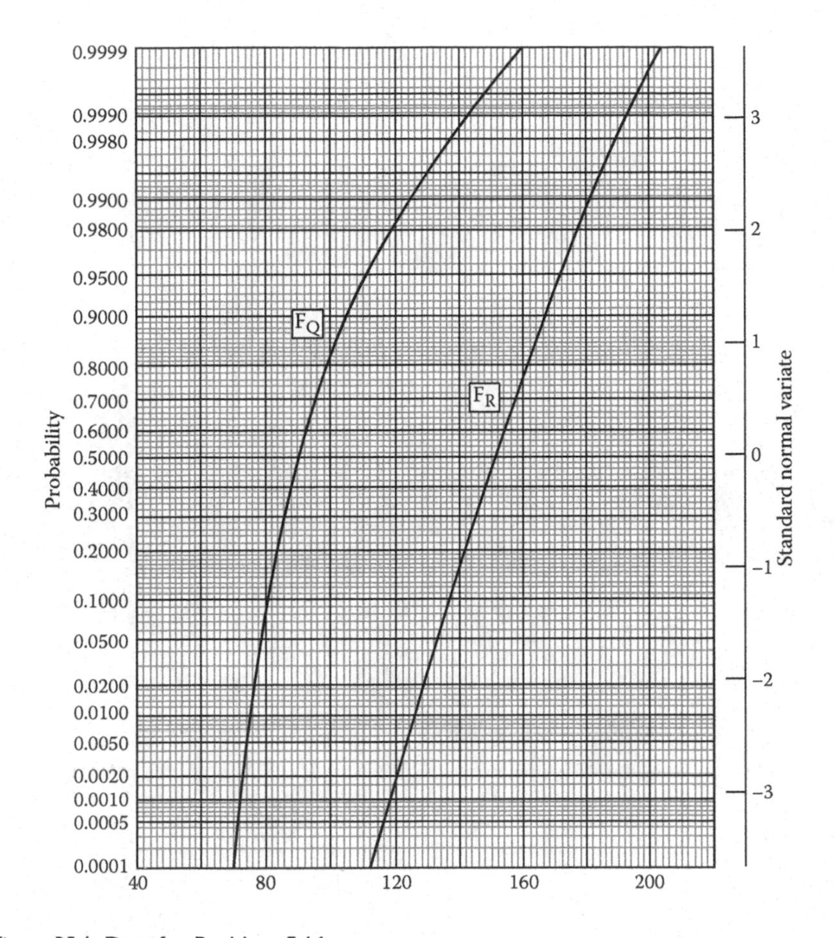

Figure P5.1 Data for Problem 5.16.

for this case. Assume all variables are normally distributed. Compare your result to that found in Example 5.2.

5.10 Calculate the Hasofer–Lind reliability index for Example 5.3.

5.11 Find the reliability index for Example 5.4 using simulation. Assume all variables are normally distributed.

5.12 Determine the reliability index for Example 5.11 using simulation.

5.13 Solve Example 5.14 by simulation.

5.14 Solve Example 5.16 by the Rackwitz–Fiessler procedure.

5.15 Solve Problem 5.4 using the graphical Rackwitz–Fiessler procedure. (You must plot the CDF functions on normal probability paper.)

5.16 Figure P5.1 shows the CDFs for the random variables R and Q. The governing limit state equation is $g(R,Q) = R - Q$. Determine the reliability index using the graphical Rackwitz–Fiessler procedure.

Chapter 6

Structural load models

To design any structure, the designer must have an understanding of the types and magnitudes of the loads that are expected to act on the structure during its lifetime. This chapter discusses many of the types of loads commonly considered in the design of buildings and bridges and describes some probabilistic models of these loads that are used in developing reliability-based design codes.

6.1 TYPES OF LOAD

There are many types of loads that act on structures. These loads can be classified into three categories based on the types of statistical data that are available and the characteristics of the load phenomenon:

- Type I. For these loads, data are obtained by load intensity measurements without regard to the frequency of occurrence. In other words, the time dependence of the loads is not explicitly considered. Examples of loads in this category are dead and sustained live loads.
- Type II. In this category, load data are obtained from measurements at prescribed periodic time intervals. Thus, some time dependence is captured. Examples of loads in this category include severe winds, snow loads, and transient live load.
- Type III. The available data for Type III loads are obtained from infrequent measurements because the data are typically not obtainable at prescribed time intervals. These loads occur during extreme events such as earthquakes and tornadoes.

6.2 GENERAL LOAD MODELS

Consider a general load Q_i which is to be modeled for the purposes of conducting a reliability analysis. It is often convenient to express the magnitude of Q_i as

$$Q_i = A_i B_i C_i \tag{6.1}$$

where A_i represents the load itself, B_i represents the variation due to the mode in which the load is assumed to act, and C_i represents the variation due to methods of analysis. The variable C_i takes into account various approximations and idealizations used in creating the analysis model of the structure. Examples of such approximations include two-dimensional idealizations of three-dimensional structures and fixed-base versus pinned-base assumptions. The variable B_i accounts for assumptions about how the loading is applied to the structure. For example, Figure 6.1 shows two beams. On the left-hand beam, the loading is represented by a combination of concentrated loads and nonuniform distributed loads. On the right-hand beam, the loading is simplified for analysis purposes into a single uniformly distributed load acting over the entire beam.

To proceed with a reliability analysis, we need at least the mean and variance [or standard deviation or coefficient of variation (COV)] for Q_i. Using the principles discussed in Chapter 3, we can linearize the function for Q_i about the mean values, and then we can apply the formulas presented in Chapter 3 to calculate the mean, bias factor, variance, and COV of a linear function. The resulting expressions are as follows:

$$\mu_Q \approx \mu_A \mu_B \mu_C \tag{6.2a}$$

$$\lambda_Q \approx \lambda_A \lambda_B \lambda_C \tag{6.2b}$$

$$\sigma_Q^2 \approx (\mu_B \mu_C)^2 \sigma_A^2 + (\mu_A \mu_C)^2 \sigma_B^2 + (\mu_A \mu_B)^2 \sigma_C^2 \tag{6.2c}$$

$$V_Q \approx \sqrt{V_A^2 + V_B^2 + V_C^2} \tag{6.2d}$$

where μ denotes mean value, λ denotes bias factor, σ^2 denotes variance, and V denotes COV. The derivation of these expressions is outlined in the example below.

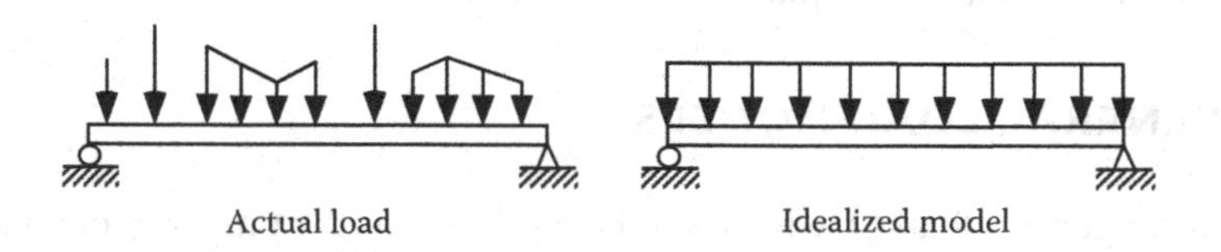

Figure 6.1 Idealization of loads on a structure as reflected in the variable B_i in Equation 6.1.

Example 6.1

Estimate the mean, bias factor, and COV of Q_i (given in Equation 6.1) as a function of the means and COVs of the variables A_i, B_i, and C_i.

First, we need to linearize the function for Q_i. Linearizing about the mean values, we get

$$\begin{aligned}Q_i &= \mu_A\mu_B\mu_C + \left.\frac{\partial Q_i}{\partial A_i}\right|_{\text{mean values}}(A_i-\mu_A)+\left.\frac{\partial Q_i}{\partial B_i}\right|_{\text{mean values}}(B_i-\mu_B)+\left.\frac{\partial Q_i}{\partial C_i}\right|_{\text{mean values}} \\ &\quad\times(C_i-\mu_C) \\ &= \mu_A\mu_B\mu_C+\mu_B\mu_C(A_i-\mu_A)+\mu_A\mu_C(B_i-\mu_B)+\mu_A\mu_B(C_i-\mu_C) \\ &= \mu_B\mu_C A_i+\mu_A\mu_C B_i+\mu_A\mu_B C_i-2\mu_A\mu_B\mu_C\end{aligned}$$

Then, we apply Equations 3.2 and 3.4 to the linearized version of Q_i to get

$$\begin{aligned}\mu_Q &\approx \mu_B\mu_C\mu_A+\mu_A\mu_C\mu_B+\mu_A\mu_B\mu_C-2\mu_A\mu_B\mu_C \\ &\approx \mu_A\mu_B\mu_C\end{aligned}$$

$$\sigma_Q^2 \approx (\mu_B\mu_C)^2\sigma_A^2+(\mu_A\mu_C)^2\sigma_B^2+(\mu_A\mu_B)^2\sigma_C^2$$

Note that in the expression for σ_Q^2, we have assumed that the variables are all uncorrelated. Now, we can obtain an expression for the COV, V_Q, as follows:

$$\begin{aligned}V_Q^2 &= \frac{\sigma_Q^2}{\mu_Q^2} \approx \frac{(\mu_B\mu_C)^2\sigma_A^2+(\mu_A\mu_C)^2\sigma_B^2+(\mu_A\mu_B)^2\sigma_C^2}{(\mu_A\mu_B\mu_C)^2} = \frac{\sigma_A^2}{\mu_A^2}+\frac{\sigma_B^2}{\mu_B^2}+\frac{\sigma_C^2}{\mu_C^2} \\ &= V_A^2+V_B^2+V_C^2\end{aligned}$$

$$V_Q \approx \sqrt{V_A^2+V_B^2+V_C^2}$$

To relate the bias factors, we first recognize that the nominal value of Q, Q_n, is simply the product of the nominal values of A, B, and C. Then, starting with the above relationship for the mean values, we can relate the bias factors as follows:

$$\mu_Q \approx \mu_A\mu_B\mu_C$$

$$\lambda_Q Q_n \approx (\lambda_A A_n)(\lambda_B B_n)(\lambda_C C_n)$$

$$\lambda_Q Q_n \approx (\lambda_A \lambda_B \lambda_C) A_n B_n C_n$$
$$\Downarrow$$
$$\lambda_Q \approx \lambda_A \lambda_B \lambda_C$$

In most cases, we need to consider load cases involving several different types of loads. When several loads are acting together (e.g., $Q_1 + Q_2 + \ldots + Q_n$), then the total load can be modeled by using Equation 6.1 for each load to get

$$Q = c\,(A_1 B_1 C_1 + A_2 B_2 C_2 + \ldots\ldots + A_n B_n C_n) \tag{6.3}$$

where c is an additional factor that is common for all loads. It can also be thought of as a load combination factor.

6.3 DEAD LOAD

The dead load considered in design is usually the gravity load due to the self-weight of the structural and nonstructural elements permanently connected to the structure. Because of different degrees of variation in different structural and nonstructural elements, it is convenient to break up the total dead load into two components: weight of factory-made elements (steel, precast concrete members) and weight of cast-in-place concrete members. Also, for bridges, a third component of dead load is the weight of the wearing surface (asphalt). All components of dead load are typically treated as normal random variables. Usually, it is assumed that the total dead load, D, remains constant throughout the life of the structure. Table 6.1 shows some representative statistical parameters of dead load.

Often, there is a tendency on the part of designers to underestimate the total dead load. Therefore, to partially account for this, it is recommended to use a bias factor of 1.05 instead of the lower values shown in Table 6.1.

Table 6.1 Representative statistical parameters of dead load

	Bias factor $\lambda_D = \frac{\mu_D}{D_n}$	$\sqrt{V_A^2 + V_B^2}$	*COV* V_D
Bridges—factory-made elements	1.03	0.04	0.08
Bridges—cast-in-place concrete	1.05	0.06	0.10

6.4 LIVE LOAD IN BUILDINGS

6.4.1 Design (nominal) live load

Live load represents the weight of people and their possessions, furniture, movable partitions, and other portable fixtures and equipment. Usually, live load is idealized as a uniformly distributed load. The design live load is specified in psf (pounds per square foot) or kN/m^2 (kilonewtons per square meter). The magnitude of live load depends on the type of occupancy. For example, live loads specified by ASCE/SEI 7-10 (ASCE, 2010), *Minimum Design Loads for Buildings and Other Structures*, range from 10 psf (0.48 kN/m^2) for uninhabited attics not used for storage to 250 psf (11.97 kN/m^2) for storage areas above ceilings. The value of live load also depends on the expected number of people using the structure and the effects of possible crowding as shown in Figure 6.2.

The statistical parameters of live load depend on the area under consideration. The larger the area that contributes to the live load, the smaller the magnitude of the load intensity. ASCE/SEI 7-10 specifies the reduction factors for live load intensity as a function of the influence area. It is important to distinguish between influence area and tributary area. The tributary area is used to calculate the live load (or load effect) in beams and columns. The influence area is used to determine the reduction factors for live load intensity.

It is possible to relate influence area and tributary area by the following formulas:

For beams: Influence area = 2 × Tributary area

For columns: Influence area = 4 × Tributary area

When the *influence area*, A_I, is larger than 400 ft^2 (37.16 m^2), the design (nominal) live load, L_n, is calculated using

$$L_n = L_0\left(0.25 + \frac{15}{\sqrt{K_{LL}A_T}}\right)\text{(psf)} \qquad (6.4a)$$

$$L_n = L_0\left(0.25 + \frac{4.57}{\sqrt{K_{LL}A_T}}\right)\left(\text{kN/m}^2\right) \qquad (6.4b)$$

where K_{LL} = 2 for beams, K_{LL} = 4 for columns, A_T = tributary area,

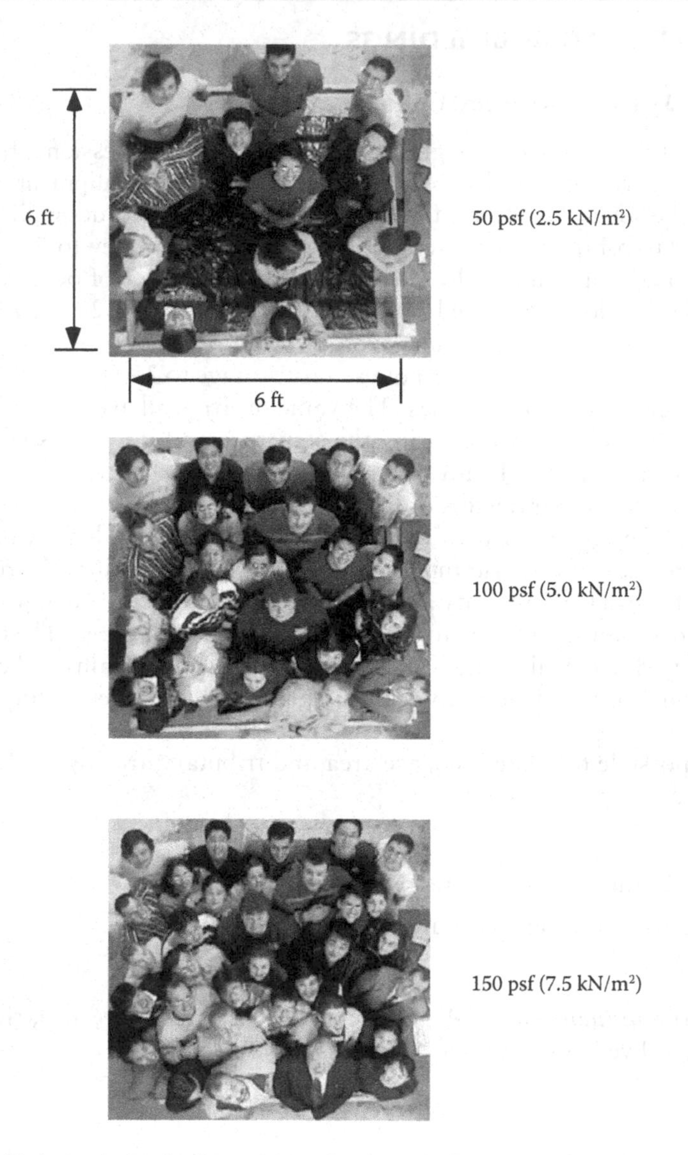

Figure 6.2 Variation in live load intensities due to crowding.

and L_0 is the unreduced design live load obtained from the code. There are some restrictions on the use of Equation 6.4; the interested reader is referred to ASCE/SEI 7-10 (ASCE, 2010) for details.

From the statistical point of view, it is convenient to consider two categories of live load: sustained live load and transient live load. These distinctions are described in the sections below.

Table 6.2 Statistical data for sustained live load as a function of influence area

Influence area (ft^2)	*Bias factor*	*COV*
200	0.24	0.59–0.89
1000	0.33	0.26–0.55
5000	0.52	0.20–0.46
10,000	0.60	0.18–0.45

Source: COVs were taken from Ellingwood, B., Galambos, T.V., MacGregor, J.G., and Cornell, C.A., *Development of a Probability Based Load Criterion for American National Standard A58*, National Bureau of Standards, NBS Special Publication 577, Washington, DC, 1980.

6.4.2 Sustained (arbitrary point-in-time) live load

Sustained live load is the typical weight of people and their possessions, furniture, movable partitions, and other portable fixtures and equipment. The term "sustained" is used to indicate that the load can be expected to exist as a usual situation (nothing extraordinary). Sustained live load is also called an *arbitrary point-in-time* live load, L_{apt}. It is the live load that you would most likely find in a typical office, apartment, school, hotel, etc.

Live load surveys have been performed by many researchers to obtain statistical data on the sustained live load. Some statistical values are shown in Table 6.2. Previous investigations (Corotis and Doshi, 1977; Ellingwood et al., 1980) have found that the sustained live load can be modeled as a gamma distributed random variable. Table 6.2 presents some typical values of the bias factors and the COVs for sustained live load as a function of influence area.

6.4.3 Transient live load

Transient live load is the weight of people and their possessions that might exist during an unusual event such as an emergency, when everybody gathers in one room (as reflected in Figure 6.2), or when all the furniture is stored in one room. Since the load is infrequent and its occurrence is difficult to predict, it is called a transient load. Like sustained live load, the transient live load is also a function of the influence area rather than the tributary area.

6.4.4 Maximum live load

For design purposes, it is necessary to consider the expected combinations of sustained live load and transient loads that may occur during the

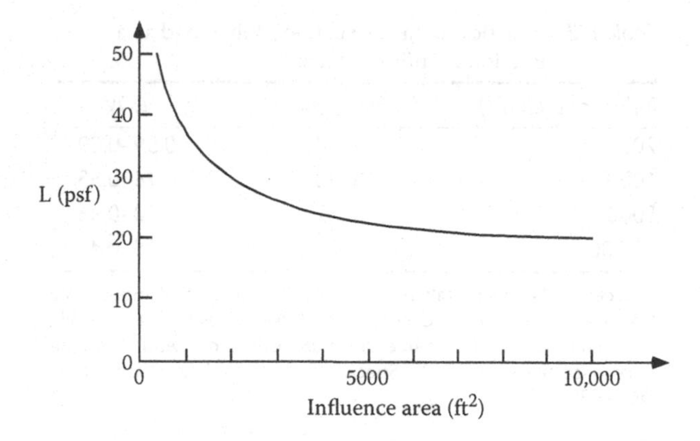

Figure 6.3 Mean maximum 50-year live load as a function of influence area.

building's design lifetime (50–100 years). The probabilistic characteristics of the maximum live load depend on the temporal variation of the transient load, the duration of the sustained load (which is related to the frequency of tenant changes or changes in use), the design lifetime, and the statistics of the random variables involved (Chalk and Corotis, 1980). The combined maximum live load can be modeled by an extreme Type I distribution (Ellingwood et al., 1980) for the range of probability values usually considered in reliability studies. More sophisticated approaches are discussed by Chalk and Corotis (1980). The mean value of the maximum 50-year live load as a function of the influence area is shown in Figure 6.3. The COVs are shown in Table 6.3.

Table 6.3 COVs of maximum 50-year live load

Influence area (ft^2)	*COV*
200	0.14–0.23
1000	0.13–0.18
5000	0.10–0.16
10,000	0.09–0.16

Source: From Ellingwood, B., Galambos, T.V., MacGregor, J.G., and Cornell, C.A., *Development of a Probability Based Load Criterion for American National Standard A58*, National Bureau of Standards, NBS Special Publication 577, Washington, DC, 1980. With permission.

6.5 LIVE LOAD FOR BRIDGES

For bridge design, the live load covers a range of forces produced by vehicles moving on the bridge. The effect of live load on the bridge depends on many parameters such as the span length, truck weight, axle loads, axle configuration, position of the vehicle on the bridge (transverse and longitudinal), number of vehicles on the bridge (multiple presence), girder spacing, and stiffness of structural members (slab and girders). Live load on bridges is characterized not only by the load itself but also by the distribution of this load to the girders. Therefore, the most important item to be considered is the load spectrum per girder.

The development of a live load model is essential for a rational bridge design and/or evaluation code. Ghosn and Moses (1985) proposed statistical parameters for truck load, including weight, axle configuration, dynamic load, and future growth. Their values were based on weigh-in-motion data (Moses et al., 1985). The live load model for AASHTO LRFD (2012) was developed by Nowak (1999) and Hong (1991) and Nowak (1993, 1999).

The design live load specified by AASHTO Standard (2002) is shown in Figure 6.4. For shorter spans, a *military load* is specified in the form of a tandem with two 24-kip axles spaced at 4 ft. The design load specified by AASHTO LRFD (2012) is shown in Figure 6.5. The *design tandem* in LRFD is based on two 25-kip axles.

The available statistical parameters of bridge live load have been determined from truck surveys and by simulations. The measurements show that the design values of bending moments and shears are lower than the actual load effects of today's heavy traffic observed on the highways. The available truck weight database is limited to selected locations and time periods of about one year. Examples of the cumulative distribution functions (CDFs) of gross vehicle weight (GVW) for trucks measured in 16 locations are shown in Figure 6.6 (on normal probability paper). The CDFs of axle loads for nine locations are presented in Figure 6.7.

The surveyed trucks were used to calculate bending moments. The CDFs of the resulting moments are plotted on normal probability paper in Figure 6.8.

The statistical parameters of bridge live load were derived in conjunction with the upgrading of the LRFD AASHTO code. An average lifetime for bridges is about 75 years, and this time period was used as the basis for calculation of loads. A statistical model was developed for the mean maximum 75-year moments and shears by extrapolation of the available truck survey data (Nowak, 1993). For longer spans, multiple presence of trucks in one lane was simulated by considering three cases: no correlation between trucks, partial correlation, and full correlation. It turned out that two fully correlated trucks governed. Figure 6.9 presents bias factor (mean maximum to HL-93 as specified in AASHTO) of live load moments for

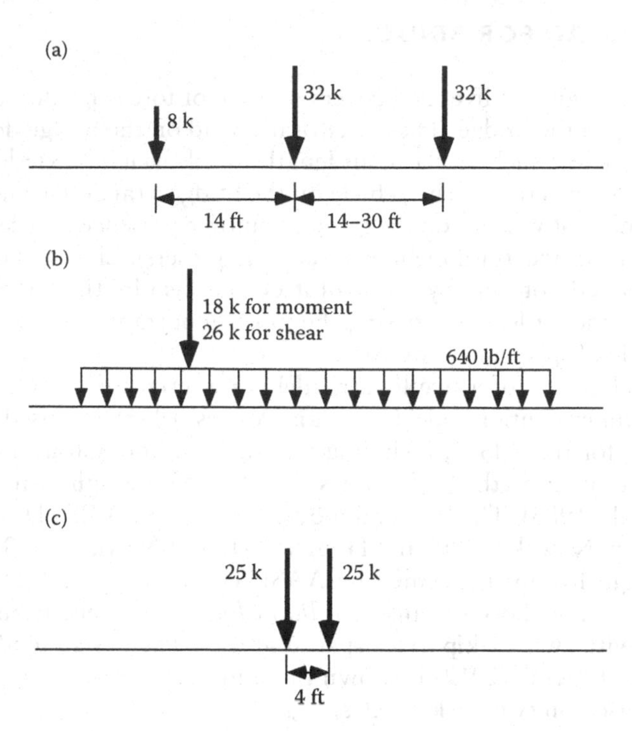

Figure 6.4 HS20 loading as defined by AASHTO (2002): (a) design truck; (b) single concentrated load with a uniform load; (c) design tandem. Conversion of units: 1 kip = 4.45 kN, 1 lb/ft = 15 N/m, 1 ft = 0.3 m.

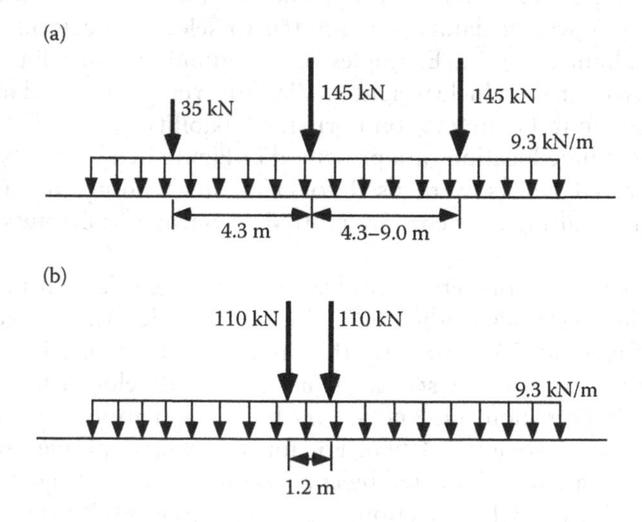

Figure 6.5 HL-93 loading specification by AASHTO LRFD (2012): (a) design truck with lane load; (b) design tandem with lane load.

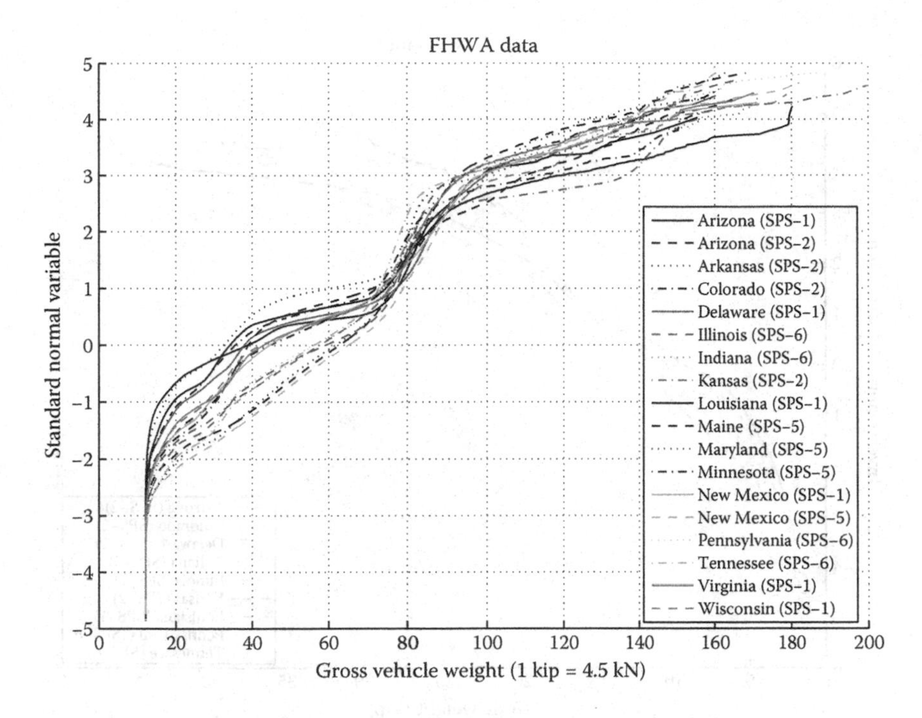

Figure 6.6 GVW of trucks surveyed on roads from 16 different states.

different average daily truck traffic (ADTT) and time intervals. For the maximum 75-year moment, the COV of truck load is 0.12. It is larger for shorter periods of time (for example, 0.20–0.25 for the maximum daily truck).

For multiple lanes, the load analysis involved not only truck loads but also girder distribution factors (GDFs). The girder moment or shear can be the resultant of more than one truck load. As in the case of a single lane, three cases of correlation between side-by-side vehicles were considered. It was determined by simulations that for two-lane bridges, trucks with fully correlated weights govern. The probability of having two very heavy trucks simultaneously on the bridge is lower than the occurrence of a single heavily loaded truck. It was calculated that in a side-by-side occurrence, each truck is about 85% of the mean maximum 75-year truck, which corresponds to the maximum 2-month truck.

Traffic frequency is very important in the statistical analysis of the heaviest trucks. ADTT varies depending on local conditions. The calculated multi-lane factors (for ADTT equal to 100, 1000, and 5000 trucks in one direction) are presented in Table 6.4.

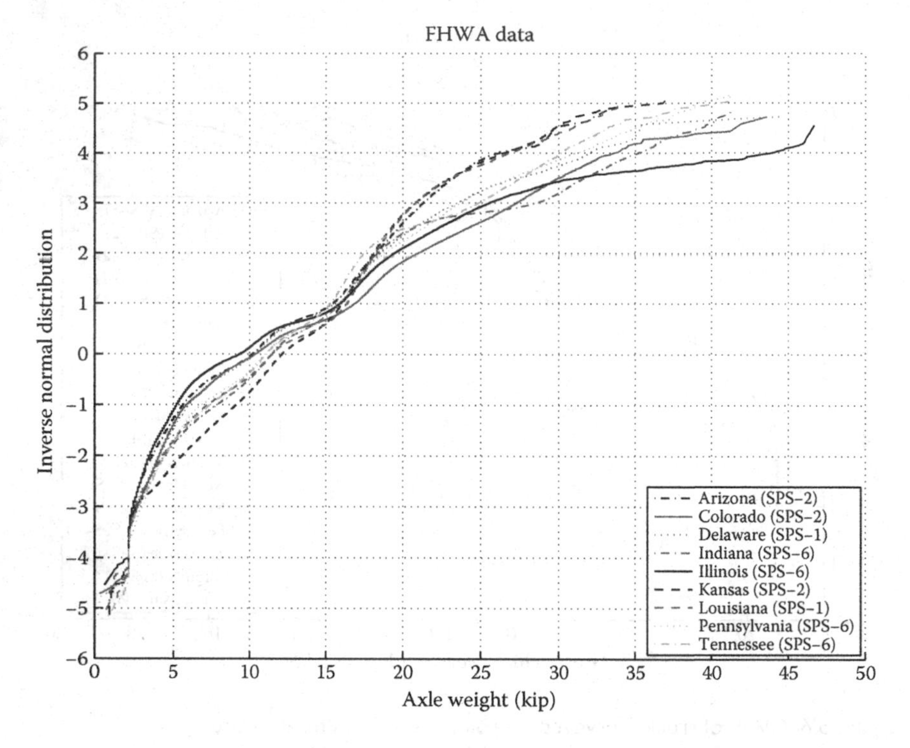

Figure 6.7 CDFs of axle loads for trucks surveyed on roads from nine different states.

Dynamic load is defined as the ratio of dynamic deflection and static deflection. The AASHTO Standard (2002) specifies impact, I, as a function of span length only, by the equation

$$I = \frac{50}{125 + L} \tag{6.5}$$

where L is the span length in feet (1 ft = 0.305 m). Actual dynamic load depends on three major factors: road roughness, bridge dynamics (natural period of vibration) and vehicle dynamics (type and condition of suspension system). The derivation of the statistical model for the dynamic behavior of bridges is presented by Hwang and Nowak (1991) and Naasif and Nowak (1995, 1996). The simulations and tests indicated that the dynamic load decreases for heavier trucks (as a percentage of static live load). The mean dynamic load is less than about 0.15 for a single truck and less than 0.10 for two trucks, for all spans. The COV of dynamic load is 0.80. The COV of a joint effect of live load and dynamic load is 0.18.

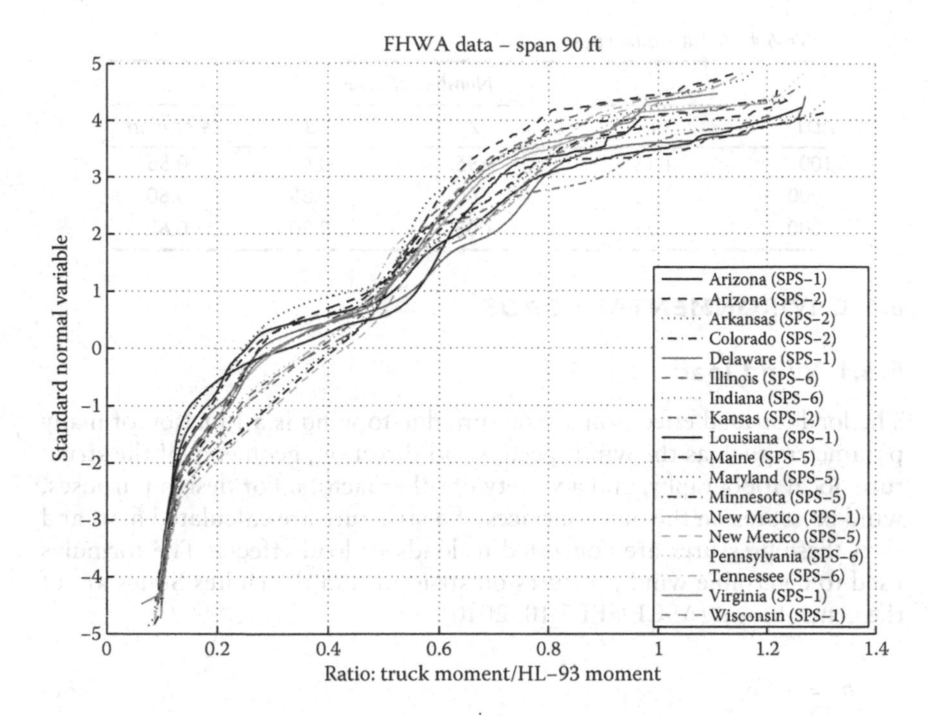

Figure 6.8 CDFs of calculated bending moments for trucks surveyed on roads from 16 different states.

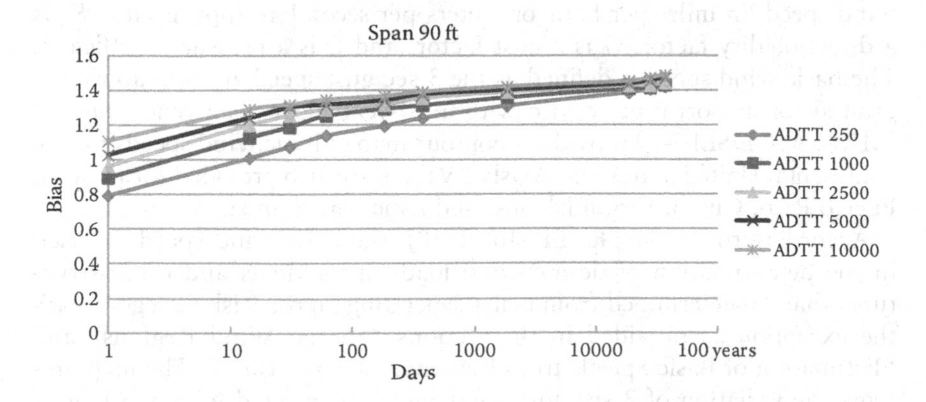

Figure 6.9 Bias factors of live load moments for different ADTT and time intervals.

Table 6.4 Multilane factors

	Number of lanes			
ADTT	*1*	*2*	*3*	*4 or more*
100	1.15	0.95	0.65	0.55
1000	1.20	1.00	0.85	0.60
5000	1.25	1.05	0.90	0.65

6.6 ENVIRONMENTAL LOADS

6.6.1 Wind load

The load (or load effect) on a structure due to wind is a function of many parameters such as the wind speed, wind direction, geometry of the structure, local topography, and a variety of other factors. For design purposes, wind pressures on the outer surfaces of a structure are calculated first, and then these pressures are converted to loads or load effects. The formulas used to determine wind pressures on structures in the Unites States are of the general form (ASCE/SEI 7-10, 2010):

$$p_Z = q_Z GC \tag{6.6a}$$

$$q_Z = 0.00256\ K_Z K_{Zt} K_d V^2 \text{ (U.S. units psf)} \tag{6.6b}$$

$$q_Z = 0.613\ K_Z K_{Zt} K_d V^2 \text{ (SI units Pa)} \tag{6.6c}$$

where p_Z is the design pressure, q_Z is the velocity pressure, K_Z is a velocity pressure exposure coefficient, K_{Zt} is a topographic factor, V is the basic wind speed (in miles per hour or meters per second as appropriate), K_d is a directionality factor, G is a gust factor, and C is a pressure coefficient. The basic wind speed is defined as the 3 sec gust speed at 10 m above the ground for airport-type terrain with a 50-year mean recurrence interval (MRI). ASCE/SEI 7-10 provides a contour map of basic wind speeds for the continental United States and Alaska. Values are also provided for Hawaii, Puerto Rico, Guam, Virgin Islands, and American Samoa.

According to the ASCE/SEI7-10 (2010), the basic wind speed, V, used in the determination of design wind load on buildings and other structures shall be determined from maps depending on the Risk Category with the exception as provided in the sections "Special Wind Regions" and "Estimation of Basic Speeds from Regional Climatic Data." The map presents the variation of 3 sec gust wind speeds associated with a height of 33 ft (10 m) for open terrain (Exposure C). Three-second gust wind speeds are used because most national weather service stations currently record and archive peak gust wind.

For Risk Category I buildings and structures, it is required to use map of wind speed corresponding to approximately a 15% probability of exceedance in 50 years (annual exceedance probability = 0.00333, MRI = 300 years). For Risk Category II buildings and structures, it is required to use map of wind speed corresponding to approximately a 7% probability of exceedance in 50 years (annual exceedance probability = 0.00143, MRI = 700 years). For Risk Category III and IV buildings and structures, it is required to use map of wind speed corresponding to approximately a 3% probability of exceedance in 50 years (annual exceedance probability = 0.000588, MRI = 1700 years).

The non-hurricane wind speed is based on 3 sec peak gust data collected at 485 weather stations where at least 5 years of data were available (Peterka, 1992; Peterka and Shahid, 1998). For non-hurricane regions, measured gust data were assembled from a number of stations in state-sized areas to decrease sampling error, and the assembled data were fit using a Fisher–Tippett Type I extreme value distribution. The hurricane wind speeds on the United States Gulf and Atlantic Coasts are based on the results of a Monte Carlo simulation model described in Applied Research Associates (2001); Vickery and Wadhera (2008); and Vickery et al. (2009a, 2009b, 2010).

Analyses of wind speed data suggest that the largest annual wind speed at a particular site tends to follow a Type I extreme value distribution (Simiu, 1979), and this distribution is the most commonly used model for wind speed (Ellingwood and Tekie, 1999). However, hurricane wind speeds tend to follow a Weibull distribution (Commentary to ASCE/SEI 7-10, 2010). Although the probability distributions for V are available, the probability distribution of the wind load (or load effect) is the one of interest for structural reliability calculations. The distribution of wind load is not necessarily Type I since the wind pressure is proportional to V^2 instead of V. Furthermore, the other parameters such K_Z, K_{Zt}, G, and C are random in nature as well. Thus, it is difficult to determine the distribution of the wind load (or load effect). However, studies (Ellingwood, 1981) have indicated that the uncertainty in wind load is dominated by the uncertainty in V^2, and the CDF for wind load can be represented by a Type I distribution. An extensive study on statistical parameters of wind speed was recently conducted by Rakoczy and Nowak (2012). They plotted the CDFs of peak gust wind speed on the normal probability paper as the best fit to the wind speed from the maps in ASCE/SEI (2010). The distribution for 50-year recurrence interval was defined as Fisher–Tippett Type I extreme value distribution (Peterka and Shahid, 1993). Based on the type of distribution and probability of exceedance in 50 years, they determined the statistical parameters of wind speed. For each location, five distributions were plotted. Table 6.5 summarizes some of the statistical parameters for wind speed in Central United States. These results were obtained in conjunction with NCHRP Project 10-80 by Rakoczy and Nowak (2012).

Table 6.5 Summary of statistical parameters of peak gust wind speeds for Central United States

MRI	*Bias factor*	*COV*
10 years	1.00	0.19
50 years	1.00	0.15
300 years	1.01	0.12
700 years	0.99	0.11
1700 years	1.02	0.10

Source: Rakoczy, A.M. and Nowak, A.S., "Statistical Parameters for Wind Speed," Progress Report for NCHRP 10-80, Report No. UNL-CE-3-2012, March 2012.

For additional information on wind loads, the interested reader is referred to the papers and reports by Ellingwood et al. (1980, 1981, 1982, 1999), Wen (1983), Simiu (1979), Galambos et al. (1982), and ASCE/SEI 7-10 (2010).

6.6.2 Ice load

Atmospheric ice load due to freezing rain, snow, and in-cloud icing shall be considered in the design of ice-sensitive structures. In areas where records or experience indicates that snow or in-cloud icing produces larger loads than freezing rain, site-specific studies shall be used. Structural loads due to hoarfrost are not a design consideration.

The design ice thickness shall be calculated using (ASCE/SEI 7-10)

$$t_d = 2.0\, tI_i\, fZ(K_{Zt})^{0.35} \tag{6.7}$$

where t is nominal ice thickness due to freezing rain at a height of 33 ft (10 m), I_i is the importance factor, f_Z is the height factor, and K_{Zt} is the topographic factor.

The projected area shall be increased by adding t_d to all free edges of the projected area. Wind loads on the increased projected area shall be used in the design of ice-sensitive structures.

The statistical parameters of load components are necessary to develop load factors and conduct reliability analysis (see Table 6.6). The design minimum load from ASCE/SEI (2010) is based on 25-year, 50-year, and 100-year events

Table 6.6 Summary of statistical parameters of ice thickness

MRI	*Bias factor*	*COV*
25 years	0.80	0.32
50 years	0.76	0.24
100 years	0.16	0.68

Source: Rakoczy, A.M. and Nowak, A.S., "Statistical Parameters for Ice Thickness," Progress Report for NCHRP 10-80, Report No. UNL-CE-8-2011, August 2011.

depending on Risk Category. Studies conducted by Rakoczy and Nowak (2011) show that the 25-year, 50-year, and 100-year event can be represented by normal distribution even though annual events have generalized Pareto distribution.

6.6.3 Snow load

The weight of snow on roofs can be a significant load to consider for structures in mountainous regions and snow belts. For design purposes, the snow load on a roof is often calculated based on information on the ground snow cover. For example, in the United States, the roof snow load for flat roofs (slope ≤ 5%) is calculated using (ASCE/SEI 7-10)

$$p_f = 0.7\, C_e C_t I_s\, p_g \tag{6.8}$$

where C_e is the exposure coefficient, C_t is the thermal factor, I_s is the snow importance factor, and p_g is the ground snow load (psf or kN/m²). For sloping roofs, the snow load is calculated using (ASCE/SEI 7-10)

$$p_s = C_s\, p_f \tag{6.9}$$

where C_s is the roof slope factor and p_f is the flat-roof snow load computed using Equation 6.8. The figure included in ASCE/SEI 7-10 shows a contour map of the United States that provides design values of the ground snow load based on a 50-year MRI.

As reflected by Equations 6.8 and 6.9, the probability distribution for the snow load on a structure will depend on the probability distributions of the ground snow load and the conversion factors C_e, C_s, and C_t. Statistical analyses of meteorological data suggest that the ground snow load can be modeled using either a lognormal or Type I distribution (Thom, 1966; Boyd, 1961; Ellingwood and Redfield, 1983). However, studies of snow data for the northeast quadrant of the United States indicate that the lognormal distribution is the preferred distribution for that region (Ellingwood and Redfield, 1983). The design values are based on a lognormal distribution model (Commentary to ASCE/SEI 7-10, 2010). The ground-to-roof conversion factor C_s is often modeled as a lognormal random variable (Ellingwood and O'Rourke, 1985) or normal random variable (Ellingwood, 1981). The distribution of the roof snow load is not obvious based on the distributions of ground snow load and C_s. Statistical studies (Ellingwood, 1981) suggest that a Type II extreme value distribution is appropriate for values of the CDF above 0.90, which is the region of interest in structural reliability studies. If all the random variables in Equations 6.8 and 6.9 are assumed to be independent lognormal variates, then a lognormal distribution for roof snow load is appropriate; this distribution was used by Ellingwood and Rosowsky (1996) in their investigation of combinations of snow and earthquake loads.

For additional details on the probabilistic treatment of snow load, the interested reader is referred to the papers and reports by Ellingwood et al. (1980, 1981, 1982), Ellingwood and Redfield (1983), Ellingwood and O'Rourke (1985), Ellingwood and Rosowky (1996), Galambos et al. (1982), Thom (1966), and Boyd (1961).

6.6.4 Earthquake

The basic design philosophy behind earthquake-resistant design is different than it is for most other environmental loads. For loads such as live and snow loads, the design event is typically defined with the expectation that a structure will suffer little or no damage if the design event occurs. However, for earthquake loading, the "design event" is defined with the understanding that some damage to the structure is expected and considered acceptable if this event occurs. Furthermore, the earthquake design limit state is defined in terms of overall system performance whereas the limit states for other environmental loads are defined for member performance (Commentary to ASCE/SEI 7-10, 2010). For economic and other reasons, it is usually not feasible to design structures so that they will suffer no damage during an earthquake event comparable to the design event. There are exceptions to this; for example, nuclear power plants and hospitals are usually designed for more stringent criteria because of the potential impact of structural damage on the functionality of the structure following an earthquake.

To understand the logic behind the design event for earthquake loading, it is important to understand two terms: hazard and risk. Hazard refers to some measure of an earthquake's potential to cause loss or damage, whereas risk focuses on the probability of occurrence of some level of loss or damage. In other words, seismic risk is associated with the consequences of exposure to the seismic hazard. For example, consider a large earthquake that may pose a significant hazard to the built environment. If it occurs in a densely populated area with poorly constructed buildings, then the risk is high. However, if it occurs in a region with well-designed and well-constructed buildings, then the risk is lower. McGuire (2004) provides a detailed discussion of the distinction between hazard and risk and provides an overview of the necessary calculations for each.

In the past, the design earthquake event has been defined in terms of hazard. In ASCE/SEI 7-10, a change was made to use a definition based on risk. The risk-targeted maximum considered earthquake (MCE_R) is defined as the ground motion input corresponding to a 1% chance of collapse in 50 years. The use of a risk-targeted input as opposed to a hazard-targeted input is due to geographic differences. As described in BSSC/FEMA P-750 (2009), "The change to risk-targeted ground motions uses the different shapes of hazard curves to adjust the uniform-hazard (2-percent-in-50-years) ground motions such that they are expected to result in a uniform annual frequency of collapse, or risk level, when used in design." The technical basis for this adjustment is described in the paper by Luco et al. (2007). Parameters for the design event earthquake are obtained by multiplying the MCE_R by a factor of 2/3. Specific values of the ground motion parameters pertaining to the MCE_R can be found online at http://earthquake.usgs.gov/designmaps.

Although an earthquake is a dynamic (time-varying) load, a static analysis procedure is commonly used to design structures for earthquakes. Furthermore, although damage and significant nonlinear material behavior is expected during the design event, the static analyses are typically based on linear elastic models of the structure. The fundamental premise behind this approach (labeled as the "equivalent lateral force procedure" in ASCE/SEI 7-10) is that a structure can be built and detailed to have a significant capacity to absorb and dissipate energy during a moderate to severe earthquake. Therefore, the intent is to design the structure so that damage is repairable following a moderate earthquake comparable to the design event. For a severe earthquake, the goal of the code provisions is to avoid structural collapse and to ensure life safety of the occupants of the structure.

In the context of the equivalent lateral force procedure applied to buildings, the code provides formulas for the design base shear force (V) at the base of the structure. To conduct the static analysis, this total base shear force is converted to a set of lateral forces distributed along the height of the structure in accordance with distribution formulas provided in the code. Figure 6.10 illustrates this process of distributing the base shear along the height of a building structure.

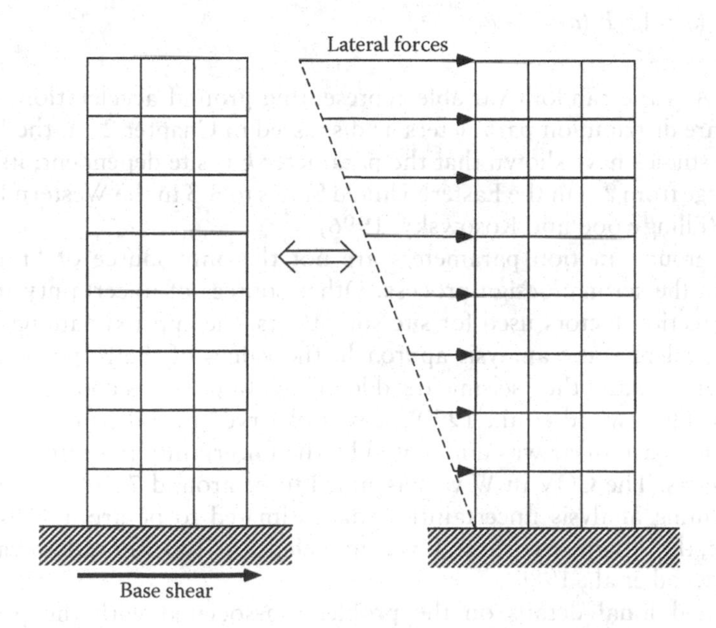

Figure 6.10 Illustration of how the design base shear is converted to lateral forces applied along the height of a building.

The seismic base shear force, V, is calculated as a percentage of the effective seismic weight W using the formula

$$V = C_S W \tag{6.10}$$

where the coefficient C_S is determined using formulas that incorporate information specific to the MCE_R, site soil coefficients, the fundamental period of the structure, a response modification factor R, and an importance factor I_e. (See ASCE/SEI 7-10 for the specific formulas.) The response modification R depends on the structural system being considered and provides an indirect measure of the ability of a particular structural system to absorb and dissipate energy during an earthquake. Values of R for various types of structural systems are provided in the code.

In early studies of structural reliability for earthquake loads, a quantity that was used to quantify the severity of ground motion was the peak ground acceleration (PGA). This acceleration was modeled as a random variable, and a Type II extreme value distribution (suggested by Cornell, 1968) was commonly used. In the seismic hazard literature, the complement of the Type II CDF was often used instead of the CDF. Mathematically, the complement $G_A(a)$ is represented as

$$G_A(a) = 1 - F_A(a) = 1 - e^{\left[-\left(\frac{u}{a}\right)^k\right]}, \tag{6.11}$$

where A is the random variable representing ground acceleration, and u and k are distribution parameters as discussed in Chapter 2. In the United States, studies have shown that the parameter k is site dependent; its value can range from 2.3 in the Eastern United States to 3.3 in the Western United States (Ellingwood and Rosowsky, 1996).

The ground motion parameters are not the only source of "randomness" in the seismic design process. Other sources of uncertainty include the correction factors used for site soil effects, the approximate nature of the equivalent static analysis approach, the values of the response reduction factor R, and the "seismic dead load" W. In previous code calibration studies (Ellingwood et al., 1980), it was observed that the uncertainty in the design base shear was dominated by the uncertainty in ground motion parameters. The COV in W was assumed to be around 7–10%. The COV representing analysis uncertainties was estimated to be around 20%. In the end, the design base shear V was modeled as a Type II random variable (Ellingwood et al., 1980).

For additional details on the problems associated with the probabilistic modeling of earthquake loads, the interested reader is referred to papers and reports by Collins et al. (1996), Cornell (1968), Ellingwood et al. (1980, 1982, 1996), Han and Wen (1994), Galambos et al. (1982), McGuire (2004), and BSSC (2009).

6.7 LOAD COMBINATIONS

6.7.1 Time variation

The time variation of loads and the possibility of simultaneous occurrence of loads are extremely important in structural reliability analysis.

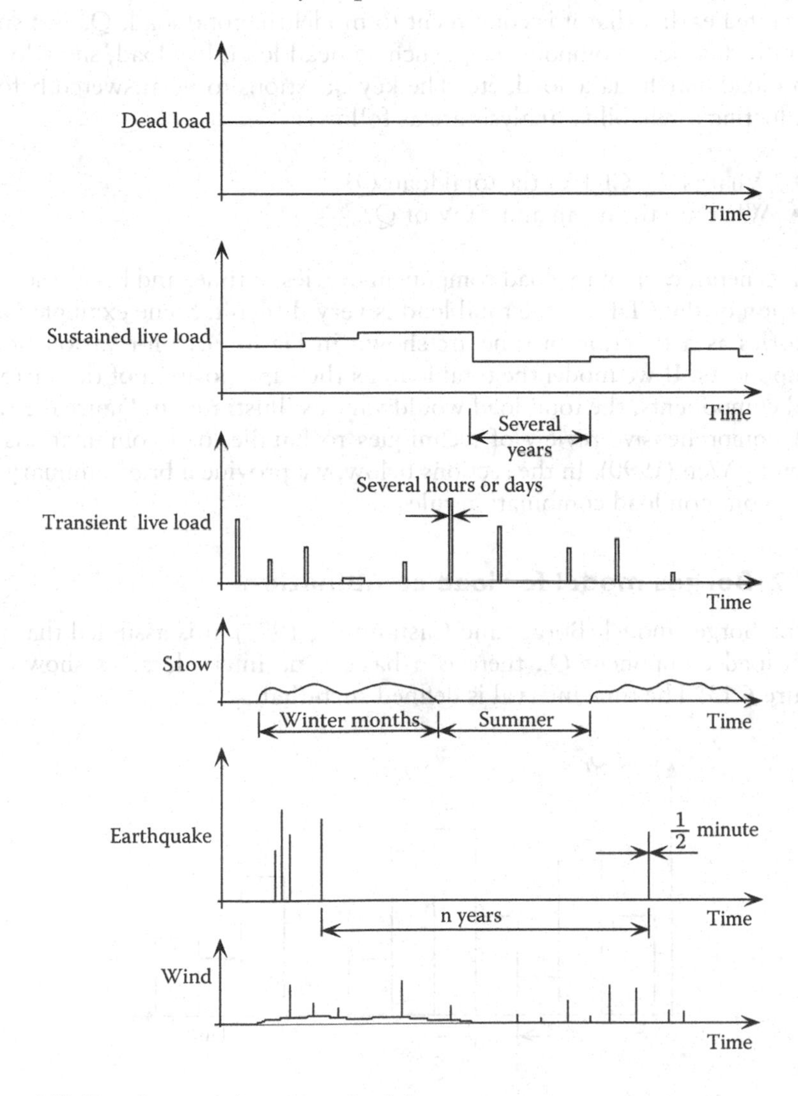

Figure 6.11 Time histories for various load components.

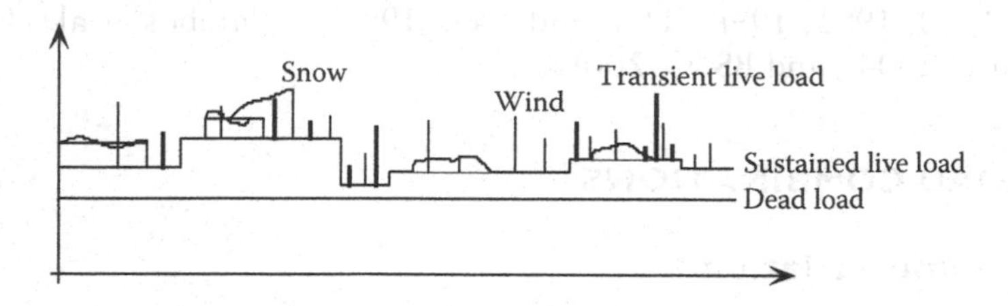

Figure 6.12 Superposition of loads.

We stated earlier that it is convenient to model the total load, Q, as a sum of individual load components Q_i such as dead load, live load, snow load, wind load, earthquake load, etc. The key questions to be answered before conducting a reliability analysis are as follows:

- What is the CDF of the total load Q?
- What are the mean and COV of Q?

In general, each of the load components varies in time, and hence the calculation of the CDF for the total load is very difficult. Some example load histories as a function of time are shown in Figure 6.11 for various load components. If we model the total load as the superposition of the various load components, the total load would vary as illustrated in Figure 6.12.

A comprehensive review of techniques to handle load combinations is given by Wen (1990). In the sections below, we provide a brief summary of some common load combination rules.

6.7.2 Borges model for load combination

In the Borges model (Borges and Castanheta, 1971), it is assumed that for each load component Q_i, there is a basic time interval, τ_i, as shown in Figure 6.13. The time interval is defined such that

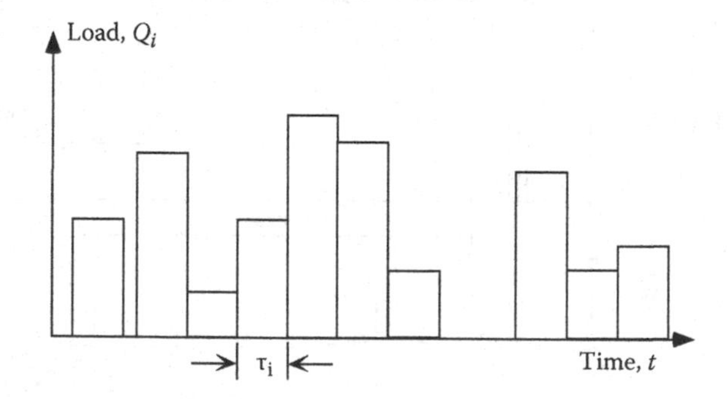

Figure 6.13 Time-variant load model.

1. The magnitude of Q_i can be considered as constant during this time period.
2. The occurrence or nonoccurrence of Q_i in each time interval corresponds to repeated independent trials with probability of occurrence p.

Given the occurrence of a particular load, the probability distribution of its amplitude is $F_i(q)$. This can be interpreted as a conditional CDF for Q_i. We assume based on physical grounds that the load must be nonnegative; hence, $F_i(q)$ is zero for $q \le 0$.

For the basic time interval τ_i, we can obtain the unconditional CDF of Q (in an interval τ_i) as

$$\begin{aligned} F(q) = P\left(Q_i \le q \text{ in interval } \tau_i\right) &= P(\text{load occurs})P\left(Q_i \le q \middle| \text{load occurs}\right) \\ &\quad + P(\text{no load occurs})P\left(Q_i \le q \middle| \text{no load occurs}\right) \\ &= pF_i(q) + (1-p)1 \\ &= 1 - p\left[1 - F_i(q)\right] \end{aligned} \tag{6.12}$$

The logic behind the last term in Equation 6.12 is as follows. The load can either occur or not occur. If p is the probability of occurrence, then $(1 - p)$ is the probability of nonoccurrence. Now, if the load does not occur, then Q_i must be zero by default since positive values of Q_i imply that the load occurs. If $Q_i = 0$, then the probability of the load being less than or equal to q must be equal to 1 (i.e., a certain event) since q must be a positive number.

Now, let us consider two basic time intervals. For two intervals, the unconditional probability can be expressed as

$$\begin{aligned} &F(q) = P(Q_i \le q \text{ in interval } 2\tau_i) = \\ &P(Q_i \le q \text{ in the first interval} \cap Q_i \le q \text{ in the second interval}) \end{aligned} \tag{6.13}$$

where $\cap$ denotes the intersection of the two events. If we assume that the values of Q_i are independent from interval to interval, the probability of the intersection is simply the product of the probabilities, or

$$F(q) = \{1 - p\,[1 - F_i(q)]\}^2 \tag{6.14}$$

If we extend this basic idea to n basic time intervals, we can conclude that

$$F(q) = \{1 - p[1 - F_i(q)]\}^n \tag{6.15}$$

Now, consider the total load Q as being composed of two components Q_1 and Q_2, i.e., $Q = Q_1 + Q_2$. The CDF of Q is difficult to determine, but the parameters (mean and variance) are relatively straightforward. Let τ_1 be the basic time interval for load component Q_1 and let τ_2 be the basic time interval for Q_2. The probabilities of occurrence in the basic time interval for each load component are p_1 and p_2, respectively. If we define k as the ratio τ_2/τ_1 where k is an integer, then the CDF of Q_1, corresponding to the time period τ_2, is

$$F(q_1) = \{1 - p_1 [1 - F_1(q_1)]\}^k \tag{6.16}$$

and for Q_2 it is

$$F(q_2) = \{1 - p_2 [1 - F_2(q_2)]\} \tag{6.17}$$

Since the variables defined by the CDFs in Equations 6.16 and 6.17 are defined for the same time period (τ_2), the sum of the variables is consistently defined. Therefore, the mean and variance can be calculated using the formulas presented in Chapter 3 for the sum of random variables. The results are

$$\mu_Q = \mu_1 + \mu_2 \tag{6.18}$$

$$\sigma_Q = \sqrt{\sigma_1^2 + \sigma_2^2} \tag{6.19}$$

where μ_1 and σ_1 are calculated for the CDF in Equation 6.16 and μ_2 and σ_2 are calculated for the CDF in Equation 6.17.

Example 6.2

Consider the load component, X, with CDF $F_X(x)$, corresponding to the basic time interval, τ, shown in Figure 6.14. If $p = 1$ and $k = 4$, then the CDF for the time interval 4τ is as shown in Figure 6.14 by the curve labeled $F^4{}_X(x)$.

6.7.3 Turkstra's Rule

The Borges approach for load combination can become cumbersome for several loads added together, and the definition of the basic time interval can be very difficult. Turkstra's Rule (Turkstra, 1970; Turkstra and Madsen, 1980) is a practical approach to modeling load combinations. It is based on the following observation: when one load component reaches an extreme value, the other load components are often acting at their average

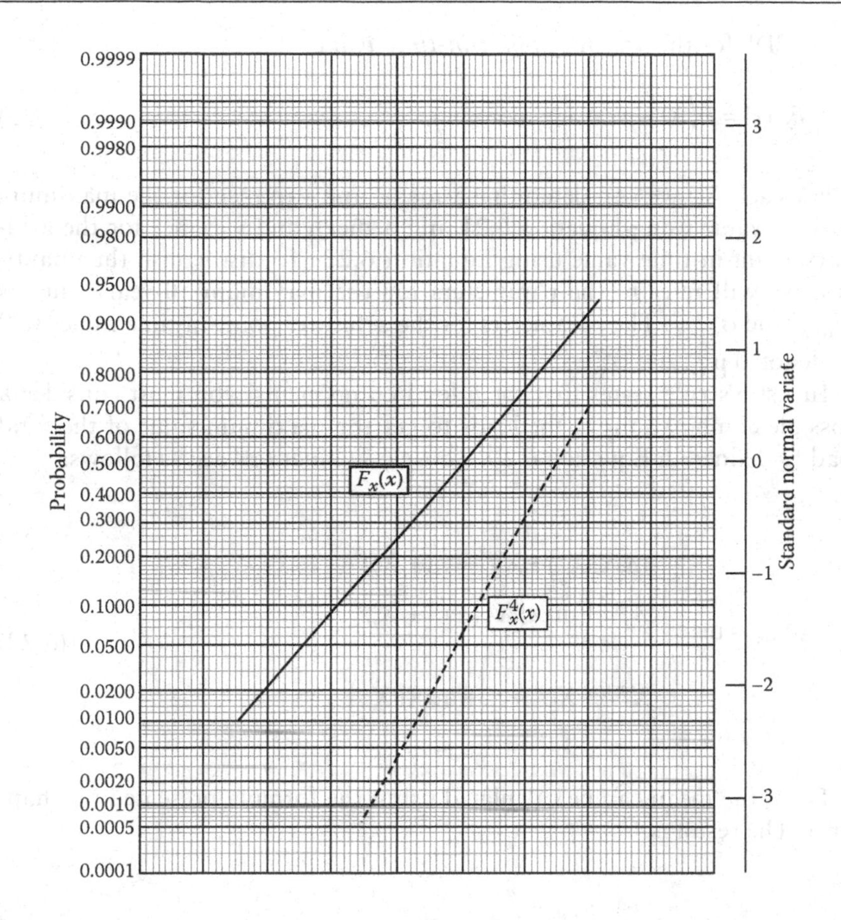

Figure 6.14 CDFs of load component considered in Example 6.2.

values. In other words, the possibility of two or more load components acting at their extreme values simultaneously is so remote that it is negligible.

Let $X_1, X_2 \ldots, X_n$ be load components considered in defining a total load Q defined as

$$Q = X_1 + X_2 + \ldots\ldots + X_n \tag{6.20}$$

Our objective is to determine the mean and variance of the *maximum* value of Q in 50 years. We will apply Turkstra's rule to obtain these quantities.

For each load component X_i, two distributions must be considered:

CDF for the *maximum* 50-year value

$$F_{X_{i-50}}(x) = P\left(\max X_i \le x \text{ in 50 years}\right) \tag{6.21}$$

CDF for the *arbitrary point-in-time value*

$$F_{X_i}(x) = P\left(X_i \leq x \text{ at any moment}\right). \tag{6.22}$$

For each X_i, we can determine a mean and variance for the maximum 50-year value using Equation 6.21 and a mean and variance for the arbitrary point-in-time value using Equation 6.22. To distinguish the quantities, we will refer to the parameters for the maximum 50-year value as $\mu_{\max X_i}$ and $\sigma_{\max X_i}$. The parameters for the arbitrary point-in-time values will be denoted $\mu_{X_i}^{\text{apt}}$ and $\sigma_{X_1}^{\text{apt}}$.

Turkstra's rule states that for n load components, you must consider n possible combinations of the loads to get the maximum value of the total load Q defined in Equation 6.20. Those combinations are as follows:

$$Q_{\max} = \max\begin{cases} \max\left(X_1\right) + X_2^{\text{apt}} + \ldots + X_n^{\text{apt}} \\ X_1^{\text{apt}} + \max\left(X_2\right) + \ldots + X_n^{\text{apt}} \\ \vdots \\ X_1^{\text{apt}} + X_2^{\text{apt}} + \ldots + \max\left(X_n\right) \end{cases} \tag{6.23}$$

Then, the mean can be calculated using the formula presented in Chapter 3. The result is

$$\mu_{Q_{\max}} = \max\begin{cases} \mu_{\max X_1} + \mu_{X_2}^{\text{apt}} + \ldots + \mu_{X_n}^{\text{apt}} \\ \mu_{X_1}^{\text{apt}} + \mu_{\max X_2} + \ldots + \mu_{X_n}^{\text{apt}} \\ \vdots \\ \mu_{X_1}^{\text{apt}} + \mu_{X_2}^{\text{apt}} + \ldots + \mu_{\max X_n} \end{cases} \tag{6.24}$$

For the combination that results in the largest mean value, we can also calculate the variance for that combination using the formula presented in Chapter 3. If the kth combination is the largest, then the variance is (assuming that the various components are uncorrelated)

$$\sigma_{Q_{\max}}^2 = \sigma_{\max X_k}^2 + \sum_{\text{other components}} \left(\sigma_{X_i}^{\text{apt}}\right)^2 \tag{6.25}$$

Example 6.3

Consider a combination of dead load, live load, and wind load. For each load component, the following information is known:

- Dead load is time invariant. Therefore, D_{max} and D^{apt} are the same. Both are normally distributed with

  $\mu_D = 20;\ V_D = 10\% \Rightarrow \sigma_D = 2.$

- For the live load, L_{max} follows an extreme Type I distribution with

  $\mu_{L_{max}} = 30;\ V_{L_{max}} = 12\% \Rightarrow \sigma_{L_{max}} = 3.6.$

  The arbitrary point-in-time value, L_{apt}, follows a gamma distribution with

  $\mu_L = 9;\ V_L = 31\% \Rightarrow \sigma_L = 2.8.$

- For the wind load, W_{max} follows an extreme Type I distribution with

  $\mu_{W_{max}} = 24;\ V_{W_{max}} = 20\% \Rightarrow \sigma_{W_{max}} = 4.8.$

  The arbitrary point-in-time value is lognormally distributed with

  $\mu_W = 1;\ V_W = 60\% \Rightarrow \sigma_W = 0.6.$

Calculate the parameters of the combined effect of these loads using Turkstra's rule.

The total load, Q, is $D + L + W$. By Turkstra's rule, we must consider three possible combinations to find Q_{max}:

$$Q_{max} = \max\begin{cases} D_{max} + L^{apt} + W^{apt} \\ D^{apt} + L_{max} + W^{apt} \\ D^{apt} + L^{apt} + W_{max} \end{cases}$$

However, in this case, since D_{max} and D^{apt} are equal, we really only need to consider two combinations:

$$Q_{max} = D^{apt} + \max\begin{cases} L_{max} + W^{apt} \\ L^{apt} + W_{max} \end{cases}$$

To find the mean value of Q_{max}, we must apply Equation 6.24.

$$\mu_{Q_{max}} = 20 + \max\begin{cases} 30+1 \\ 9+24 \end{cases} = 20 + \max\begin{cases} 31 \\ 33 \end{cases} = 53$$

Since the second load combination controls, we calculate the variance based on that combination. Applying Equation 6.25, we find

$$\sigma^2_{Q_{max}} = \left(\sigma_D^{apt}\right)^2 + \left(\sigma_L^{apt}\right)^2 + \left(\sigma_{W_{max}}\right)^2 = (2)^2 + (2.8)^2 + (4.8)^2$$
$$= 34.9 \Rightarrow \sigma_{Q_{max}} = 5.9$$

6.7.4 Load coincidence method

6.7.4.1 *Poisson pulse processes*

Before considering the load coincidence method, it is necessary to introduce the concept of a Poisson pulse process. A Poisson pulse process is a useful mathematical model that can be used to represent the temporal variation and intensity of loadings that act on structures.

The basic idea behind the Poisson pulse process is to represent a load $Q(t)$ as a sequence of pulses that occur during some reference time interval T. Two examples are shown in Figure 6.15. Both the intensity, X, and the duration, τ, of each pulse are treated as random variables, and their values

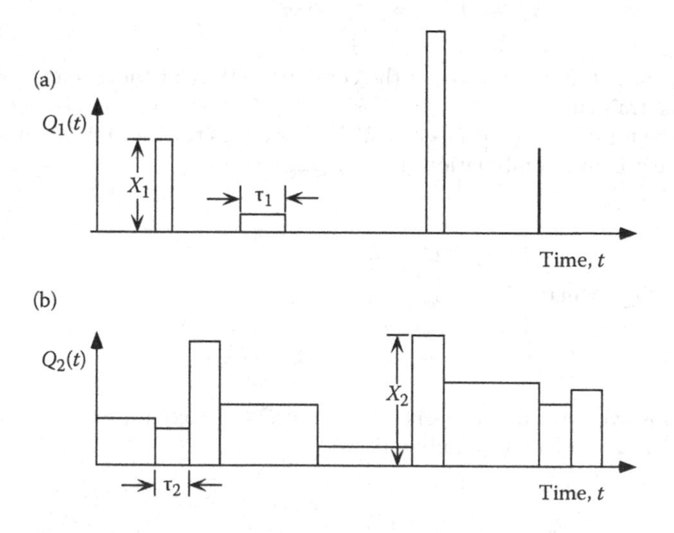

Figure 6.15 Examples of rectangular Poisson pulse processes. (a) A sparse process. (b) A full process.

are independent from occurrence to occurrence. (In other words, the intensity and duration do not depend on previous intensities and durations.) In general, the pulses can have arbitrary shapes, but a rectangular shape is considered here for simplicity. In Figure 6.15, the load $Q_1(t)$ is representative of a *sparse* process, which means that there are significant periods of time during which the load is zero. On the other hand, load $Q_2(t)$ is representative of a *full* process, which means that the load is never zero, but the intensity can vary in a stepwise manner as shown.

Suppose that a load $Q(t)$ can be represented as a Poisson pulse process with an occurrence rate λ and a mean duration μ_τ. Furthermore, assume that the intensity X of each pulse follows a CDF $F_X(x)$. For design purposes, we may be interested in the distribution of the maximum lifetime value Q_{max} of the load for a particular time period $(0,T)$. It can be shown (Wen, 1977, 1990; Melchers, 1987) that the CDF for Q_{max} is of the form

$$F_{Q_{max}}(q,T) = \exp\left\{-\lambda T\left[1 - F_X(q)\right]\right\} \tag{6.26}$$

6.7.4.2 Combinations of Poisson pulse processes

In its simplest form, the load coincidence model is a load combination model in which Poisson pulse processes are combined. For example, suppose that we are interested in determining the distribution of a combined total load (or load effect) $Q(t)$ defined as

$$Q(t) = Q_1(t) + Q_2(t) \tag{6.27}$$

in which $Q_1(t)$ and $Q_2(t)$ are independent loads that are modeled as Poisson pulse processes. The relevant parameters for the load models are the mean durations μ_{τ_1} and μ_{τ_2}, the occurrence rates λ_1 and λ_2, and the intensity distributions F_{X_1} and F_{X_2} of the pulses. Figure 6.16 schematically illustrates this combination problem. The maximum values of Q_1 and Q_2 in the reference interval $(0,T)$ are identified as $Q_{1,max}$ and $Q_{2,max}$. For a linear combination of pulse processes such as Equation 6.27, the possibility of simultaneous occurrence (overlap) of pulses exists as shown by the shaded regions in Figure 6.16c. The occurrence of these overlapping regions is referred to as the coincidence process (Wen, 1990) and is identified as $Q_{12}(t)$. The maximum of the coincidence process is identified as $Q_{12,max}$. Wen (1977, 1990) has shown that the coincidence process is also a Poisson pulse process with occurrence rate λ_{12} and mean duration $\mu_{\tau_{12}}$. Approximate relationships for λ_{12} and $\mu_{\tau_{12}}$ are (Wen, 1977, 1990)

$$\lambda_{12} = \lambda_1\lambda_2(\mu_{\tau_1} + \mu_{\tau_2}) \tag{6.28a}$$

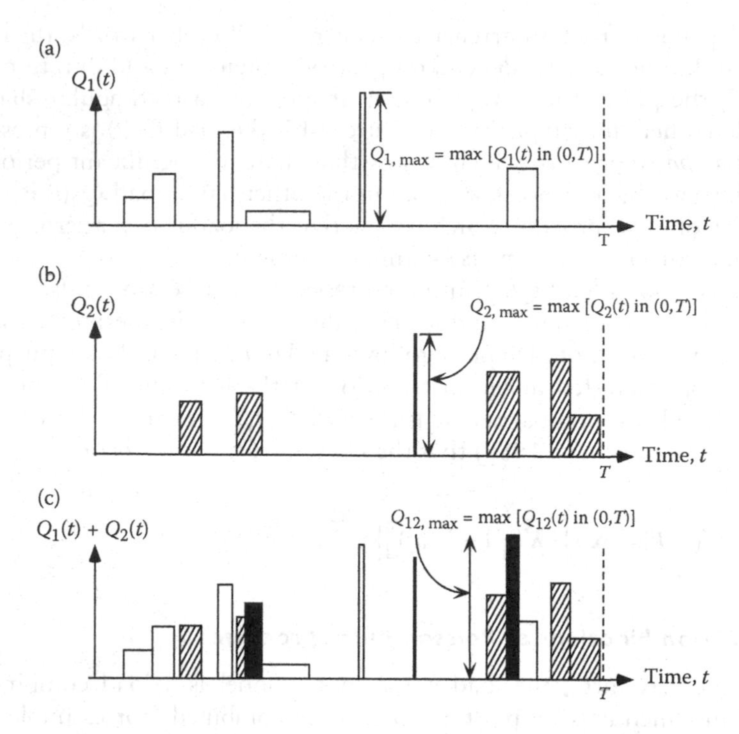

Figure 6.16 Illustration of the sum of two rectangular pulse processes. (a) Pulse process $Q_1(t)$ acting alone. (b) Pulse process $Q_2(t)$ acting alone. (c) Combined process $Q(t) = Q_1(t) + Q_2(t)$.

$$\mu_{\tau_{12}} = \frac{\mu_{\tau_1}\mu_{\tau_2}}{\mu_{\tau_1} + \mu_{\tau_2}} \tag{6.28b}$$

Now, consider the determination of the CDF of the maximum value of $Q(t)$ as defined in Equation 6.27. Referring to Figure 6.16c, the CDF of $Q_{\max}$ can be expressed as

$$F_{Q_{\max}}(q) = P\left[\left(Q_{1,\max} \le q\right) \cap \left(Q_{2,\max} \le q\right) \cap \left(Q_{12,\max} \le q\right)\right] \tag{6.29}$$

According to Wen (1990), a conservative approximation to Equation 6.29 is

$$F_{Q_{\max}}(q) \approx P\left(Q_{1,\max} \le q\right) P\left(Q_{2,\max} \le q\right) P\left(Q_{12,\max} \le q\right) \tag{6.30}$$

Since $Q_{1,max}$, $Q_{2,max}$, and $Q_{12,max}$ are random variables representing maximum values of Poisson pulse processes, Equation 6.26 can be used to model each one. Making the appropriate substitutions of Equation 6.26 into Equation 6.30, we get

$$F_{Q_{max}}(q) \approx \exp\left\{-\lambda_1 T\left[1-F_{X_1}(q)\right]\right\}\exp\left\{-\lambda_2 T\left[1-F_{X_2}(q)\right]\right\}$$
$$\times\exp\left\{-\lambda_{12} T\left[1-F_{X_{12}}(q)\right]\right\}$$
$$\approx \exp\left\{-\lambda_1 T\left[1-F_{X_1}(q)\right]-\lambda_2 T\left[1-F_{X_2}(q)\right]-\lambda_{12} T\left[1-F_{X_{12}}(q)\right]\right\}$$
$$(6.31)$$

There are two things to be noted. First, Equation 6.31 requires some knowledge of the CDF of the intensity, X_{12}, of the coincidence process; for rectangular pulses, this intensity is simply $X_1 + X_2$. Second, the approximation in Equations 6.30 and 6.31 is based on the assumption that $Q_{1,max}$, $Q_{2,max}$, and $Q_{12,max}$ are mutually independent of each other. $Q_{1,max}$ and $Q_{2,max}$ are independent since $Q_1(t)$ and $Q_2(t)$ were assumed to be independent loads. However, $Q_{12,max}$ is positively correlated with both $Q_{1,max}$ and $Q_{2,max}$. The approximation is conservative because it tends to overestimate the probability of exceedance of a particular value of q (Wen, 1990).

Example 6.4

Consider a combination of the form shown by Equation 6.27. Assume that the parameters describing the two processes $Q_1(t)$ and $Q_2(t)$ are as follows:

$$\lambda_1 = 2/\text{year} \quad \mu_{\tau_1} = 1 \text{ day} = \frac{1}{365} \text{ year} \quad X_1 \text{ is normal}$$

with $\mu = 1.2$ and $\sigma = 0.3$

$$\lambda_2 = 5/\text{year} \quad \mu_{\tau_2} = 2 \text{ days} = \frac{2}{365} \text{ year} \quad X_2 \text{ is normal}$$

with $\mu = 1.5$ and $\sigma = 0.4$

A time interval of 50 years ($T = 50$) is considered.

A. Determine the probability of exceeding a load value of 2.7 in 50 years.
B. Determine the value of combined load that has a probability of exceedance of 10% in 50 years.

First, we need to determine λ_{12} using Equation 6.28a:

$$\begin{aligned}\lambda_{12} &= \lambda_1\lambda_2\left(\mu_{\tau_1} + \mu_{\tau_2}\right)\\ &= (2)(5)\left(\frac{1}{365} + \frac{2}{365}\right)\\ &= 8.219\times10^{-2}/\text{year.}\end{aligned}$$

Next, we need to determine the CDF for the coincidence process. Since the pulse processes are rectangular, the resulting intensity of the coincidence process is simply

$$X_{12} = X_1 + X_2.$$

Since X_1 and X_2 are both normal random variables and since they are assumed to be independent, their sum is also a random variable. Thus, using the results of Chapter 3, we can show that

$$\begin{aligned}\mu_{12} &= \mu_1 + \mu_2 = 2.7\\ \sigma_{12} &= \sqrt{\sigma_1^2 + \sigma_2^2} = \sqrt{0.25} = 0.5.\end{aligned}$$

To solve part A, we substitute the given information into Equation 6.31 and determine the value of the CDF for a value of 2.7. The result is

$$F_{Q_{max}}(q = 2.7) = 9.142\times10^{-2}.$$

The probability of exceedance is simply $1 - (9.142 \times 10^{-2}) = 0.9086$. Hence, there is about a 91% chance that the maximum load will exceed 2.7 in 50 years.

To solve part B, we have to find the value of q such that

$$P(Q_{max} > q) = 1 - F_{Q_{max}}(q) = 0.10$$

or

$$F_{Q_{max}}(q) = 0.90.$$

By trial and error (or by plotting the function), the value of q is about 3.67.

The load coincidence model can be extended to include more than two load components, and other enhancements and modifications are also possible. The interested reader is referred to Wen (1977, 1990) for additional information and discussion.

Table P6.1 Load parameters for Problem 6.1

Load	*Nominal value*	*50-year extreme*		*Arbitrary point in time*	
		Bias	*COV*	*Bias*	*COV*
Dead	40	1.03	0.10	1.03	0.10
Live	60	0.85	0.12	0.25	0.60
Wind	25	0.80	0.20	0.08	0.85
Earthquake	30	0.30	0.95	0.01	3.00

PROBLEMS

6.1 Consider a steel beam. There are four load components: dead load, live load, wind load, and earthquake load. The load data are summarized in Table P6.1. The load-carrying capacity, R, is normally distributed. The bias factor for R is 1.15 and the COV is 11.5%. Determine the required R so that $\beta = 3.5$. Use Turkstra's rule to calculate the mean and standard deviation of the total load.

6.2 Consider a reinforced concrete column. The total load, Q, consists of three components: dead, live, and wind. The nominal value of the dead load effect is 100 kN. The nominal value of the live load effect is 125 kN. The nominal value of the wind load effect is 150 kN. The load-carrying capacity, R, is the sum of the capacity of the longitudinal reinforcement steel and concrete. This can be expressed as

$$R = R_S + R_C,$$

where the nominal values of R_S and R_C are 250 kN and 225 kN, respectively. It is assumed that all variables are uncorrelated. Calculate the reliability index for the column using the statistical data given in

Table P6.2 Statistical data for Problem 6.2

Item	*Arbitrary point in time*		*50-year extreme*	
	Bias	*COV*	*Bias*	*COV*
D	1.04	0.09	1.04	0.09
L	0.23	0.55	0.89	0.145
W	0.30	0.45	0.85	0.175
R_S	1.075	0.125	1.075	0.125
R_C	1.065	0.14	1.065	0.14

Table P6.2. HINT: Use Turkstra's rule to calculate the parameters of the total load, Q.

Chapter 7

Models of resistance

In Chapter 6, we discussed the subject of modeling the uncertainty in loads acting on structures. In this chapter, we look at some of the models used to represent the uncertainty in the capacity (i.e., resistance) of structural elements.

7.1 PARAMETERS OF RESISTANCE

The load-carrying capacity of a structure depends on the resistance of its components and connections. The resistance of a component, commonly denoted by R, is typically a function of material strength, section geometry, and dimensions. Although in design these quantities are often considered as deterministic, in reality, there is some uncertainty associated with each quantity. Therefore, the resistance R is a random variable.

The possible sources or causes of uncertainty in resistance can be put into three categories:

- Material properties: uncertainty in the strength of the material, the modulus of elasticity (MOE), cracking stresses, and chemical composition
- Fabrication: uncertainty in the overall dimensions of the component, which can affect the cross-section area, moment of inertia, and section modulus
- Analysis: uncertainty resulting from approximate methods of analysis and idealized stress/strain distribution models

The variability of the resistance of components has been quantified (to some extent) by tests, observations of existing structures, and engineering judgment. Most of this information is available for the basic structural materials and components. However, structural members are often made of several materials (composite members) that require special methods of analysis. Since information on the variability of the resistance of such members is not always available, it is often necessary to develop resistance models using the available material test data and numerical simulations.

In reliability analysis, one popular way to model the resistance R is to consider the resistance as a product of the nominal resistance, R_n, used in design and three parameters that account for some of the sources of uncertainty mentioned above. Mathematically, this model of resistance is of the form

$$R = R_n MFP \tag{7.1}$$

where M is a parameter reflecting variation in the strength of the material, F is a variable reflecting uncertainties in fabrication (dimensions), and P is an analysis factor (also known as a professional factor), which accounts for uncertainties due to the methods of analysis used. The material factor is defined as the ratio of the actual to nominal material properties (e.g., compressive strength of concrete or yield strength of steel). The fabrication factor is defined as the ratio of actual to nominal cross-sectional properties (i.e., dimensions, plastic modulus Z, section modulus S, moment of inertia I). The professional factor is defined as the ratio of test capacity (representing the actual *in situ* performance) to the predicted capacity (according to the model used in calculations).

The design (nominal) resistance, R_n, is the value of resistance specified by the code. For example, for a compact steel beam, the nominal resistance in bending (for plastic analysis) is

$$R_n = F_y Z \tag{7.2}$$

where F_y is the yield stress and Z is the plastic section modulus. Similarly, for a reinforced concrete (R/C) beam, the nominal resistance is

$$R_n = A_s f_y \left[d - 0.59 A_s f_y / (f_c' b) \right] \tag{7.3}$$

where A_s is the area of reinforcing steel, f_y is the yield stress of the reinforcing steel, d is the effective depth of the member, f_c' is the compressive strength of concrete, and b is the width of the member.

For the resistance model shown in Equation 7.1, the mean value of the resistance, μ_R, is (based on a first-order approximation)

$$\mu_R = R_n \mu_M \mu_F \mu_P \tag{7.4}$$

where μ_R, μ_F, and μ_P are the mean values of M, F, and P, respectively. The first-order estimates of the bias factor, λ_R, and the coefficient of variation of R, V_R, are

$$\lambda_R = \lambda_M \lambda_F \lambda_P \tag{7.5}$$

$$V_R = \sqrt{(V_M)^2 + (V_F)^2 + (V_P)^2} \qquad (7.6)$$

where λ_M, λ_F, and λ_P are the bias factors and V_M, V_F, and V_P are the coefficients of variation of M, F, and P, respectively. (The derivation of Equations 7.4 through 7.6 parallels the derivation outlined in Example 6.1.)

The statistical parameters for M, F, and P (i.e., bias factors, mean values, and coefficients of variation) are available in the literature. A detailed summary of the relevant parameters for building structures is provided in Ellingwood et al. (1980, 1982) and Galambos et al. (1982). For bridges, the parameters have been developed for steel girders (composite and noncomposite), R/C T-beams, and prestressed concrete AASHTO-type girders (Tabsh and Nowak, 1991; Nowak et al., 1994). The following sections provide a brief summary of some representative information that has been used in past reliability studies. The interested reader is encouraged to refer to the references cited for additional details (and limitations) and to review the literature as more up-to-date information becomes available.

7.2 STEEL COMPONENTS

7.2.1 Hot-rolled steel beams (noncomposite behavior)

The behavior of noncomposite steel girders depends on the yield strength of the steel (F_y) and on the "compactness" of the cross section. The dimensions of hot-rolled steel beams have low coefficients of variation (typically less than 0.03) and hence can be treated as deterministic values for practical purposes. The linear and nonlinear flexural behavior of a cross section is described by the moment–curvature relationship. Using a moment–curvature diagram, the elastic and plastic flexural rigidities and the level of ductility can be determined. The shape of the moment–curvature relationship depends on the shape factor of the steel section. (Recall that the shape factor is defined as the ratio of the plastic section modulus to the elastic section modulus.)

In a simple bending test on a section, a plastic hinge will form when the bending moment reaches a value of $M_p = F_y Z$ where M_p is the moment causing yielding of the whole section and Z is the plastic section modulus. The strength benefit derived from exploiting the plastic range is small for I-sections since the shape is already efficient under elastic conditions. Most of the material in the section is positioned furthest away from the neutral axis and is therefore fully stressed.

Some representative statistical parameters of M, F, and P for hot-rolled steel elements used in buildings are summarized in Table 7.1. It should be

Table 7.1 Representative statistics for the resistance of structures produced from hot-rolled steel elements

Element type	μ_P	V_P	μ_M	V_M	μ_F	V_F	λ_R^a	V_R
Tension member (yielding)	1.00	0	1.05	0.10	1.00	0.05	1.05	0.11
Tension member (ultimate)	1.00	0	1.10	0.10	1.00	0.05	1.10	0.11
Compact beam (uniform moment)	1.02	0.06	1.05	0.10	1.00	0.05	1.07	0.13
Compact beam (continuous)	1.06	0.07	1.05	0.10	1.00	0.05	1.11	0.13
Elastic beam (lat.-torsional buckling)	1.03	0.09	1.00	0.06	1.00	0.05	1.03	0.12
Inelastic beam (lat.-torsional buckling)	1.06	0.09	1.05	0.10	1.00	0.05	1.11	0.14
Plate girders (flexure)	1.03	0.05	1.05	0.10	1.00	0.05	1.08	0.12
Beam-columns	1.02	0.10	1.05	0.10	1.00	0.05	1.07	0.15

Source: Adapted from Ellingwood, B., Galambos, T.V., MacGregor, J.G., and Cornell C.A., *Development of a Probability Based Load Criterion for American National Standard A58*, National Bureau of Standards, NBS Special Publication 577, Washington, DC, 1980. With permission.

[a] The nominal resistance used to determine the bias factor is based on 2005 version of the AISC steel design code. These values may change slightly if nominal resistances based on 2011 versions of the AISC code (or other steel design codes) are used.

noted that the bias factors shown in Table 7.1 are based on nominal design values obtained from an older version of the American Institute of Steel Construction (AISC, 2011) Manual of Steel Construction; slightly different values are possible when the nominal design values are based on current code provisions.

In Table 7.1, note that the parameters for the fabrication factor, F, are $\lambda_F = 1.0$ and $V_F = 0.05$ for all cases. However, if a component contains fillet welds, it is recommended that $V_F = 0.15$ be used. The larger coefficient of variation is intended to account for the variability of the throat area of the weld (Ellingwood et al., 1980).

For noncomposite steel girders used in bridges, the response to bending moment has been evaluated for representative sizes using a computer procedure developed by Tabsh (1990). The resulting moment–curvature relationships are shown in Figures 7.1 through 7.4. The middle lines correspond to the average. Also shown are curves corresponding to one standard deviation above and one standard deviation below the average.

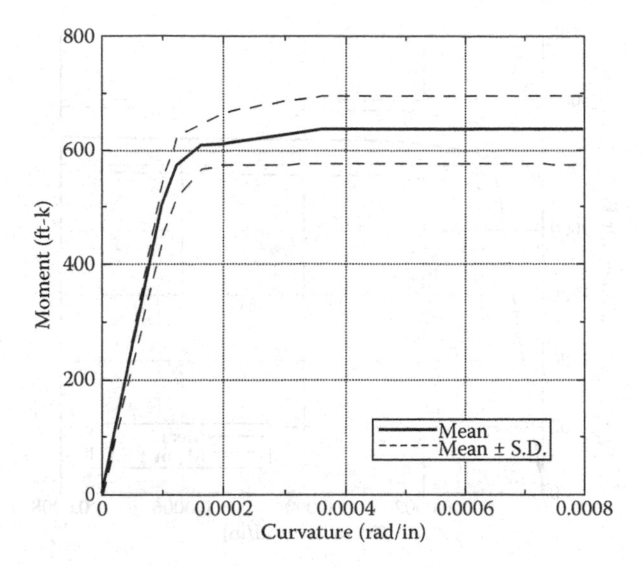

Figure 7.1 Moment–curvature curves for a noncomposite W24 × 76 steel section.

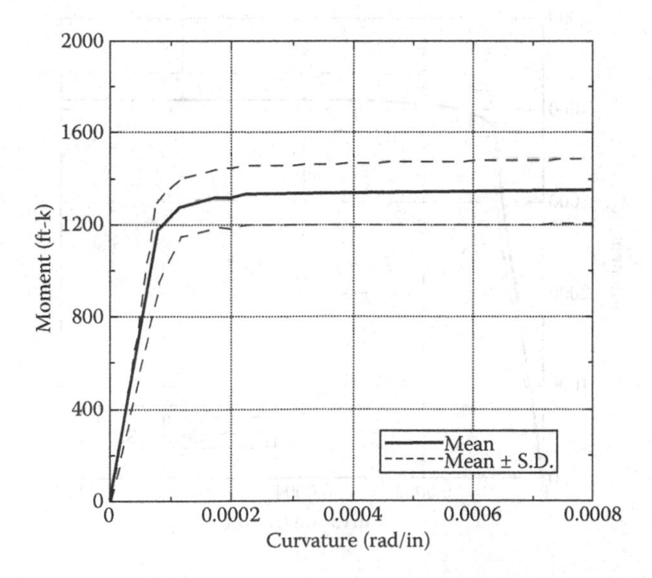

Figure 7.2 Moment–curvature curves for a noncomposite W33 × 118 steel section.

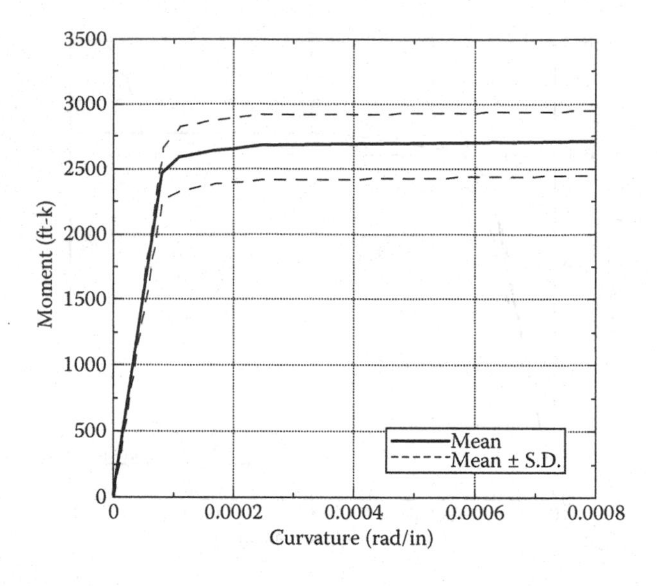

Figure 7.3 Moment–curvature curves for a noncomposite W36 × 210 steel section.

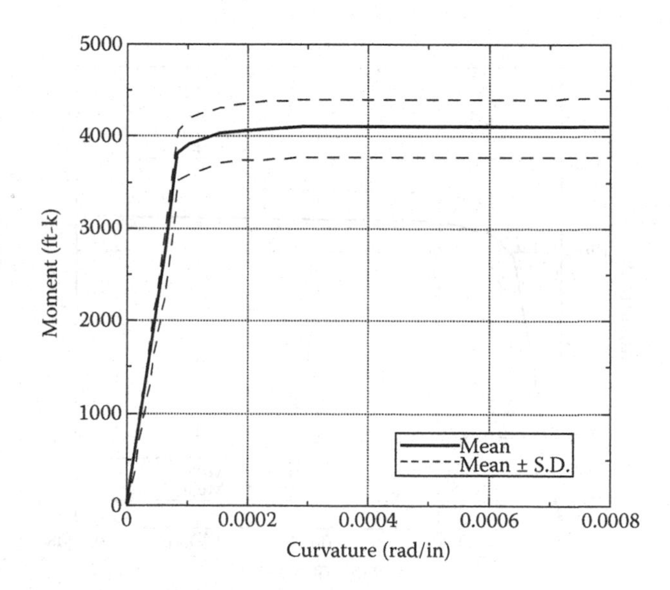

Figure 7.4 Moment–curvature curves for a noncomposite W36 × 300 steel section.

Based on data provided by the steel industry [American Iron and Steel Institute (AISI, 2008)], the observed bias factor (for M and F together) is $\lambda_{MF} = 1.095$ and the coefficient of variation is $V_{MF} = 0.075$. The parameters of the professional factor, P, are $\lambda_P = 1.02$ and $V_P = 0.06$ (Ellingwood et al., 1980). Therefore, for the resistance, R, the parameters are $\lambda_R = (1.095)(1.02) = 1.12$ and $V_R = \sqrt{(0.075)^2 + (0.06)^2} = 0.10$.

7.2.2 Composite steel girders

The behavior of composite steel and concrete cross sections has been summarized by Tantawi (1986). In that study, the major stresses considered were flexural, torsional, and shear. The ultimate torsional capacity of the cross section was also considered. Material properties (strength and dimensions) were modeled using the data provided by Kennedy (1982) and Ellingwood et al. (1980). Crushing of concrete in the positive moment region was the dominant failure mode, provided the longitudinal reinforcement in the cross section was at the minimum level.

The moment–curvature relationship in a composite beam depends on the stress–strain relationship of the structural steel, concrete, reinforcing steel, and the effective flange width of the cross section. A computer procedure was developed by Tabsh (1990) to calculate the moment–curvature relationship under monotonically increasing loading. The procedure was based on the following assumptions:

- A complete composite action between concrete and steel section was assumed. The effect of slip was neglected based on experimental and theoretical work done by Kurata and Shodo (1967).
- The typical stress–strain curves for concrete, reinforcing steel, and structural steel were used. In the analysis, the curves were generated by Monte Carlo simulations.
- The tensile strength of concrete was neglected.
- In the case of unshored construction, the effect of existing stresses and strains in the cross section before composite action takes place was not considered.

An iterative method was used for the development of the nonlinear moment–curvature relationship (Tantawi, 1986). The section was idealized as a set of uniform rectangular layers. The strain was increased gradually by increments. At each strain level, the corresponding moment was calculated using the nonlinear stress–strain relationships for the materials. The strain throughout the section was assumed constant during the analysis.

A closed-form expression for the moment–curvature relationship was developed by Zhou (1987) and Zhou and Nowak (1988). The formula is fairly flexible and accurate for most engineering purposes. Moreover, it can be used for a wide variety of cross sections. The basic equation is:

$$\phi = \frac{M}{EI_e} + C_1\left(\frac{M}{M_y}\right)^{C_2} \tag{7.7}$$

where ϕ is the curvature, EI_e is the elastic bending rigidity, M_y is the yield moment, M is the internal moment due to applied load, and C_1 and C_2 are constants controlling the shape of the curve. These constants can be determined from the conditions at yield and at ultimate. For composite girders, C_2 ranges between 16 and 24 whereas C_1 ranges between 0.00015/ft and 0.0003/ft.

Moment–curvature relationships for some typical composite steel bridge girders were obtained by Tabsh and Nowak (1991) using Monte Carlo simulation techniques. The moment–curvature relationships at the mean, mean plus one standard deviation, and mean minus one standard deviation for typical sections are shown in Figures 7.5 through 7.8. The concrete slab width considered is 6 ft, and the slab thickness is assumed to be 7 in. Based on data from the AISI (2008), the statistical parameters for MF are λ_{MF} = 1.07 and V_{MF} = 0.08. For the analysis factor, P, λ_P = 1.05 and V_P = 0.06 is assumed. Hence, for the ultimate moment, λ_R = 1.12 and V_R = 0.10.

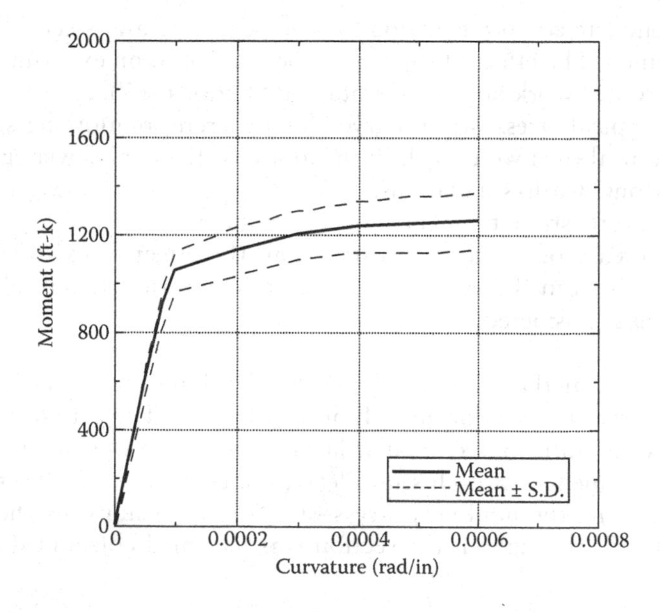

Figure 7.5 Moment–curvature relationship for a composite W24 × 76 steel section.

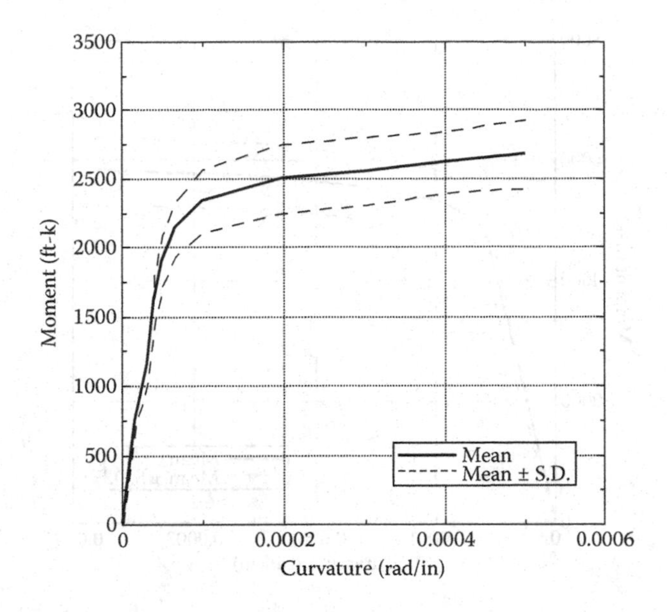

Figure 7.6 Moment–curvature relationship for a composite W33 × 130 steel section.

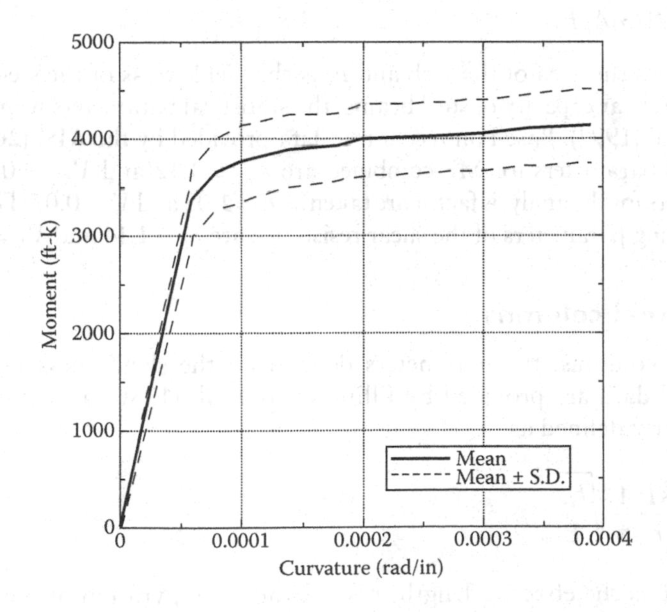

Figure 7.7 Moment–curvature relationship for a composite W36 × 210 steel section.

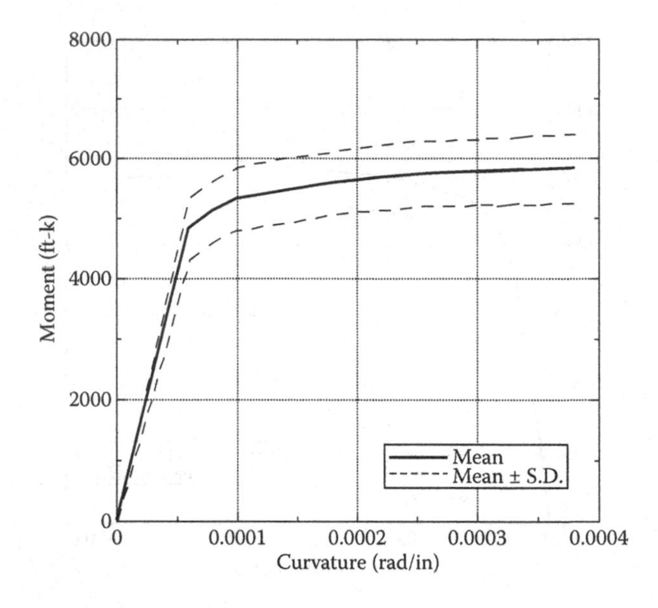

Figure 7.8 Moment–curvature relationship for a composite W36 × 300 steel section.

7.2.3 Shear capacity of steel beams

The ultimate shear capacity of steel sections, V_u, can be computed using

$$V_u = (1/3)A_w F_y \tag{7.8}$$

where A_w is the area of the web and F_y is the yield stress of the steel.

For the shear capacity of steel beams, the statistical parameters were derived by Yamani (1992). Based on recent test data provided by the AISI (2008), the statistical parameters for MF combined are $\lambda_{MF} = 1.12$ and $V_{MF} = 0.08$. The parameters for the analysis factor are taken as $\lambda_P = 1.02$ and $V_P = 0.07$. Therefore, the resulting parameters of the shear resistance are $\lambda_R = 1.14$ and $V_R = 0.106$.

7.2.4 Steel columns

For steel columns, the parameters depend on the slenderness ratio. The statistical data are provided by Ellingwood et al. (1980) as a function of parameter χ defined as

$$\chi = \frac{KL}{r}\frac{1}{\pi}\sqrt{\frac{F_{ys}}{E}} \tag{7.9}$$

where KL is the effective length, r is the radius of gyration of the section, F_{ys} is the static yield stress, and E is the MOE.

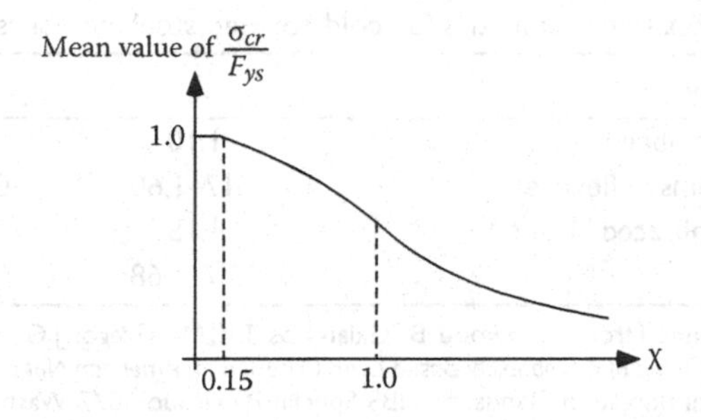

Figure 7.9 Critical stress for steel columns based on Equation 7.11.

The data presented in Ellingwood et al. (1980) is based on an older formulation of the mean column strength (Johnston, 1976; Bjorhovde, 1972). Their definition of mean strength is expressed in terms of the critical stress divided by F_{ys}. The critical stress is defined as

$$\sigma_{cr} = \frac{P_{cr}}{A} \tag{7.10}$$

where P_{cr} = critical load. The mean column strength is given by the following equations:

$$\text{mean value } \frac{\sigma_{cr}}{F_{ys}} = \begin{cases} 1 & \chi \leq 0.15 \\ 1.035 - 0.202\,\chi - 0.222\,\chi^2 & 0.15 \leq \chi \leq 1.0 \\ -0.111 + \dfrac{0.636}{\chi} + \dfrac{0.087}{\chi^2} & 1.0 \leq \chi \leq 2.0 \end{cases} \tag{7.11}$$

The variation of mean column strength with χ is shown in Figure 7.9.

The mean resistance is calculated using

$$\mu_R = \left(\frac{\sigma_{cr}}{F_{ys}}\right)_{\text{mean}} \mu_P \mu_M \mu_F \tag{7.12}$$

According to Ellingwood et al. (1980), the product $\mu_M\mu_P\mu_F$ is 1.08 for values of χ ranging from 0.3 to 1.9 For the same range of χ values, the coefficient of variation, V_R, varies between 0.12 and 0.15.

7.2.5 Cold-formed members

For cold-formed steel members, the statistical parameters are given in Table 7.2. These data are based on research conducted at the University of

Table 7.2 Resistance statistics for cold-formed steel members

Member type	λ_R	V_R
Tension member	1.10	0.11
Braced beams in flexure[a]	1.17–1.60	0.17–0.28
Laterally unbraced beams	1.15	0.17
Columns[b]	0.97–1.68	0.09–0.26

Source: Adapted from Ellingwood, B., Galambos, T.V., MacGregor, J.G., and Cornell C.A., *Development of a Probability Based Load Criterion for American National Standard A58*, National Bureau of Standards, NBS Special Publication 577, Washington, DC, 1980. With permission.

[a] The values of bias factor and coefficient of variation depend on whether the flanges are stiffened or unstiffened.

[b] The values of bias factor and coefficient of variation depend on whether the mode of buckling is flexural or torsional-flexural and whether the behavior is elastic or inelastic. For additional details, see Ellingwood et al. (1980).

Missouri-Rolla (Ellingwood et al., 1980). Note that the statistical parameters are given for the resistance explicitly; information on M, F, and P is not needed. The values of nominal resistance used in determining the bias factor values are based on the 1968 AISI Specification for cold-formed structures and include the factor of safety specified in that code. The factor of safety values range from 5/3 to 23/12.

7.3 ALUMINUM STRUCTURES

Table 7.3 presents some statistics for aluminum structures based on research conducted at Washington University in St. Louis (Ellingwood et al., 1980).

Table 7.3 Resistance statistics for aluminum structures

Member type	*Factor of safety*	λ_R	V_R
Tension member—yield limit state	1.65	1.10	0.08
Tension member—ultimate limit state	1.95	1.10	0.08
Beams—yield limit state	1.65	1.10	0.08
Beams—lateral buckling limit state	1.65	1.03	0.13
Beams—inelastic local buckling limit state	1.65	1.00	0.09
Columns[a]	1.82–1.95	0.87–1.10	0.08–0.14

Source: Adapted from Ellingwood, B., Galambos, T.V., MacGregor, J.G., and Cornell C.A., *Development of a Probability Based Load Criterion for American National Standard A58*, National Bureau of Standards, NBS Special Publication 577, Washington, DC, 1980. With permission.

[a] The values of bias factor and coefficient of variation depend on the specific limit state of yielding or buckling being considered. For additional details, see Ellingwood et al. (1980).

The nominal resistance used to determine the bias factor is the allowable resistance based on design specifications multiplied by the factor of safety shown.

7.4 REINFORCED AND PRESTRESSED CONCRETE COMPONENTS

7.4.1 Concrete elements in buildings

For R/C components, the basic variables that influence the resistance of the component are concrete strength in compression and tension, yield strength of the reinforcement, and dimensions of the cross section. In quantifying the variability of the resistance, there are several important assumptions that must be noted:

- The variability of the material properties and dimensions correspond to the "average" quality of construction expected in practice.
- Material strengths are assumed to be representative of relatively slow loading rates for dead, live, and snow loads.
- Long-term strength changes of the concrete and steel (due to the increasing maturity of the concrete, reduction of strength due to fatigue, and possible future corrosion of the reinforcement) are ignored.

To illustrate how the long-term aging of concrete can affect its strength, consider this real-world example from Gardiner and Hatcher (1970). Concrete strength was tested at 99 points (locations) in a 22-year-old building. The average strength from the tests was 8050 psi (56 N/mm^2) and the standard deviation was 500 psi (3.5 N/mm^2). The average 28-day cylinder strength for the same project was 3780 psi (26 N/mm^2), and the specified design strength (nominal strength) was 3000 psi (21 N/mm^2). If we compute the ratio of 22-year strength and 28-day strength, we get 8050/3780 = 2.13. This is clearly a significant increase, and the fact that it is ignored (as noted in the third bullet above) must be kept in mind.

For compressive strength of concrete, the statistical parameters are summarized in Table 7.4, and for yield strength of reinforcing bars in Table 7.5.

Fabrication factor, F, represents the variation in dimensions and geometry. The recommended statistical parameters are based on study by Ellingwood et al. (1980). For the dimensions of concrete components, the recommended parameters are listed in Table 7.6. For steel components, reinforcing bars and stirrups, the bias factor of dimensions is $\lambda = 1.0$ and

Table 7.4 Statistical parameters for compressive strength of concrete

	Compressive strength		Shear strength	
Property	*Bias factor, λ*	*Coefficient of variation, V*	*Bias factor, λ*	*Coefficient of variation, V*
Lightweight concrete				
$f'_c = 3000$ psi	1.38	0.155	1.38	0.185
$f'_c = 3500$ psi	1.33	0.145	1.33	0.175
$f'_c = 4000$ psi	1.29	0.140	1.29	0.170
$f'_c = 4500$ psi	1.25	0.135	1.25	0.160
$f'_c = 5000$ psi	1.22	0.130	1.22	0.155
$f'_c = 5500$ psi	1.20	0.125	1.20	0.150
$f'_c = 6000$ psi	1.18	0.120	1.18	0.145
$f'_c = 6500$ psi	1.16	0.120	1.16	0.145
Ordinary concrete				
$f'_c = 3000$ psi	1.31	0.170	1.31	0.205
$f'_c = 3500$ psi	1.27	0.160	1.27	0.190
$f'_c = 4000$ psi	1.24	0.150	1.24	0.180
$f'_c = 4500$ psi	1.21	0.140	1.21	0.170
$f'_c = 5000$ psi	1.19	0.135	1.19	0.160
$f'_c = 5500$ psi	1.17	0.130	1.17	0.155
$f'_c = 6000$ psi	1.15	0.125	1.15	0.150
$f'_c = 6500$ psi	1.14	0.120	1.14	0.145
High strength				
$f'_c = 7000$ psi	1.13	0.115	1.13	0.140
$f'_c = 8000$ psi	1.11	0.110	1.11	0.135
$f'_c = 9000$ psi	1.10	0.110	1.10	0.135
$f'_c = 10000$ psi	1.09	0.110	1.09	0.135
$f'_c = 12000$ psi	1.08	0.110	1.08	0.135

Source: Adapted from Rakoczy, A.M. and A.S. Nowak, "Resistance Model of Lightweight Concrete Members," *ACI Materials Journal*, American Concrete Institute, (accepted); and Rakoczy, A.M. and A.S. Nowak, "Resistance Factors for Lightweight Concrete Members," *ACI Materials Journal*, American Concrete Institute (under review).

Note: 1 ksi = 6900 Pa.

$V = 0.01$. The area of reinforcing steel, A_s, is also treated as a practically deterministic value, with $\lambda = 1.0$ and $V = 0.015$.

Professional (analysis) factor, P, represents the variation in the ratio of the actual resistance and what can be analytically predicted using accurate material strength and dimension values. Most of the statistical parameters of P are based on the previous study by Ellingwood et al. (1980). The

Table 7.5 Statistical parameters for reinforcing steel bars

Bar size	*Bias factor, λ*	*Coefficient of variation, V*
#3	1.18	0.04
#4	1.13	0.03
#5	1.12	0.02
#6	1.12	0.02
#7	1.14	0.03
#8	1.13	0.025
#9	1.14	0.02
#10	1.13	0.02
#11	1.13	0.02
#14	1.14	0.02

Source: Adapted from Nowak, A.S., Rakoczy, A.M., and Szeliga, E., "Revised Statistical Resistance Models for R/C Structural Components," ACI SP-284-6, American Concrete Institute, March 2012, pp. 1–16.

Table 7.6 Statistical parameters for fabrication factor

Item	*Bias factor, λ*	*Coefficient of variation, V*
Width of beam, cast in place	1.01	0.04
Effective depth of a R/C beam	0.99	0.04
Effective depth of prestressed concrete beam	1.00	0.025
Effective depth of slab, cast in place	0.92	0.12
Effective depth of slab, plant cast	1.00	0.06
Column width and breadth	1.005	0.04
Area of reinforcement, A_s, A_v	1.00	0.015
Spacing of shear reinforcement, s	1.00	0.04

Source: Adapted from Ellingwood, B., Galambos, T.V., MacGregor, J.G., and Cornell C.A., *Development of a Probability Based Load Criterion for American National Standard A58*, National Bureau of Standards, NBS Special Publication 577, Washington, DC, 1980. With permission.

professional factor for shear in slabs without shear reinforcement is based on the recent results of shear tests in slabs (Collins and Kuchma, 1999; Reineck et al., 2003; Teng et al., 2004). The recommended values are listed in Table 7.7. The statistical parameters of resistance (load carrying capacity) of reinforced concrete members are given in Table 7.8.

There are two items to note. First, Table 7.4 shows that the mean and coefficient of variation of the compressive strength of concrete generally

Table 7.7 Statistical parameters for professional factor

Item	*Bias factor, λ*	*Coefficient of variation, V*
Ordinary concrete		
R/C beams, flexure	1.02	0.06
R/C beams, shear	1.075	0.10
Slab, flexure and shear with shear reinforcement	1.02	0.06
Slab, shear without shear reinforcement	1.16	0.11
Column, tied	1.00	0.08
Column, spiral	1.05	0.06
Bearing strength	1.02	0.06
Lightweight concrete		
R/C beams, flexure	0.98	0.06
R/C beams, shear	0.975	0.10
Slab, flexure and shear with shear reinforcement	0.98	0.06
Slab, shear without shear reinforcement	1.10	0.11

Source: Nowak, A.S., Rakoczy, A.M., and Szeliga, E., "Revised Statistical Resistance Models for R/C Structural Components," ACI SP-284-6, American Concrete Institute, March 2012, pp. 1–16; Ellingwood, B., Galambos, T.V., MacGregor, J.G., and Cornell C.A., *Development of a Probability Based Load Criterion for American National Standard A58*, National Bureau of Standards, NBS Special Publication 577, Washington, DC, 1980.

Table 7.8 Statistical parameters of R/C members

Structural type and limit state	*Design cases*	*NWC*		*LWC*	
		λ_R	V_R	λ_R	V_R
(R/C)	ρ = 0.6%	1.140	0.080	1.100	0.080
Beam—flexure	ρ = 1.6%	1.130	0.085	1.110	0.085
R/C beam—shear	No shear reinforcement	1.265	0.215	1.230	0.210
f'_c = 4000 psi	Minimum code shear reinforcement	1.250	0.170	1.165	0.165
(27.6 MPa)	Minimum practical shear reinforcement	1.230	0.135	1.135	0.135
	Average shear reinforcement	1.225	0.130	1.124	0.125
Concrete—bearing	f'_c = 4000 psi (27.6 MPa)	1.275	0.175	1.275	0.165
R/C slab	Slab depth: 4, 6, 8 in; one-way flexure	1.055	0.145	1.020	0.145
f'_c = 4000 psi	Slab depth: 4, 6, 8 in; one-way shear	1.260	0.245	1.240	0.235
(27.6 MPa)	Slab depth: 4, 6, 8 in; two-way shear	1.430	0.245	1.420	0.242
R/C column	f'_c = 4000 psi (27.6 MPa)	1.16-	0.115	–	–
Tied and spiral		1.28	0.155	–	–

Source: Adapted from Nowak, A.S., Rakoczy, A.M., and Szeliga, E., "Revised Statistical Resistance Models for R/C Structural Components," ACI SP-284-6, American Concrete Institute, March 2012, pp. 1–16.

Note: 1 ksi = 6900 Pa and 1 in = 25.4 mm. Various reinforcing details were considered. See Nowak et al. (2012) for details.

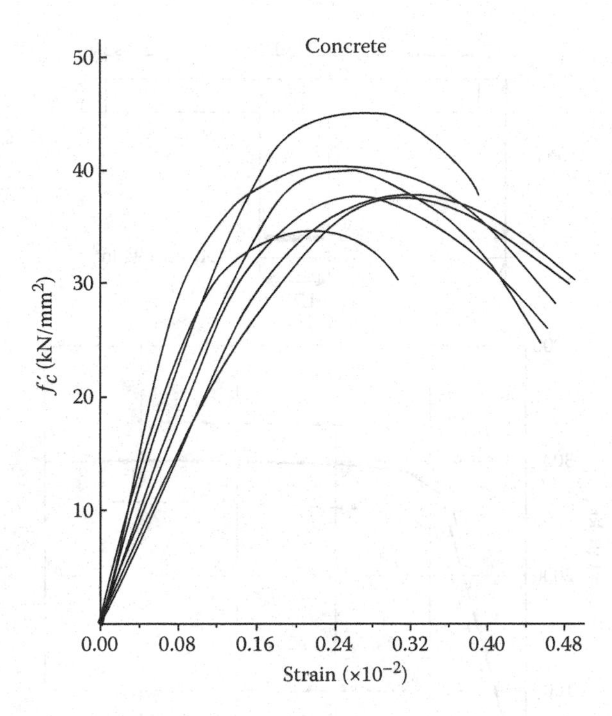

Figure 7.10 Examples of the stress–strain relationship for concrete in compression.

depend on the specified compressive strength f_c' of the mix. Figure 7.10 shows a typical stress–strain relationship of concrete in compression. The tensile strength of concrete is significantly less than its compressive strength. Second, it turns out that the standard deviation for dimensions is roughly independent of the beam size. Therefore, the coefficient of variation tends to be smaller for larger beam sizes.

7.4.2 Concrete elements in bridges

7.4.2.1 Moment capacity

For R/C T-beams in bridges, the statistical parameters of bending resistance were derived by Nowak et al. (1994). Sample moment–curvature diagrams for typical bridge T-beams are shown in Figures 7.11 through 7.13. These curves were generated using the numerical procedures developed by Ting (1989). Three sections are considered; a flange width of 7 ft and a slab

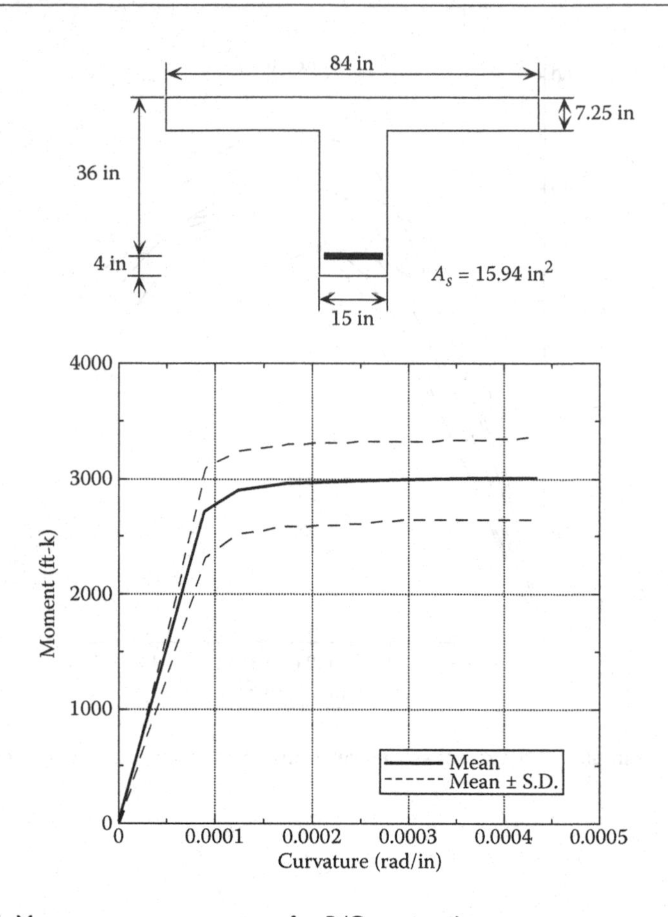

Figure 7.11 Moment–curvature curves for R/C section A.

thickness of 7.25 in are assumed for all three sections. These beams are used for spans ranging from 40 to 80 ft. The major parameters that determine the structural performance include the amount of reinforcement, steel yield stress, and concrete strength.

The bias factor and coefficient of variation of MF (materials and fabrication) for lightly R/C T-beams were assumed to be 1.12 and 0.12, respectively. The parameters for analysis factors were taken as $\lambda_P = 1.00$ and $V_P = 0.06$. Therefore, for the resistance R, the parameters were $\lambda_R = 1.12$ and $V_R = 0.13$.

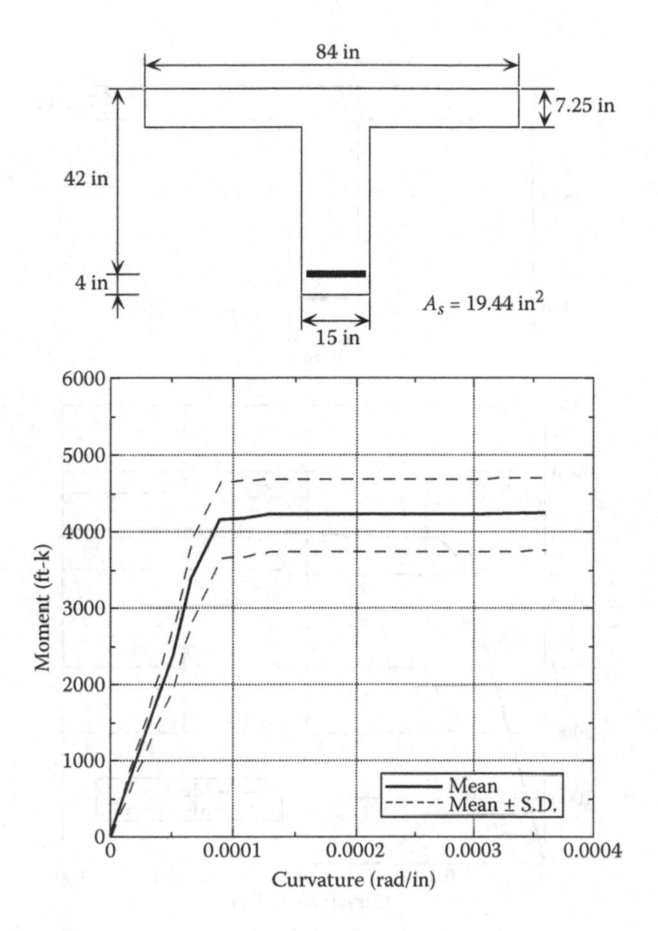

Figure 7.12 Moment–curvature curves for R/C section B.

The parameters of resistance for prestressed concrete bridge girders are derived on the basis of the statistical data from Ellingwood et al. (1980) and Siriaksorn and Naaman (1980). The results are based on simulation studies using the computer program developed by Ting (1989). Within a given cross section, the strains are assumed to be linearly distributed. Material properties are assumed to be uniform throughout the section. The section is divided into a number of rectangular horizontal strips of a small depth. For given strains, stresses are calculated using material stress–strain curves. The bending moment is calculated as the resultant of the internal stresses.

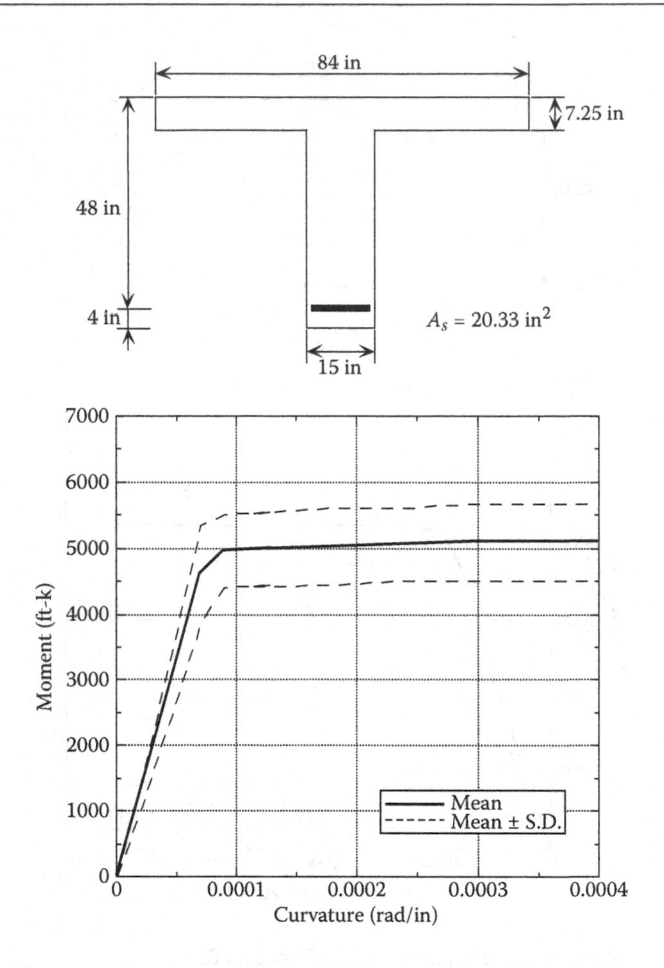

Figure 7.13 **Moment–curvature curves for R/C section C.**

Uncracked and cracked sections are considered. The section is uncracked until the tension in concrete exceeds the tensile strength. In a cracked section, all tension is resisted by the steel. The ultimate stiffness corresponds to the prefailure part of the moment–curvature plot. It is important to note that the moment–curvature relationship changes under a cyclic loading pattern such as that caused by passing trucks. If the total bending moment, M_Q, exceeds the cracking moment, M_{cr}, the section cracks and the tensile strength of the affected concrete are reduced to zero. After the first cracking, the crack stays open any time M_Q exceeds the decompression moment,

M_d. In other words, if $M_Q < M_d$, then all concrete is compressed; if $M_Q > M_d$, then the crack opens. For typical bridge girders, the ultimate moment is about twice the decompression moment. The section cracks for the first time when the moment reaches approximately 1.15 M_d.

Figures 7.14 through 7.16 show some representative moment–curvature relationships for typical AASHTO girders. The solid line corresponds to the average, whereas the dashed lines correspond to the average plus one and minus one standard deviation.

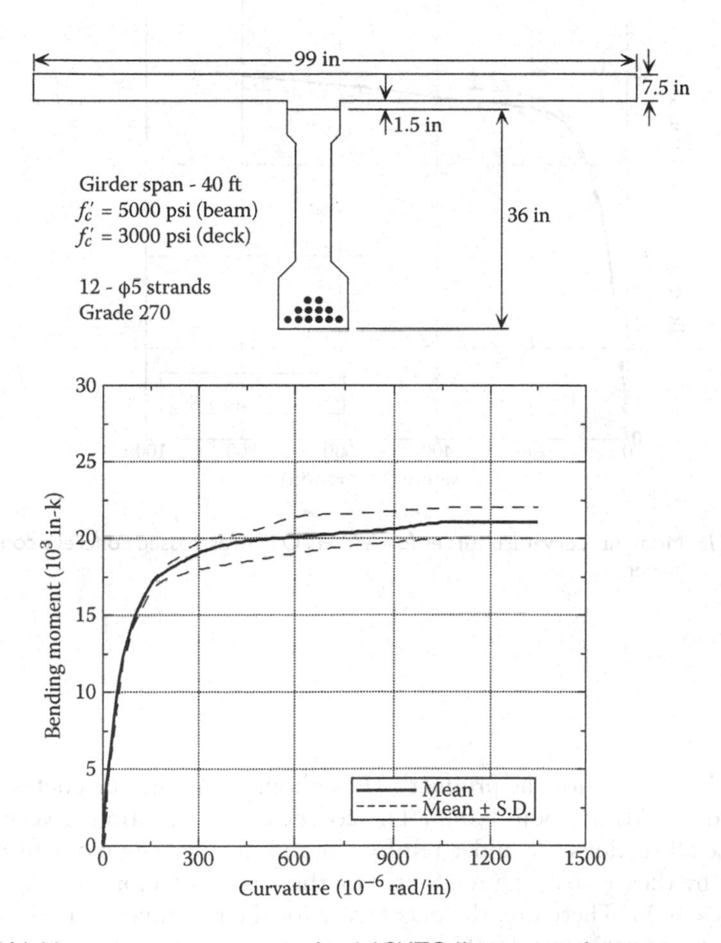

Figure 7.14 Moment–curvature curves for AASHTO II prestressed concrete composite girder.

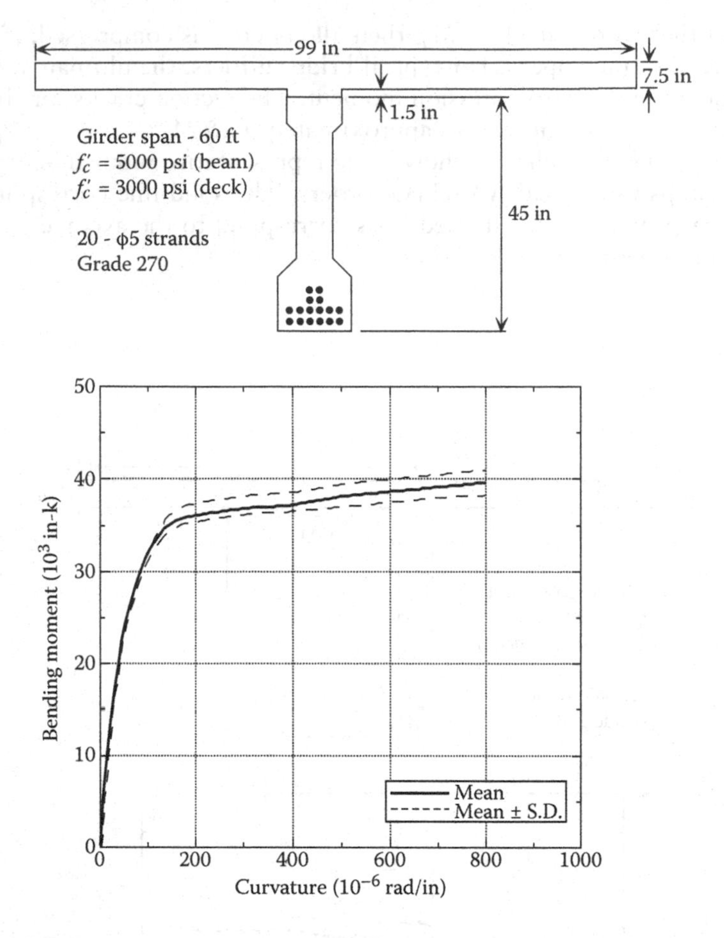

Figure 7.15 **Moment–curvature curves for AASHTO III prestressed concrete composite girder.**

The bias factor for the product MF is about 1.04, and the coefficient of variation of MF is about 0.045. The coefficient of variation is very small because all sections are under-reinforced and the ultimate moment is controlled by the prestressing tendons. For the analysis factor bias, $\lambda_P = 1.01$ and $V_P = 0.06$. Therefore, the bias factor for the resistance is 1.05 and the coefficient of variation is 0.075.

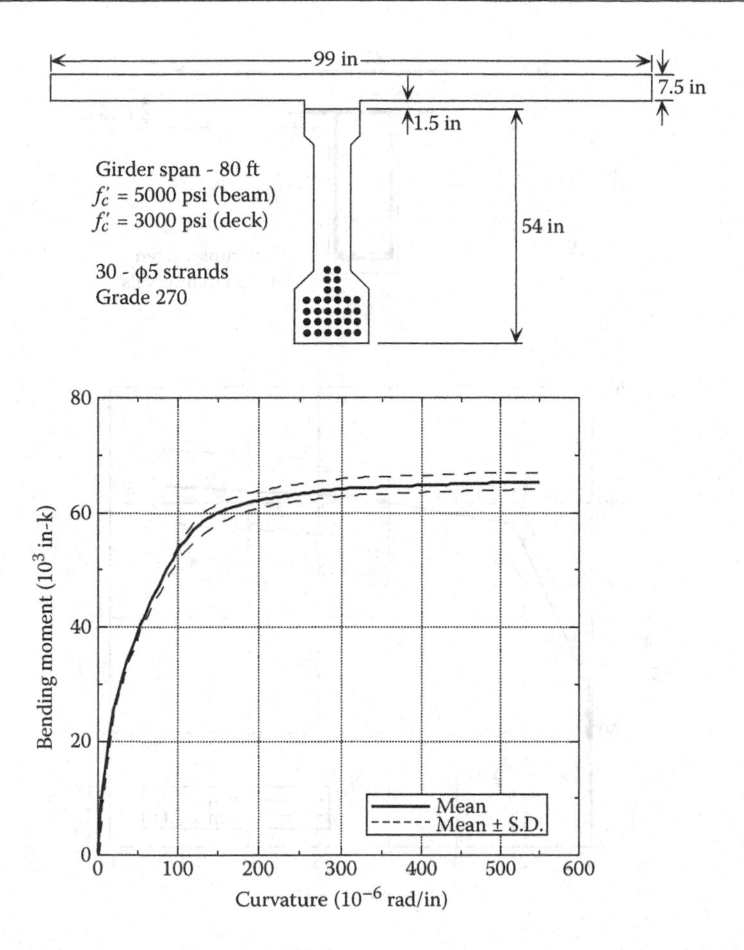

Figure 7.16 Moment–curvature curves for AASHTO IV prestressed concrete composite girder.

7.4.2.2 Shear capacity

The shear capacity of R/C girders can be calculated using the modified compression field theory (Vecchio and Collins, 1982, 1986). The statistical parameters were determined on the basis of simulations performed by Yamani (1992). The relationship between shear force and shear strain was established for representative T-beams. The results are shown in Figures 7.17 through 7.19. The nominal (design) value of shear capacity was calculated according to current AASHTO (2012) provisions.

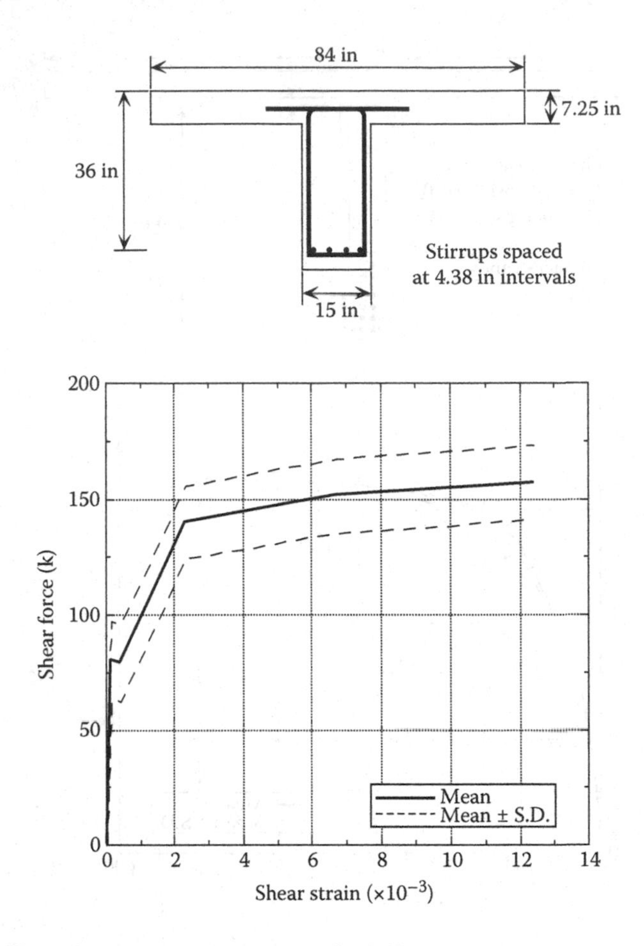

Figure 7.17 Shear force–shear strain curves for R/C section A.

The parameters of the nominal shear capacity, V_n, depend on the amount of shear reinforcement. If shear reinforcement is used, λ_{MF} = 1.13 and V_{MF} = 0.12. For the analysis factor, λ_P = 1.075 and the coefficient of variation is V_P = 0.10. Therefore, for the shear resistance, λ_R = 1.20 and V_R = 0.155. If no shear reinforcement is used, then λ_R = 1.40 and V_R = 0.17.

M. Collins (yet unpublished) observed that, in most cases, failure in flexure occurs before failure in shear. The nominal flexural capacity, M_n, and nominal shear capacity, V_n, are correlated in the statistical sense. An

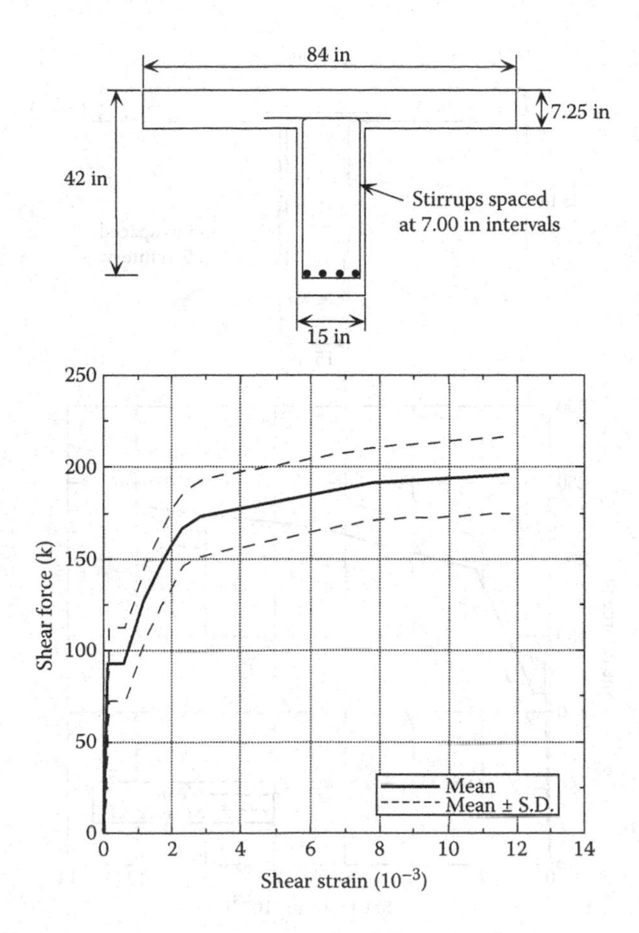

Figure 7.18 Shear force–shear strain curves for R/C section B.

increase in M_n causes an increase in V_n. In practice, shear governs only in cross sections with zero bending moment and large shear forces (e.g., some sections in box culverts).

The shear capacity of prestressed concrete girders can be calculated on the basis of the modified compression field theory (Vecchio and Collins, 1982, 1986). The parameters for the shear resistance are simulated using available test data and a computer procedure developed by Yamani (1992). The nominal (design) value of the shear capacity is calculated using the current AASHTO (2012) provisions.

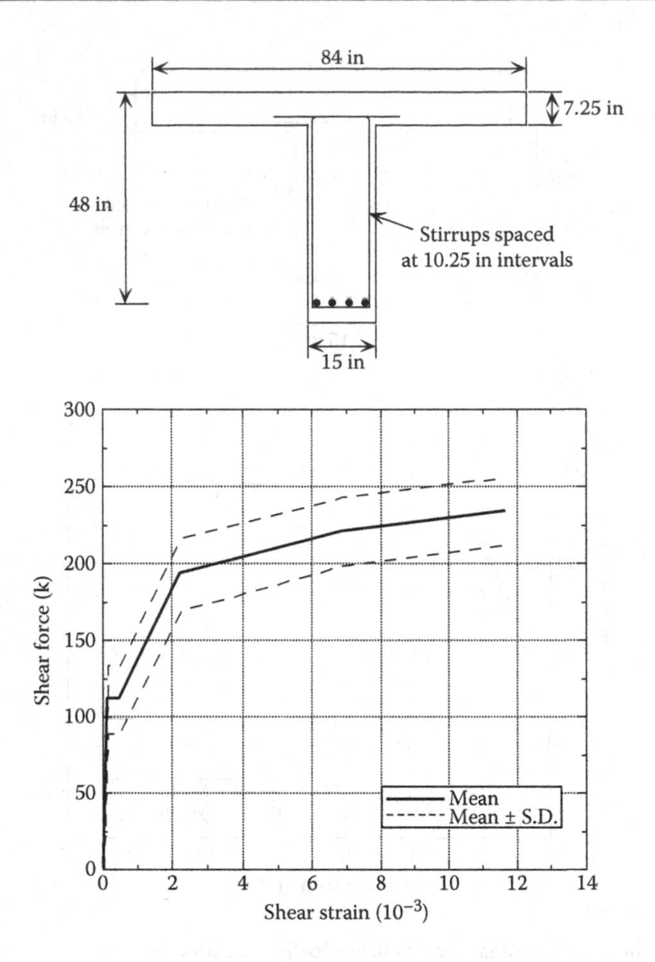

Figure 7.19 Shear force–shear strain curves for R/C section C.

For typical AASHTO type girders, the resulting relationships between the shear force and shear strain are shown in Figures 7.20 through 7.22. The curves correspond to the mean, mean plus one standard deviation, and mean minus one standard deviation.

The parameters of *FM* (fabrication and materials) are λ_{MF} = 1.07 and V_{MF} = 0.10. For the analysis factor *P*, λ_P = 1.075 and V_P = 0.10. Therefore, for the shear resistance, λ_R = 1.15 and V_R = 0.14.

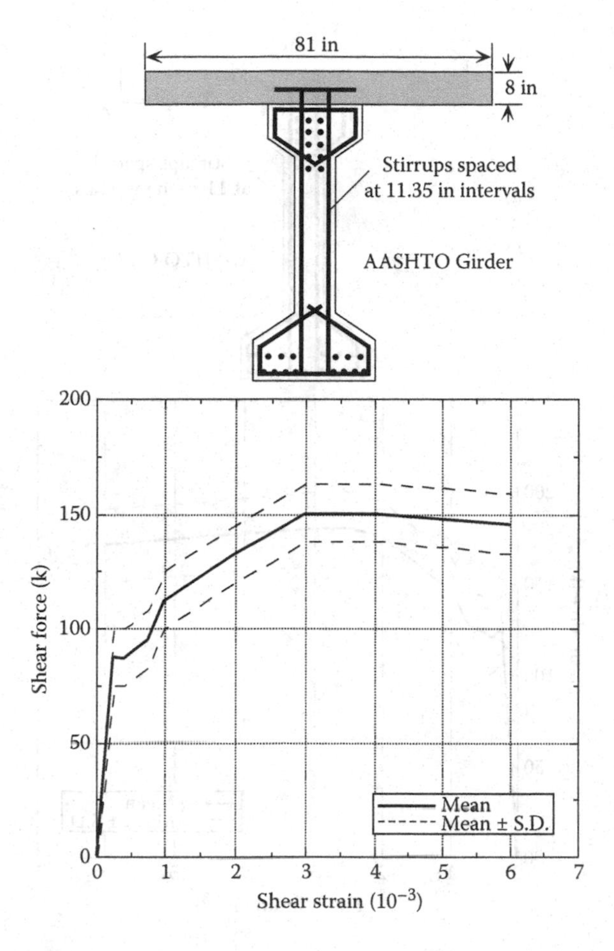

Figure 7.20 Shear force–shear strain curves for AASHTO II prestressed concrete composite girder.

7.4.3 Resistance of components with high-strength prestressing bars

The resistance of components made with high-strength prestressing bars is determined by the mechanical properties of the prestressing bars. Figure 7.23 presents the results of tests on 30 samples conducted by DYWIDAG Systems International. The tests were conducted to determine the yield

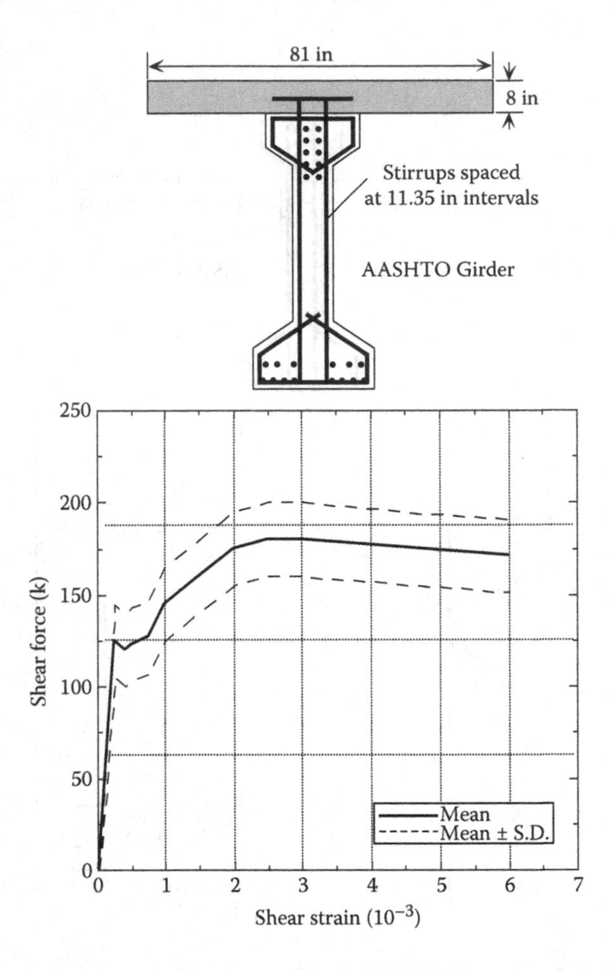

Figure 7.21 **Shear force–shear strain curves for AASHTO III prestressed concrete composite girder.**

stress, F_y, and the tensile strength (ultimate stress), F_u. In Figure 7.23, the data are plotted on normal probability paper. The calculated coefficients of variation are 0.03 for F_y and 0.01 for F_u. However, the lower tails of the cumulative distribution functions (CDFs) show a higher variation, which is important in reliability analysis. The statistical parameters of the resistance can be assumed to be the same as for R/C T-beams.

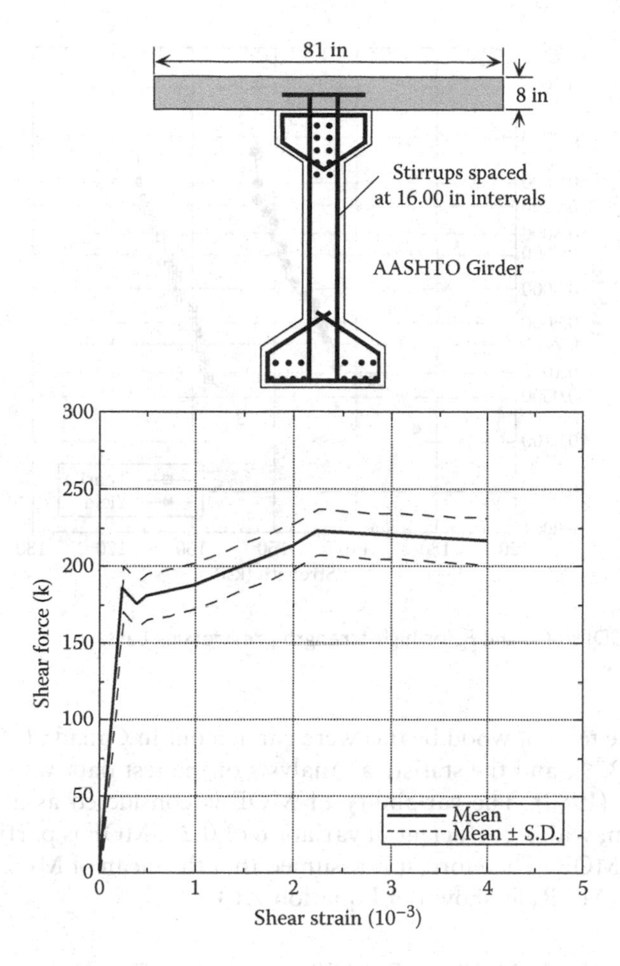

Figure 7.22 Shear force–shear strain curves for AASHTO IV prestressed concrete composite girder.

7.5 WOOD COMPONENTS

7.5.1 Basic strength of material

The major material parameters governing the resistance of structural members made of wood are the modulus of rupture (MOR) and the MOE. Statistical parameters of MOR and MOE have been determined in tests; a typical test configuration is shown Figure 7.24. A typical load–deflection curve is shown in Figure 7.25 (Sexsmith et al., 1979).

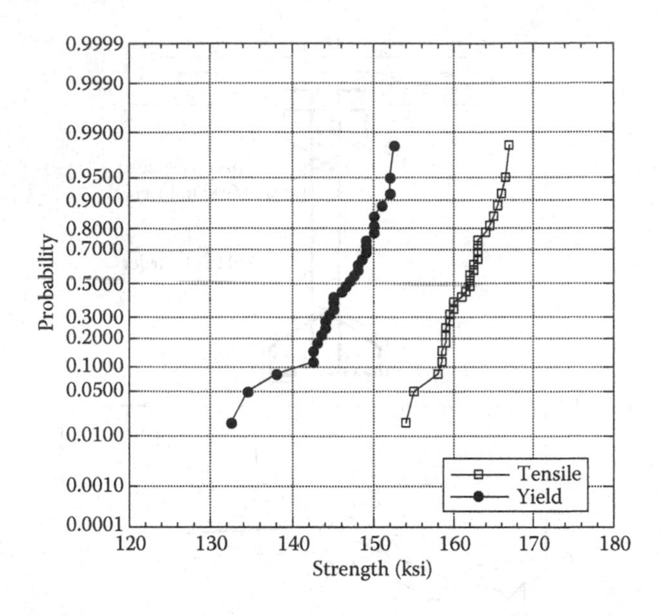

Figure 7.23 CDFs of F_y and F_u for high-strength prestressing bars.

Extensive tests of wood beams were carried out in Canada (Madsen and Nielsen, 1978), and the statistical analysis of the test data was performed by Nowak (1983). The variability of MOE is considered as a lognormal distribution, with a coefficient of variation of 0.20. MOE is partially correlated with MOR. Therefore, it is assumed that the mean of MOE is a linear function of MOR, as shown in Equation 7.13.

$$\text{MOE} = (0.15 \cdot \text{MOR} + 0.7) \cdot 1000 \tag{7.13}$$

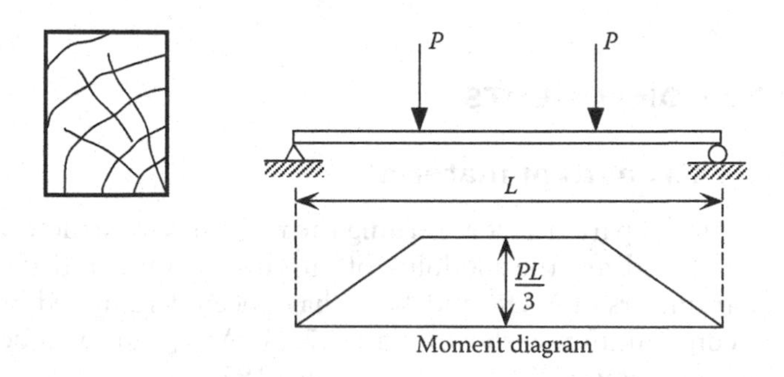

Figure 7.24 Tests of wood beams to determine MOR and MOE.

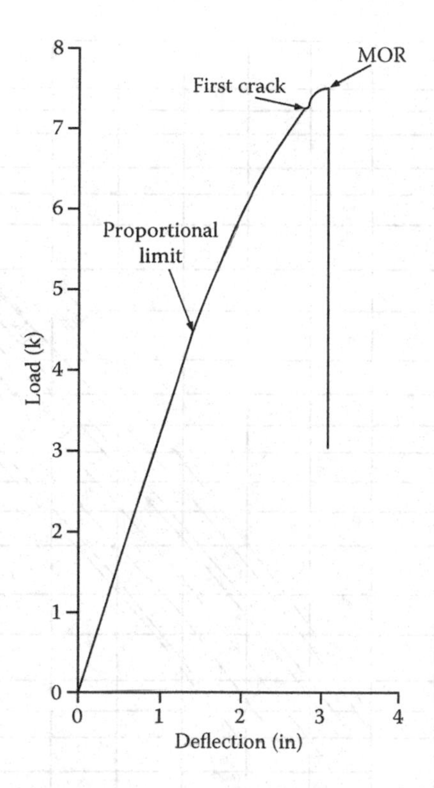

Figure 7.25 Typical load–deflection curve for a wood beam.

Typical CDFs for the MOR for Douglas Fir (Select Structural Grade) are shown in Figure 7.26. Similar curves are shown in Figure 7.27 for Grades 1 and 2 Douglas Fir. Scatterplots of MOR vs. MOE for Douglas Fir are presented in Figures 7.28 and 7.29 for two beam sizes.

The dimensions of wood beams were measured and compared with design values (Madsen and Nielsen, 1978; Nowak, 1983). The bias factor varies depending on the nominal size, but typical values range from 0.97 to 1.04. The coefficient of variation ranges from 0.01 to 0.04.

7.5.2 Flatwise use factor

The results of flatwise versus edgewise loading on deck planks are described by Stankiewicz and Nowak (1997). Tests were performed on Red Pine, sizes 4 × 6, 4 × 8, 4 × 10, and 4 × 12. The typical resulting CDFs of MOR are shown on normal probability paper in Figure 7.30.

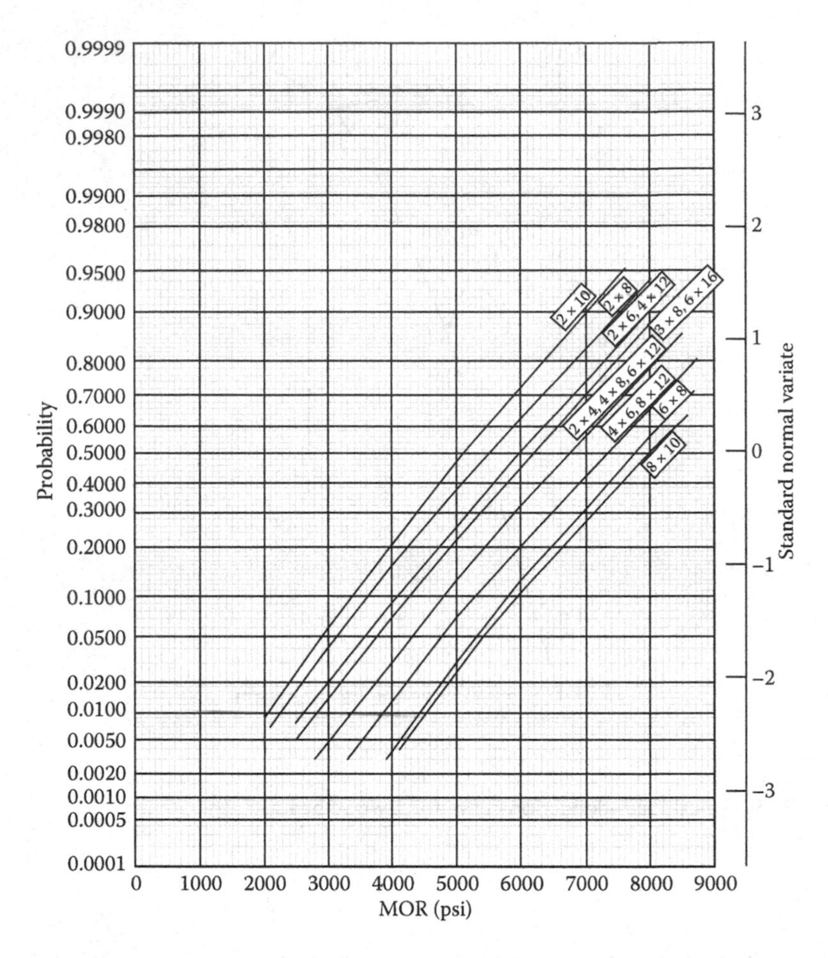

Figure 7.26 CDFs of MOR for Douglas Fir (Select Structural Grade).

The statistical parameters of MOR, the mean and coefficient of variation, V, are listed in Table 7.9 for edge-wise and flat-wise loading (for the actual dimensions). The nominal (design) strength is specified as the lower 5th or 10th percentiles. Therefore, these percentiles are also given in Table 7.9.

The flatwise use factor is the ratio of MOR for edge-wise and flat-wise loading. They are calculated for the mean, 10th percentile, and 5th percentile and listed in Table 7.10.

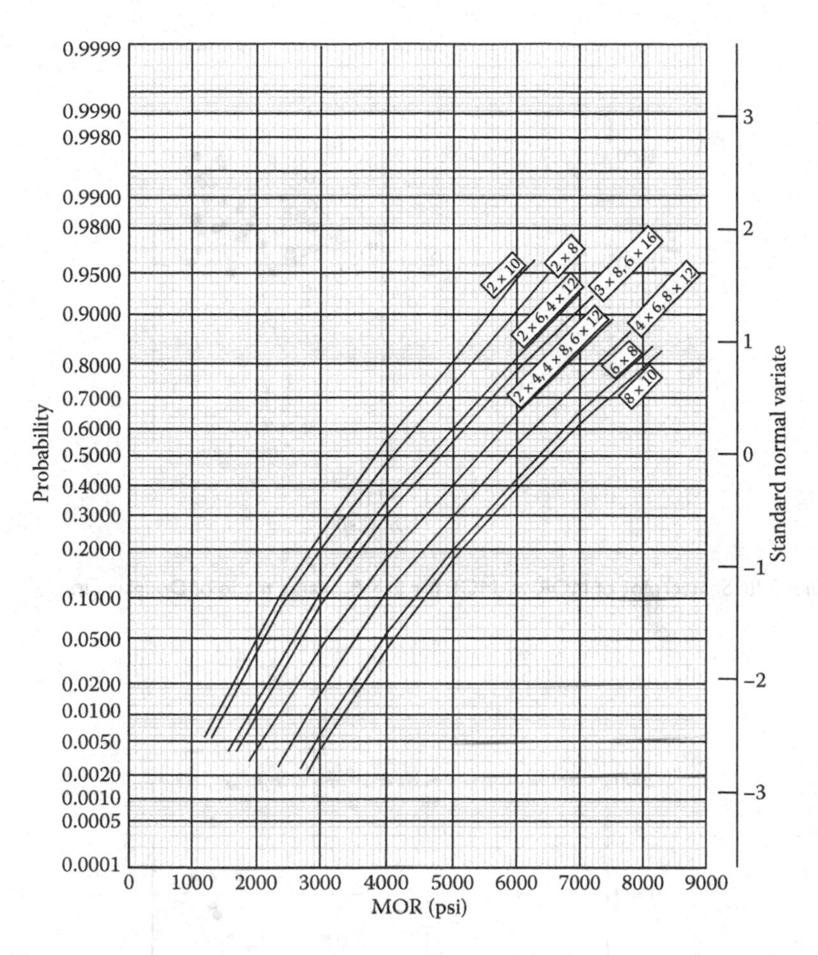

Figure 7.27 CDFs of MOR for Douglas Fir (Grades 1 and 2).

7.5.3 Resistance of structural components

In wood bridges, more than in the case of steel and concrete structures, it is important to consider the interaction between elements. For wood nail-laminated decks, after years of service, it was observed that there is a very limited load sharing effect (Csagoly and Taylor, 1979).

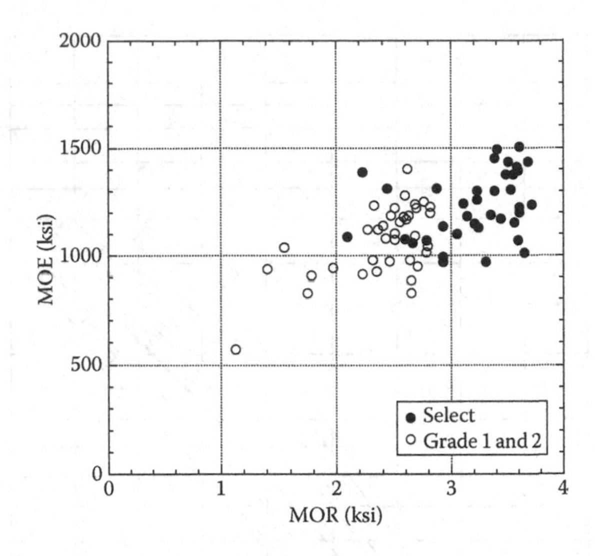

Figure 7.28 Scatterplot of MOR vs. MOE for 2 × 8 beams made of Douglas Fir.

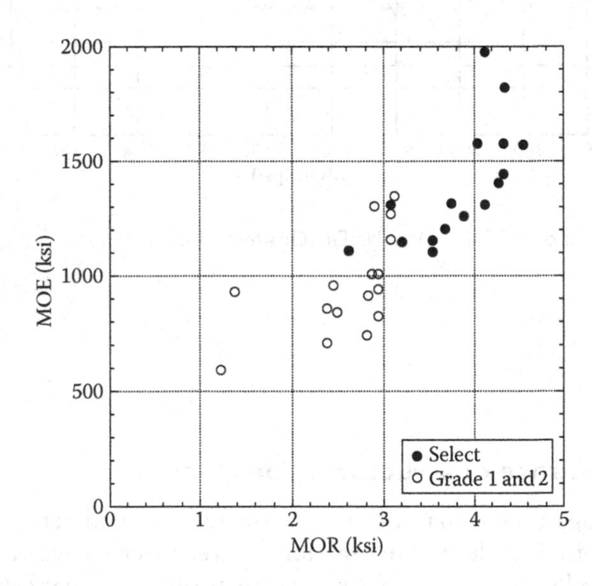

Figure 7.29 Scatterplot of MOR vs. MOE for 3 × 8 beams made of Douglas Fir.

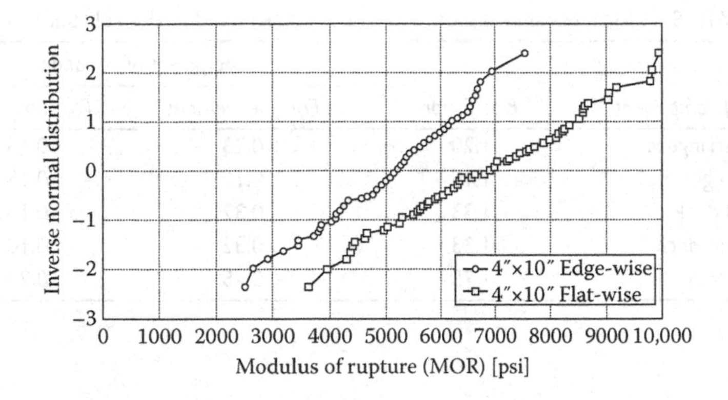

Figure 7.30 Typical MOR probability plot of Red Pine, edge-wise versus flat-wise.

The resistance of the deck subsystem involves the summation of component (i.e., individual laminations) resistances, and hence, the resulting statistical distribution of resistance for the system is assumed to be normal. In the case of plank decks, however, the number of participating components is small, and thus, the distribution is treated as lognormal. The statistical parameters for resistance that can be used in the reliability analysis are summarized in Table 7.11.

Table 7.9 Statistical parameters of MOR for tested planks

	Flat-wise MOR (psi)				*Edge-wise MOR (psi)*			
Size	*Mean*	*10th*	*5th*	*V*	*Mean*	*10th*	*5th*	*V*
4 × 6	4350	2800	2250	0.31	3800	2300	1950	0.30
4 × 8	3750	1800	1250	0.44	3300	1500	1000	0.45
4 × 10	5100	3750	3200	0.22	4100	2900	2400	0.24
4 × 12	5100	3350	2900	0.25	3400	2050	1750	0.32

Table 7.10 Statistical parameters of flat use factor from test results

Size	*Mean*	*10th Percentile*	*5th Percentile*	*Coefficient of variation, V*	*Recommended value*
4 × 6	1.14	1.22	1.15	0.31	1.10
4 × 8	1.14	1.20	1.25	0.30	1.15
4 × 10	1.24	1.29	1.33	0.275	1.25
4 × 12	1.50	1.63	1.66	0.25	1.50

Table 7.11 Statistical parameters of resistance (MOR) used in the reliability analysis

		Coefficient of variation	
Type of component	*Bias factor*	*For one element*	*For subsystem*
Sawn stringers	1.20	0.23	0.15
Glulam girders	1.15	0.15	0.15
Nailed deck	1.33	0.32	0.15
Stressed deck	1.33	0.32	0.10
Plank deck	1.75	0.25	0.20

Chapter 8

Design codes

The objective of this chapter is to present the basic philosophy behind the development of reliability-based design codes. The concepts presented in Chapter 5 are used to calibrate design codes so that designers can create designs with a consistent level of reliability without doing a detailed reliability analysis for each design.

8.1 OVERVIEW

Design codes play a central role in the building process because they specify the requirements that the designer must satisfy so that the minimum acceptable safety level is provided. Probability-based (reliability-based) codes have been developed in the United States for steel building structures [American Institute of Steel Construction (AISC) LRFD, 2011], loads for buildings [American Society of Civil Engineers (ASCE)/SEI 7-10], highway bridges [American Association of State Highway and Transportation Officials (AASHTO) LRFD, 2012], wood building structures [American National Standards Institute (ANSI)/AF&PA NDS-2005], and offshore oil platform structures (American Petroleum Institute, 1989), as well as in Europe (Eurocodes 1–9) and Canada (CAN/CSA-S6, 2006; CSA S16-09). The common feature of current codes is that they provide guidance on determining design loads and design load-carrying capacities of structural members. Safety reserve is implemented through conservative load and resistance factors applied to the design loads and design load-carrying capacities.

In this chapter, the major steps in the development of a probability-based code are reviewed. These steps include determining the scope and objective of the code, formulating the demand function, determining the target safety level, identifying a format for presenting the code requirements, and developing design-checking formulas. The steps are demonstrated by considering the development of an LRFD (load and resistance factor design) code for girder bridges in Section 8.6 and reinforced concrete building structures in Section 8.7. The approach presented herein was successfully applied in the development of bridge codes in the United States and Canada (Nowak and Lind, 1979; Grouni and Nowak, 1984; Nowak and Grouni, 1994; Nowak, 1995).

8.2 ROLE OF A CODE IN THE BUILDING PROCESS

The building process includes planning, design, manufacturing of materials, transportation, construction, operation/use, and demolition. It involves many different trades and professions, and the major parties can be put into four categories: the owner/investor, the designer, the contractor, and the user/operator. These parties often have conflicting interests. In particular, the owner is usually interested in maximum profits, which means reduced costs, and the user/operator would like to have a comfortable and safe structure, which means higher costs. The designer provides the documentation, including calculations and drawings, and the contractor is expected to build the physical structure according to the documentation. Both the designer and the contractor are under pressure from the owner to keep the costs down. The role of a design code is to establish the requirements needed for a structure so that the reliability can be at an acceptable level. The central role of a code is shown in Figure 8.1.

Structural failures are always undesirable events. They occur because of ignorance, negligence, greed, physical barriers, and sometimes act of God. The probability of failure is often higher for projects involving new materials, technology, and extreme parameters (such as, for example, span, height, thickness, and weight) for which there is little or no prior experience. Therefore, the specified design provisions include built-in safety

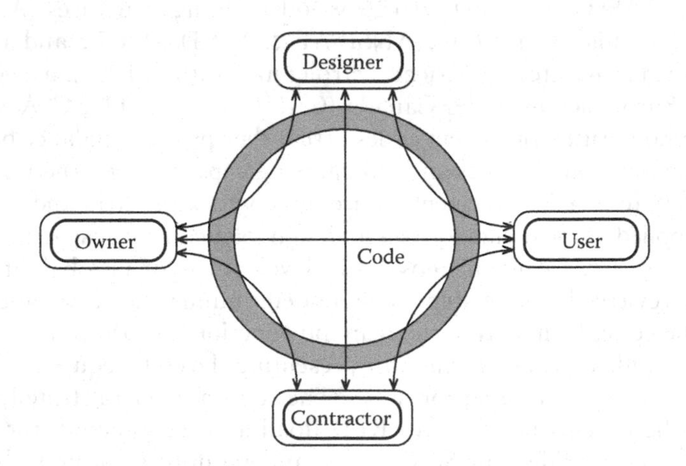

Figure 8.1 Central role of the code in the design and construction of structures.

margins: load effects are usually overestimated and resistances are underestimated. However, this safety reserve cannot cover all possible causes of structural failure. From the legalistic point of view, the code defines the acceptable practice. In case of failure, the designer can be found responsible if he or she does not satisfy the code requirements.

The acceptable reliability levels implied in current building codes depend on the systems of values assigned by the society to human life, material loss, disruption of services, and so on. The actual rate of failure in the building industry can be used as an indication of how important this sector of the national economy is in the eyes of our society. The number of fatalities and injuries due to a structural collapse can be compared with loss of life in car accidents, airplane accidents, and others. The comparison points out that there is a considerable variation in acceptable failure rate, depending on industry, geographical region, subjective aversion to risk, and tradition.

The meaning of "acceptable safety level" has evolved through the centuries. Historically, the notion of a safe structure was associated with a structure surviving a number of years without collapse. It was a builder's responsibility to build the structure so that failure would not occur. The oldest preserved building code is Hammurabi's code from ancient Babylonia. A translation of a portion of this code is shown in Figure 8.2. It dates back almost 4000 years and is on display in the Louvre Museum in Paris. In this code, the responsibilities were clearly determined. If the building collapsed and killed the owner, the builder was put to death. If the son of the owner died, the son of the builder was put to death, and so on.

In the middle ages, the construction of large structures (churches, towers) was done by skilled craftsmen. The "margin of safety" was provided by comparison to other existing successful structures. The learning process was based on trial and error. For example, when a new cathedral was planned, taller than any other structure at that time, the wall thickness was increased compared to other existing cathedrals. The amount of the increase was determined by judgment. If the structure collapsed, then it would mean that the required wall thickness was underestimated, and it would have to be increased in the next trial.

As mentioned earlier, current design codes specify design and detailing requirements that are intended to achieve a minimum acceptable safety level. The code provisions are expressed in terms of formulas and procedures. In the allowable stress design approach, the strength (resistance) of the material is divided by a safety factor to obtain the allowable strength (stress, deflection, force). In LRFD or limit state design, load components are multiplied by load factors and resistance is multiplied by resistance factor. The basic form of the LRFD equation is

$$\phi R \geq \sum \gamma_i Q_i \tag{8.1}$$

A. If a builder build a house for a man and do not make its construction firm and the house which he has built collapse and cause the death of the owner of the house – that builder shall be put to death.

B. If it cause the death of the son of the owner of the house – they shall put to death a son of that builder.

C. If it cause the death of a slave of the owner of the house – he shall give to the owner of the house a slave of equal value.

D. If it destroy property, he shall restore whatever it destroyed, and because he did not make the house which he built firm and it collapsed, he shall rebuild the house which collapsed at his own expense.

E. If a builder build a house for a man and do not make its construction meet the requirements and a wall fall in, that builder shall strengthen the wall at his own expense.

Figure 8.2 Schematic diagram of a portion of Hammurabi's code along with English translation. Helander, M.: *Human Factors/Ergonomics for Building and Construction*. 1981. Copyright Wiley-VCH Verlag GmbH & Co. KGaA. English translation taken from Harper, R.F., *Hammurabi, King of Babylonia*, The University of Chicago Press, Chicago, 1904.

where γ_i is a "load factor" applied to load component (or load effect) Q_i and ϕ is a "resistance factor" applied to the resistance (measure of load-carrying capacity) R. In words, Equation 8.1 says that the capacity of the structural member (modified by the factor ϕ) must be larger than the total effect of all the loads acting on the member. The design formulas are developed by code committees with input from practicing engineers, researchers, and scientists. Given this type of design equation, it is the designer's responsibility to make sure that, for given load and resistance factors (specified in the code), the design loads and resistance satisfy the inequality expressed in Equation 8.1.

At the present time, failures still provide information that is valuable in the development of future design requirements. However, the development of structural analysis techniques, the advances in materials science, and the development of structural reliability theory provide the basis for the modern approach to structural design. In the future, it is anticipated that many code-writing organizations will make a transition toward "performance-based" design codes. Instead of simply providing requirements to achieve a minimum acceptable safety level, performance-based codes would allow the owners of a structure to choose a higher level of safety, and the codes would provide guidance on designing for that higher level. Significant research efforts are currently underway to develop these procedures for seismic design (i.e., the design of structures for earthquakes). For additional details on performance-based design, the reader is referred to the report issued by the Federal Emergency Management Agency (FEMA, 1996).

8.3 CODE LEVELS

The intent of a design code is to provide a minimum safety level. Current codes use deterministic formulas; however, the optimum design will require the consideration of structural reliability as an acceptance criterion. Depending on the approach to reliability, there are four levels (categories) of design codes (Madsen et al., 1986):

- *Level I codes:* These codes use deterministic design formulas. The safety margin is introduced through central safety factors (ratio of design resistance to design load) or partial safety factors (load and resistance factors). Equation 8.1 would be the type of equation used in a Level I code.
- *Level II codes:* These codes define the design acceptance criterion in terms of the "closeness" of the actual reliability index for a design to the target reliability index or other safety-related parameters.
- *Level III codes:* These codes require a full reliability analysis to quantify the probability of failure of the structure under various loading

scenarios. The acceptance criterion is defined in terms of the closeness of the actual reliability index to the optimum reliability level (or probability of failure).
- *Level IV codes:* These codes use the total expected cost of the design as the optimization criterion. The acceptable design maximizes the utility function that describes the difference between the benefits and costs associated with a particular design.

In practice, the current design codes are based on a Level I code philosophy. However, in the newly developed Level I codes, the design parameters (load and resistance factors) are derived using Level II methods. At present, Level III and IV methods are used mainly in advanced research or in the design of critical structures.

8.4 CODE DEVELOPMENT PROCEDURE

A structural design code is basically a set of requirements to be satisfied by a class of structures to be designed in a jurisdictional area. These requirements include values and/or procedures to determine design load and resistance. Therefore, the development of the code involves not only determination of safety factors but also verification of the nominal (design) values of load and resistance, and analytical procedures (structural analysis).

The design loads are specified depending on the use of the structure (e.g., buildings, bridges, offshore platforms, power plants). They can be given in various forms (uniform loads, concentrated forces, moving forces). The load analysis methods can be very specific (i.e., actual values of load effects can be listed) or vague, leaving the decision to the designer.

Resistance can be specified in terms of the allowable stress and/or deformation (e.g., deflection), load-carrying capacity of the component (beam, column, weld, bolt), or load-carrying capacity of the whole structure (system resistance). The acceptance criteria can be given in the form of descriptive requirements, formulas, or allowable values.

The major steps involved in the development of a design code were formulated by Lind and Davenport (1972) and are listed below:

Step 1. Define scope and data space.
Step 2. Define code objective(s).
Step 3. Establish frequency of demand.
Step 4. Select code space metric.
Step 5. Select code format(s).

The following subsections will provide additional details on these steps.

8.4.1 Scope of the code

A code is usually developed for a certain group or class of structures. It is very important to determine that class by identifying the range of parameters covered and not covered by the code. These parameters can be type of material (steel, concrete, wood, plastic), type of function (office, apartment, hotel, hospital, highway bridge, railway bridge, industrial), span length (short, medium, long), structural type (frame, beam, column, connection), thickness of components (hot-rolled steel, cold-formed), and type of connection (welded, riveted, bolted).

To avoid unintentional misuse of the code provisions, the code-writing committee should clearly specify the scope. The scope is a parametered set of structures and the set of parameters is called the data space. It can be narrow (e.g., anchor bolts used in a nuclear power plant concrete wall) or very wide (e.g., all types of bridges).

An example of a code with a specified scope is a bridge design code. For each parameter, the range can be either listed as a discrete set of values or provided in the form of an interval (specified using numerical values or formula). A parameter set for the function of a bridge might include highway, railway, transit guideway, pedestrian, and other bridges. Structural types would be girder, slab, truss, arch, frame, cantilever, cable stayed, or suspension. Materials used for bridge construction may include steel, reinforced concrete, pretensioned concrete, posttensioned concrete, wood, glulaminated wood, and stressed wood. The code may also specify the method of analysis for bridges: simple girder distribution factors (as specified by AASHTO, 2002, or AASHTO LRFD, 2012), two-dimensional analysis, three-dimensional analysis, finite element analysis, or other numerical procedures.

The code may deal with various limit states. Limit states must be clearly defined, with the major parameters identified and the acceptance criteria determined in the form of limit state functions. For example, the ultimate limit states (ULSs) may include flexural capacity, shear, compression, and tension. Serviceability limit states are dependent on the material and structural type and may include cracking, deflection, vibration, and excessive permanent deformation. Fatigue limit states include the fatigue load and resistance expressed in terms of load cycles. For example, in the case of highway bridges, this can be in terms of truck traffic (magnitude and frequency).

The major codes used in the United States vary considerably with regard to content. The following codes for building structures focus on the resistance side of Equation 8.1 for structures constructed using specific materials:

American Concrete Institute, ACI-318 Reinforced Concrete (ACI, 2011)

American Institute of Steel Construction, Steel Structures (AISC, 2011)

American Iron and Steel Institute, Cold-Formed Steel Structures (AISI, 2008)

Load components include dead load, live load (static and dynamic), environmental forces (temperature, wind, earthquake, ice pressure, water pressure), and special forces (e.g., fire and gas explosions for buildings, emergency braking, and collision for bridges). Codes that specify design values of loads to be considered in the design of buildings include the following:

Uniform Building Code (UBC)
Building Officials and Code Administration Code (BOCA)
American Society of Civil Engineers/American National Standards Institute (ASCE/ANSI 7-10)

For the design of bridges, requirements for loads and resistances are specified in a single document issued by the AASHTO (AASHTO, 2002; AASHTO LRFD, 2012).

The scope of a code is a compromise between simplicity and closeness to the objective. It is desirable to cover a wide range of structures by the same provisions. A simple code is easier for the designer to follow, and the probability of error (i.e., use of a wrong design formula) is reduced. On the other hand, very simplistic code requirements may make it difficult to achieve the target safety level for all the structures covered by the code. It may turn out that the margin of safety is close to the target value for one group of components and is either above or below the target level for another group. If the scope is narrowed down to structures with similar parameters, it is easier to satisfy the required safety criterion.

The current trend is to cover load and resistance by the same code and provide a rational basis for comparing the performance of various materials (e.g., steel, concrete, and wood). From the designer's point of view, it is convenient to use the same load and resistance factors in Equation 8.1 for all design cases regardless of material.

8.4.2 Code objective

A very general objective of a design code is to make sure that the structures designed according to the code provisions will have a minimum safety level. More specific objectives may be appropriate for some classes of structures. The code-writing organizations must define acceptability criteria for the covered structures (define failure), select the target safety level(s), and develop design provisions that satisfy the objective.

The definition of safety can be expressed in terms of the expected risk (i.e., consequences of failure), the failure probability, or the reliability index. Code objectives may vary depending on the comparison criterion and target safety value. Some examples of code objectives are as follows:

- Design a structure to have a reliability index (β) close to some specified target value (β_T).
- Design a structure to have a probability of failure lower than the predetermined maximum acceptable value for a preselected period of time (e.g., 1 year for a temporary structure).
- Maximize the total utility, which is the difference between revenues and costs. The costs usually include the initial cost, cost of maintenance, and expected cost of failure (actual cost of failure multiplied by the probability of failure). Revenues may include profits and user convenience.
- Achieve a negligible failure frequency with a reasonable material economy. This is a very vague code objective.

The approach that has been adopted in developing most reliability-based design codes is to specify the target reliability index, β_T. The target reliability index can be determined for a class of structures, components, and/or limit states. Its value can be different depending on the time period. (In general, the reliability index for a structure or component decreases with time.)

The optimum value of β_T depends on the expected cost of failure and the cost of upgrading (cost of increasing safety reserve). In current codes, there are considerable differences in β_T. For example, consider the optimum target reliability indices for a hot-rolled steel beam and a beam-to-column connection using fasteners (e.g., bolts). Should they be the same? The reliability of a hot-rolled beam section depends on the plastic modulus Z, the yield stress F_y, and thickness-to-width ratios. For a given value of F_y, to increase the reliability index β, Z must be increased, which typically means a larger cross section and a heavier beam. For fasteners, β can be increased by adding a bolt or bolts, which is usually less expensive than increasing the beam size. Therefore, in practice, it costs less to increase the reliability index for fasteners than for beams. If safety can be considered a commodity, it is cheaper in the case of bolts. Therefore, for the beams, typical values of β are between 3 and 4, but for fasteners, β can range from 6 to 8. This simple example illustrates why a *family* of prescribed target reliability indices is the only feasible and acceptable objective for a code. Collectively, the β values represent what is known as the target reliability index function.

A reliability index may be associated with any stochastic (random) system that can attain two states: failure and non-failure. Ideally, one would like to assign a reliability index to an entire structure. However, there are usually many different modes of failure which are not all equivalent. This is reflected in the safety checks required for each failure mode. The only practical alternative is to associate a prescribed target reliability index with each safety check. The target reliability index may vary with the loading condition considered (e.g., function of load ratio such as live-to-dead load ratio), type of failure mode (e.g., shear, flexure, buckling), and type of material. Moreover, it may vary within a "single" failure mode (e.g., failure by buckling for long, short, and intermediate columns). Whether such a variation is permissible and desirable is a matter for the code committee to decide. For example, the committee may consider whether the reliability of beams should be independent of the *D/L* ratio. If there is no valid reason to prescribe different reliability indices, then the target reliability index should be constant.

As a guide to the selection of a target reliability index, the past performance of structures in service is most valuable. The index can be calculated (given the appropriate statistical data) for any structural member using appropriate models of loads and resistance. From the reliability indices thus computed for designs based on existing codes, a target index is selected. This can be a function or a constant.

In particular, it is possible (but normally a rather pointless exercise) to select β values of the "old" code as target values for the new code. This makes sense, however, when the code change is meant as a change in form and not content, such as when a new analysis formula is proposed to replace an old one.

8.4.3 Demand function and frequency of demand

The third step in code development is the determination of the frequency of occurrence of a particular safety check. In general, it is difficult (if not impossible) for a design code to be extremely simplistic and yet exactly satisfy the code objective in all circumstances. Therefore, it is necessary to define the most important structural data for which the objective is to be met. For example, if the ratio of dead load to live load is ordinarily between 0.5 to 2, then it may be possible to develop a simple code procedure that will satisfy this objective (at least approximately) over this range, but it may be difficult for the procedure to satisfy the objective for all possible ratios ranging from zero to infinity. In the context of this discussion, we can say that ratios between 0.5 and 2 have a higher frequency of occurrence than other values, and we can use this information in developing the code. The frequency of occurrence is a scalar point function called the demand function.

The demand function can be defined by the analysis of past and current practice. Various design cases can be identified in terms of load components and resistance, and the occurrence frequency of each design case is estimated by statistical analysis using available data. For example, the ratio of dead load to live load varies. For beams, it is low for short spans (live load dominates) and high for longer spans (dead load dominates). The frequency of occurrence for different load ratios can be considered a demand function. The demand function can be defined by assigning "fuzzy" or qualitative values (e.g., often, sometimes, rarely, unlikely), percentage values, or probabilities of occurrence.

In the development of a new code, it is important to consider the future demand function. The code should provide a good fit to the target reliability for the most frequently expected design situations in the future. Therefore, the optimization may involve a prediction of future trends and practices.

8.4.4 Closeness to the target (space metric)

The fourth step in design code development is to select a measure (or metric) to quantify how close the code is in achieving its objective. For example, let β_T denote the desired target value of the reliability index in a particular safety check, and let β be the actual value produced by the code procedure. The difference $\beta_T - \beta$ varies over the entire data set. For some cases, it may be positive while for others it may be negative. The criterion of closeness of the code to the objective may be, for example, to minimize the expected value of $(\beta_T - \beta)^2$ over the demand space. Many other criteria of closeness could be considered appropriate. If it is desired to penalize larger deviations from the objective, then $(\beta_T - \beta)^p$ with $p > 2$ might be employed. Since under-design is more serious than over-design, a skew function may be preferred, which places a higher penalty on under-design. The specific function used is commonly referred to as the β-*metric* and is denoted by $M(\beta_T - \beta)$.

Another way to formulate the β-metric is in terms of costs. Neglecting maintenance and demolition costs, the total cost of a structure against a single limit state can be expressed as

$$C_T = C_I + C_F P_F \tag{8.2}$$

in which C_T, C_I, and C_F are the total cost, initial cost, and failure cost, respectively, and P_F is the probability of failure. The initial cost C_I can be fitted with good accuracy, at least in the neighborhood of the target reliability index β_T, by

$$C_I = a[1 + b\beta] \tag{8.3}$$

where a and b are constants and β is the reliability index. Note that C_I increases as β increases; this is reasonable because we would expect the initial cost of material and construction to be higher to achieve a safer structure. The probability of failure P_F can be often be approximated by

$$P_F = c \exp\left[\frac{-\beta}{d}\right] \tag{8.4}$$

in which c and d are constants. This is a decreasing function of β as expected since the failure probability should decrease as the reliability index increases. Assuming that the value of risk (i.e., failure cost C_F) is independent of the reliability index β, the total cost is given as

$$C_T = a[1+b\beta]+C_F\left\{c\exp\left[\frac{-\beta}{d}\right]\right\} \tag{8.5}$$

Equation 8.5 represents the sum of an increasing function of β and a decreasing function of β. Figure 8.3 provides an illustration of how the total cost varies with the value of β. The optimum design corresponds to the value of β, which minimizes the total cost. The skewness of the plotted curve indicates the difference of consequence of over/under-design on the total cost.

To formulate the β-metric in terms of cost, we need to formulate some sort of penalty function that depends on how far away the actual β is from the target (optimum) reliability index β_T. We can define the change in total cost, ΔC_T, as C_T – (C_T at β_T). At the optimum β_T, the first derivative of C_T is equal to zero; this gives an expression that relates c and C_F to the other problem parameters. Combining this result with the definition of ΔC_T above, and after doing some algebraic manipulations and normalization, we arrive at (Lind and Davenport, 1972)

$$\text{normalized } \Delta C_T = \frac{\Delta C_T}{abd} = \frac{\beta-\beta_T}{d} - 1 + \exp\left[\frac{-(\beta-\beta_T)}{d}\right] \tag{8.6}$$

Since this normalized ΔC_T depends on $(\beta - \beta_T)$ and is equal to zero when $\beta = \beta_T$, it can be used as a β-metric since it provides a measure of closeness.

Given a β-metric function, the variation of the measure of closeness over the data space can be accounted for by formulating a *weighted average measure of closeness*, T, such as

$$T = \int M(\beta_T - \beta)\, D_f(\beta)\, d\beta \tag{8.7}$$

where $M()$ is the β-metric and $D_f()$ is the demand function that describes the frequency of occurrence of values of β.

The appropriate β-metric depends on how the structure is utilized and the difference in consequences for over- and under-design (Lind and Davenport, 1972). It is important for a code committee to have an idea of how the code is likely to be used. If the target reliability cannot be met exactly, and if the value of the β-metric varies, then the structural data that occur most frequently must be known so that the target reliability can be met, as closely as possible, for these data. The end result of the calibration is not very sensitive to details of demand function variation, and rather crude

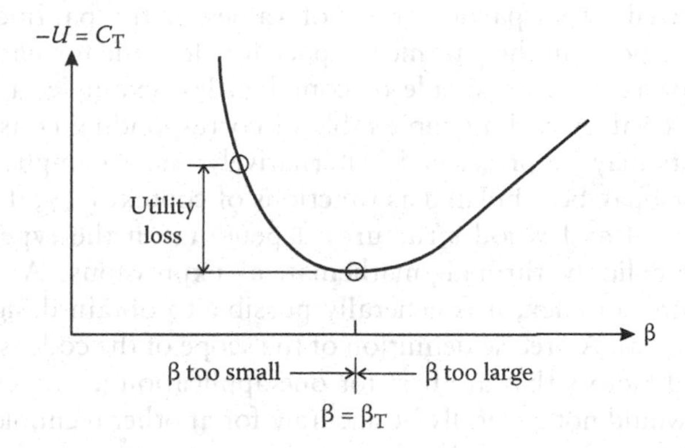

Figure 8.3 Total cost C_T versus reliability index β.

estimates, based on sampling from past designs, would normally suffice in practice. In general, the importance of the demand function decreases as the distance to the target decreases.

In summary, the optimality criterion (defined in terms of how close the actual and target reliability indices are) can be taken as minimizing a weighted average of closeness over all safety checks. However, a code committee may choose to replace this criterion with another criterion or add constraints to it. In any event, without a closeness criterion, it may not be possible to determine which of several codes is better, and it may not be possible to make a rational selection of final code parameters or the code format.

8.4.5 Code format

The fifth step is the selection of a sequence of trial code formats, arranged in order of increasing complexity. Even the simplest conceivable objective cannot be met exactly by a Level I code except at a practically unacceptable level of complexity. It is therefore necessary to confine the search to a set of formats that lead to sufficiently simple design procedures. In each format, there exists, generally, an optimal solution that comes closest to the objective. Concerning closeness criterion, the best solution is the simplest one, which also meets the criterion.

It is convenient to use the notion of a code format associated with a code or a *code grid*. Apart from well-defined physical or mathematical constants (such as the density of water or other stored goods), the numerical constants contained in a code may be considered as variables. As these variables (parameters) take on various different values, a set of different codes is generated. This set is called the code format of the code; the actual code adopted for use is one of many realizations of the format. Each realization is characterized by its particular set of values of the parameters, corresponding to a point in the parameter space for the code format.

A code format may be simple or complex. For example, a fixed set of characteristic values and a simple table of corresponding constant partial safety factors may be prescribed. Alternatively, more complicated partial safety factors may be tabulated as functions of context (e.g., different load factors for steel and wood structures depending on the type of risk) or prescribed implicitly through mathematical expressions. As the format becomes more complex, it is generally possible to obtain designs that are closer to optimal. A precise definition of the scope of the code is a necessity. A set of load factors that are best for one application (e.g., design of steel structures) would not generally be accurate for another technology (such as design of concrete structures).

A popular code format is LRFD mentioned earlier. In LRFD codes, the design formula is expressed as

$$\text{factored nominal resistance} \geq \text{total factored nominal load} \tag{8.8a}$$

or

$$\phi R_n \geq \sum \gamma_i Q_{ni} \tag{8.8b}$$

where Q_{ni} is the nominal (design) value of load component i, γ_i is the load factor for load component i, R_n is the nominal (design) value of resistance or capacity, and ϕ is the resistance factor.

An example of how the nominal (design) load, the mean load, and the factored load are related is shown in Figure 8.4. In some recent developments

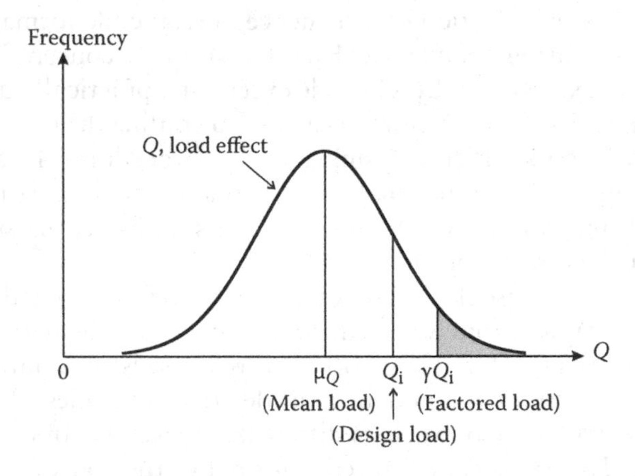

Figure 8.4 Relationships among nominal load, mean load, and factored load.

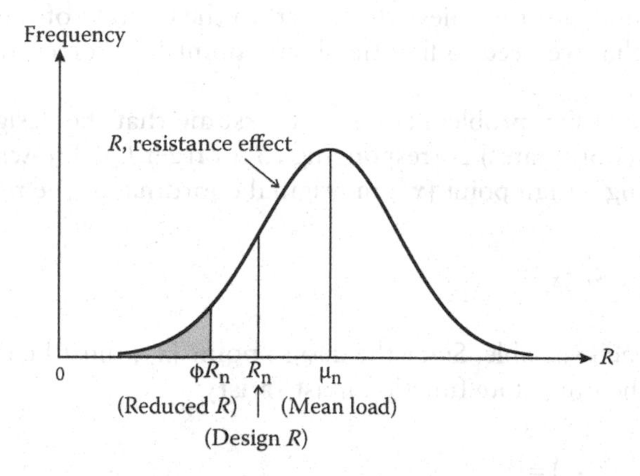

Figure 8.5 Relationships among nominal resistance, mean resistance, and factored resistance.

of design codes, the factored load was taken so that the shaded area (see Figure 8.4) was the same for all load components (e.g., dead load and live load). A similar example showing the relationship among nominal resistance, mean resistance, and factored resistance is presented in Figure 8.5. The actual value of the resistance factor is determined by calibration with the objective to obtain $\beta = \beta_T$.

8.5 CALIBRATION OF PARTIAL SAFETY FACTORS FOR A LEVEL I CODE

As noted in the previous section, one possible format for Level I codes is the LRFD format presented in Equations 8.1 and 8.8b. Design equations following this format contain partial safety factors ϕ and γ that must be calibrated based on the target reliability index adopted by the code. This section describes the rationale behind these factors and presents one technique for determining their values.

To understand these coefficients, it is necessary to modify the thought process behind the algorithms presented in Chapter 5. In Chapter 5, two algorithms were presented to determine β given a limit state function (or performance function) and the mean and variance [or standard deviation or coefficient of variation (COV)] for each of the random variables used in the limit state function. These algorithms are based on finding a design point $\{z^*\}$ in reduced variates that is then used to calculate β based on the "shortest distance" interpretation proposed by Hasofer and Lind. In determining partial safety factors, the problem is reversed. A target β is specified, and it is then necessary to determine the required mean values of the resistance and loads to achieve the target. In the context of the algorithms, this means that we need to find the design point $\{z^*\}$ corresponding to the target β.

To put the design problem in context, assume that the design point $\{z^*\}$ (in reduced coordinates) corresponding to a target β is known. To get the corresponding design point $\{x^*\}$ in original coordinates, the relation

$$x_i^* = \mu_{X_i} + z_i^* \sigma_{X_i} \tag{8.9}$$

is used for each variable. Since the design point $\{x^*\}$ must be on the failure boundary, the limit state function must satisfy

$$g(x_1^*, x_2^*, \ldots, x_n^*) = 0 \tag{8.10}$$

For design purposes, it is necessary to relate each design point value, x_1^*, to a value of the variable used in design (e.g., a nominal design value specified by the code). If the nominal design value of X_i is denoted by $\tilde{x}_i$, then the partial safety factor γ_i is defined as

$$\gamma_i = \frac{x_i^*}{\tilde{x}_i} \tag{8.11}$$

The partial safety factor is nothing more than a scaling factor that allows the designer to convert a nominal design value of a variable to the value needed to satisfy Equation 8.10 for the target β. Using Equation 8.11, the limit state condition in Equation 8.10 can be expressed as

$$g(\gamma_1 \tilde{x}_1, \gamma_2 \tilde{x}_2, \ldots, \gamma_n \tilde{x}_n) = 0 \tag{8.12}$$

Observe that Equation 8.12 is simply the limit state equation expressed in terms of nominal design values and partial safety factors. As discussed later in this section, the partial safety factors can be calibrated to achieve the target β.

To further clarify this concept of partial safety factor, consider the basic case

$$g(R,Q) = R - Q \tag{8.13}$$

where R is the resistance and Q is the load. If this equation is evaluated at the design point $\{r^*, q^*\}$, and if the design points are expressed in terms of partial safety factors and nominal values, the equation becomes

$$g(\gamma_R \tilde{r}, \gamma_Q \tilde{q}) = \gamma_R \tilde{r} - \gamma_Q \tilde{q} = 0 \tag{8.14}$$

By replacing the equality in Equation 8.14 by an inequality that reflects the condition to be satisfied for a "safe" design (i.e., $g \geq 0$), we get

$$\gamma_R \tilde{r} - \gamma_Q \tilde{q} \geq 0 \text{ or } \gamma_R \tilde{r} \geq \gamma_Q \tilde{q} \tag{8.15}$$

which is in the LRFD format presented in Equation 8.1. Thus, if γ_R and γ_Q are properly calibrated, then the designer can use Equation 8.15 to determine the required nominal value of resistance ($\tilde{r}$) given a nominal value of loading ($\tilde{q}$) to achieve a target reliability index.

There are two possible ways to calibrate the partial safety factors for a given problem. One approach is to use "trial and error." In other words, one can assume some nominal values for the random variables in the problem, and then the approach described in Chapter 5 (or a similar approach) can be used to determine the reliability index and the design points for each case. (It is assumed that the COV or standard deviation is known for all random variables as well as bias factors.) If one of the assumed sets of nominal values gives the correct β, then the partial safety factors can be calculated using Equation 8.4. An alternative approach, which works well for a variety of problems, is to modify the algorithm presented in Chapter 5. The algorithm is still iterative in nature, but usually the partial safety factors can be found directly without resorting to trial and error. The modified procedure is as follows:

Procedure for calibrating partial safety factors for a given β:

1. Formulate the limit state function and the design equation. Determine the probability distributions and appropriate parameters for as many of the random variables X_i ($i = 1, 2, \ldots, n$) as possible. It is assumed that the COV or standard deviation is known for all random variables. There can be at most only *two* unknown mean values in the analysis. Typically, one unknown mean value corresponds to the resistance variable, and the other unknown mean value corresponds to a load variable. Load ratios are used to relate the mean values of the loads. For the first iteration, we can use the limit state equation $g = 0$ evaluated at the mean values to get a relationship between the two unknown means.
2. Obtain an initial design point $\{x_i^*\}$ by assuming values for $n - 1$ of the random variables X_i. (Mean values are often a reasonable initial choice.) Solve the limit state equation $g = 0$ to obtain a value for the remaining variable. This ensures that the trial design point is on the failure boundary.

3. For each of the design point values x_i^* corresponding to a non-normal distribution, determine the equivalent normal mean $\mu_{X_i}^e$ and standard deviation $\sigma_{X_i}^e$ using Equations 5.34 and 5.35. If one or more x_i^* values correspond to a normal distribution, then the equivalent normal parameters are simply the actual parameters. (Since some of the mean values are not known in advance, it may not always be possible to carry out this step.)
4. Determine the partial derivatives of the limit state function with respect to the reduced variates using Equation 5.23b. For convenience, define a column vector $\{G\}$ as the vector whose elements are these partial derivatives, i.e.,

$$\{G\} = \begin{Bmatrix} G_1 \\ G_2 \\ \vdots \\ G_n \end{Bmatrix}, \text{where } G_i = -\left.\frac{\partial g}{\partial Z_i}\right|_{\text{evaluated at design point}} \tag{8.16}$$

5. Calculate the column vector $\{\alpha\}$ using

$$\{\alpha\} = \frac{[\rho]\{G\}}{\sqrt{\{G\}^T[\rho]\{G\}}} \tag{8.17}$$

where $[\rho]$ is the matrix of correlation coefficients.
6. Determine a new design point in reduced variates for $n - 1$ of the variables using

$$z_i^* = \alpha_i \beta_{\text{target}} \tag{8.18}$$

where β_{target} is the target reliability to be achieved.
7. Determine the corresponding design point values in original coordinates for the $n - 1$ values in step 6 using

$$x_i^* = \mu_{X_i}^e + z_i^* \sigma_{X_i}^e \tag{8.19}$$

8. Determine the value of the remaining random variable (i.e., the one not found in steps 6 and 7) by solving the limit state function $g = 0$. Also update the relationship between the two unknown mean values (if applicable). This can be done by assuming $\tilde{x}_i = \mu_{X_i}$ and manipulating Equation 8.11 as follows:

$$\gamma_i = \frac{x_i^*}{\mu_{X_i}}$$

$$= \frac{\mu_{X_i} + z_i^* \sigma_{X_i}}{\mu_{X_i}} \tag{8.20}$$

$$= 1 + z_i^* V_{X_i}$$

$$= 1 + \alpha_i \beta V_{X_i}$$

Therefore,

$$\mu_{X_i} = \frac{x_i^*}{1 + \alpha_i \beta V_{X_i}} \tag{8.21}$$

9. Repeat steps 3 through 8 until $\{\alpha\}$ converges.
10. Once convergence is achieved, calculate the design factors using Equation 8.11.

Although the procedure parallels the procedure discussed in Chapter 5, it can be confusing. It is best explained by applying it to some example problems.

Example 8.1

Consider the fundamental case

$$g = R - Q$$

where R is the resistance and Q is the load. A possible design equation (in LRFD format) is

$$\gamma_R \mu_R \geq \gamma_Q \mu_Q$$

in which the nominal values are assumed to be equal to the mean values. (In practice, the symbol ϕ is often used to represent the partial safety factor of resistance instead of γ_R.) Assume that we know from a combination of experience and test data that V_R = 10% and V_Q = 12%. Determine the partial safety factors that must be used in design to achieve a β_{target} = 3.0. Since we do not have any other information, we will assume that both R and Q are normally distributed and uncorrelated.

The steps in the first cycle of iteration are as follows:

1. Formulate limit state function and design equation. This has already been done.
2. Obtain an initial design point. For this simple case, $r^* = q^*$. There are only two variables; hence, we have two unknown mean values. For the first iteration, we need to use the limit state equation evaluated at the mean values to relate the two unknown mean values. Doing this results in $\mu_R = \mu_Q$.
3. Determine equivalent normal parameters. This is not needed here.
4. Determine the $\{G\}$ vector:

$$G_1 = -\frac{\partial g}{\partial R}\sigma_R\bigg|_{\text{design point}} = -\sigma_R = -V_R\mu_R = -0.1\mu_R = -0.1\mu_Q$$

$$G_2 = -\frac{\partial g}{\partial Q}\sigma_Q\bigg|_{\text{design point}} = +\sigma_Q = V_Q\mu_Q = 0.12\mu_Q$$

Note that $\mu_R = \mu_Q$ was used to evaluate G_1.

5. Calculate $\{\alpha\}$. For this problem, $[\rho]$ is the identity matrix so the matrix math simplifies somewhat.

$$\{\alpha\} = \frac{[\rho]\{G\}}{\sqrt{\{G\}^T[\rho]\{G\}}}$$

$$= \frac{1}{\sqrt{(-0.1\mu_Q)^2 + (0.12\mu_Q)^2}}\begin{Bmatrix} -0.1\mu_Q \\ 0.12\mu_Q \end{Bmatrix}$$

$$= \frac{1}{0.156\mu_Q}\begin{Bmatrix} -0.1\mu_Q \\ 0.12\mu_Q \end{Bmatrix}$$

$$= \begin{Bmatrix} -0.641 \\ 0.769 \end{Bmatrix} = \begin{Bmatrix} \alpha_R \\ \alpha_Q \end{Bmatrix}$$

Observe that the unknown mean values of the variables eventually cancel out and the vector $\{\alpha\}$ contains only numbers.

6. Determine a new design point for $n - 1$ of the variables. We will calculate z_Q^* in this way.

$$z_Q^* = \alpha_Q\beta_{\text{target}} = 0.769(3.0) = 2.31$$

7. Determine q^* using

$$q^* = \mu_Q + z_Q^*\sigma_Q = \mu_Q(1 + z_Q^* V_Q) = \mu_Q(1 + 2.31(0.12)) = 1.28\mu_Q$$

8. Determine r^* by solving $g = 0$. Thus, $r^* = 1.28\mu_Q$.
9. Before iterating again, we need to get an improved estimate of μ_R in terms of μ_Q for use in calculating $\{G\}$ and $\{\alpha\}$. Using Equation 8.21,

$$\mu_R = \frac{r^*}{(1 + \alpha_R \beta_{\text{target}} V_R)} = \frac{1.28\mu_Q}{1 - 0.641(3.0)(0.10)} = 1.58\mu_Q$$

The results of subsequent iterations are shown in Table 8.1.

Observe that $\{\alpha\}$ has almost converged after three iterations. Therefore, assuming the mean values and the nominal design values are the same, the design factors are

$$\gamma_R = \frac{r^*}{\mu_R} = \frac{1.22\mu_Q}{1.60\mu_Q} = 0.763$$

$$\gamma_Q = \frac{q^*}{\mu_Q} = \frac{1.22\mu_Q}{\mu_Q} = 1.22$$

Table 8.1 Iteration results for Example 8.1

	Iteration #		
	1	2	3
r^* (start)	μ_Q	$1.28\,\mu_Q$	$1.22\mu_Q$
q^* (start)	μ_Q	$1.28\,\mu_Q$	$1.22\mu_Q$
μ_R (start)	μ_Q	$1.58\,\mu_Q$	$1.60\mu_Q$
α_R	−0.641	−0.796	−0.800
α_Q	0.769	0.605	0.600
r^* (end)	$1.28\,\mu_Q$	$1.22\,\mu_Q$	$1.22\mu_Q$
q^* (end)	$1.28\,\mu_Q$	$1.22\,\mu_Q$	$1.22\mu_Q$
μ_R (end)	$1.58\,\mu_Q$	$1.60\,\mu_Q$	$1.60\mu_Q$

It is important to understand how to interpret these design factors. Assume that $\tilde{q} = \mu_Q = 10$. This is a nominal design value of the load obtained from the code. From the design equation shown at the beginning of this example, the minimum required $\tilde{r} = \mu_R$ to achieve $\beta = 3$ would be

$$\mu_R = \frac{\gamma_Q \mu_Q}{\gamma_R} = \frac{(1.22)(10)}{0.763} = 16.0$$

Thus, if the partial safety factors are calibrated correctly, a nominal resistance of 16 will ensure a target $\beta = 3$ is achieved if the nominal loading is equal to 10. For this simple case of a linear limit state function, we can check our answer using Equation 5.14:

$$\beta = \frac{\mu_R - \mu_Q}{\sqrt{\sigma_R^2 + \sigma_Q^2}}$$

If we substitute all the numbers we have determined into this expression, and noting that $\sigma_R = V_R \mu_R$, we should get $\beta = 3$.

$$\beta = \frac{\mu_R - \mu_Q}{\sqrt{\sigma_R^2 + \sigma_Q^2}} = \frac{16.0 - 10}{\sqrt{[(0.1)(16.0)]^2 + [(0.12)(10)]^2}} = \frac{6.0}{2.00} = 3.0$$

Hence, the partial safety factors allow us to determine the nominal values (in this case, the mean values) of the design variables that are needed to achieve a desired reliability index.

Example 8.2

Reconsider Example 8.1 above, except that the design equation is to be written in the form

$$\gamma_R^n R_n \geq \gamma_Q^n Q_n$$

where R_n and Q_n are the nominal values of resistance and load, respectively. The bias factors (ratios of mean value to nominal value) for R and Q are known to be $\lambda_R = 1.1$ and $\lambda_Q = 1.2$. What are the partial safety factors for this case?

From the results of Example 8.1, we know the partial safety factors when mean values are used in the design equation. We also know that

$$\mu_R = \lambda_R R_n \quad \mu_Q = \lambda_Q Q_n$$

Hence, we need to relate the two versions of the design equation as follows:

$$\gamma_R \mu_R \geq \gamma_Q \mu_Q \quad \text{(from Example 6.1)}$$

$$\gamma_R (\lambda_R R_n) \geq \gamma_Q (\lambda_Q Q_n)$$

$$(\gamma_R \lambda_R) R_n \geq (\gamma_Q \lambda_Q) Q_n$$

$$\gamma_R^n R_n \geq \gamma_Q^n Q_n$$

where

$$\gamma_R^n = \gamma_R \lambda_R \quad \gamma_Q^n = \gamma_Q \lambda_Q$$

Thus, combining the results of Example 8.1 with the bias factor information provided above, we find

$$\gamma_R^n = (0.763)(1.1) = 0.839 \quad \gamma_Q^n = (1.22)(1.2) = 1.46$$

These are the partial safety factors to be used when nominal values are used in the design equation.

Example 8.3

Consider the simply supported steel beam shown in Figure 8.6. It is subjected to a concentrated load at midspan.

A plastic mechanism will form when the load is sufficient to create a plastic hinge under the load. If the limit state condition is defined in terms of forming a mechanism, then "failure" will occur when

$$M_p < \frac{PL}{4}$$

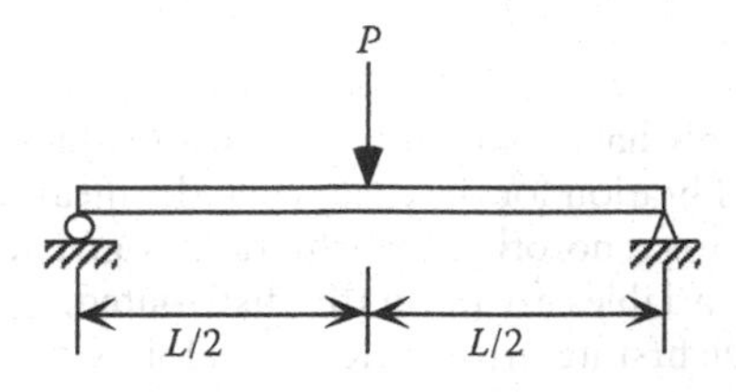

Figure 8.6 Beam considered in Example 8.3.

where M_p is the plastic moment capacity of the section. Recalling that the limit state function $g()$ indicated failure when it is less than zero, we can formulate the limit state equation for this case to rearrange the above inequality as follows:

$$M_p < \frac{PL}{4} \quad \text{at failure}$$

$$M_p - \frac{PL}{4} < 0 \iff g(M_p, P) < 0$$

$$g(M_p, P) = M_p - \frac{PL}{4}$$

Also, since $M_p = ZF_y$ (Z is the plastic section modulus and F_y is yield stress), the limit state equation can be rewritten as

$$g(F_Y, Z, P) = F_Y Z - \frac{PL}{4}$$

In this form, the limit state equation is a nonlinear equation involving three random variables. (It is assumed that L is deterministic for this example). The goal of this example is to formulate a design equation and calibrate the partial safety factors to achieve a target β of 2.5. Assume that the following information is given:

- The coefficients of variation for the variables are estimated to be $V_F = 12.5\%$, $V_Z = 5\%$, and $V_P = 15\%$.
- There is correlation between Z and F_y, and the correlation coefficient is $\rho_{FZ} = 0.25$. No other correlations are believed to exist.
- $L = 20$ ft = 240 in.

First, we need to formulate the design equation. Based on the limit state equation above, the design equation (in terms of mean values) will be

$$\gamma_F \mu_{F_Y} (\gamma_Z \mu_Z) \geq \gamma_P \mu_P \left(\frac{L}{4} \right)$$

Since we can only have two unknown mean values in each analysis, we will do the calibration for the case when the mean value of the yield stress is 50 ksi. Since no other information is given, we will assume that all random variables are normally distributed.

The steps in the first iterative cycle are as follows:

1. Formulate limit state function and design equation. This has already been done.

2. Obtain an initial design point. Assume $f_Y^* = \mu_F = 50$ ksi and assume $z^* = \mu_Z$ (which is unknown at this point). Then, from $g = 0$, $p^* = (200/L)\mu_Z = 0.8333\mu_Z$. We will also need some initial relationship between μ_Z and μ_P for the first iteration. Using the mean values in $g = 0$, we get $\mu_P = (200/L)\mu_Z = 0.8333\mu_Z$.
3. Equivalent normal parameters are not needed here.
4. Determine $\{G\}$.

$$G_1 = -\frac{\partial g}{\partial F_Y}\sigma_{F_Y}\bigg|_{\text{design point}} = -Z\sigma_{F_Y}\Big|_{\text{design point}} = -z^*\sigma_{F_Y} = -z^* V_{F_Y}\mu_{F_Y}$$

$$G_2 = -\frac{\partial g}{\partial Z}\sigma_Z\bigg|_{\text{design point}} = -F_Y\sigma_Z\Big|_{\text{design point}} = -f_y^*\sigma_Z = -f_y^* V_Z\mu_Z$$

$$G_3 = -\frac{\partial g}{\partial P}\sigma_P\bigg|_{\text{design point}} = -\left(-\frac{L}{4}\right)\sigma_P\Big|_{\text{design point}} = \frac{L}{4}V_P\mu_P$$

Substituting the appropriate numbers for the first iteration,

$$G_1 = -\mu_Z V_{F_Y}\mu_{F_Y} = -\mu_Z(0.125)(50) = -6.25\mu_Z$$

$$G_2 = -\mu_{F_Y} V_Z\mu_Z = -(50)(0.05)\mu_Z = -2.5\mu_Z$$

$$G_3 = \frac{L}{4}V_P\mu_P = \frac{L}{4}(0.15)\left(\frac{200}{L}\mu_Z\right) = 7.5\mu_Z$$

5. Calculate $\{\alpha\}$.

$$\{\alpha\} = \frac{[\rho]\{G\}}{\sqrt{\{G\}^T[\rho]\{G\}}}$$

$$= \frac{\begin{bmatrix} 1 & 0.25 & 0 \\ 0.25 & 1 & 0 \\ 0 & 0 & 1 \end{bmatrix}\begin{Bmatrix} -6.25\mu_Z \\ -2.5\mu_Z \\ 7.5\mu_Z \end{Bmatrix}}{\sqrt{\begin{Bmatrix} -6.25\mu_Z \\ -2.5\mu_Z \\ 7.5\mu_Z \end{Bmatrix}^T \begin{bmatrix} 1 & 0.25 & 0 \\ 0.25 & 1 & 0 \\ 0 & 0 & 1 \end{bmatrix}\begin{Bmatrix} -6.25\mu_Z \\ -2.5\mu_Z \\ 7.5\mu_Z \end{Bmatrix}}} = \begin{Bmatrix} -0.6574 \\ -0.3885 \\ 0.7171 \end{Bmatrix}$$

6. Determine new design point values for $n - 1 = 2$ of the variables.

$$z_{F_Y}^* = \alpha_{F_Y}\beta = (-0.6574)(2.5) = -1.644$$

$$z_Z^* = \alpha_Z\beta = (-0.3885)(2.5) = -0.9713$$

7. Determine corresponding values in original coordinates.

$$f_y^* = \mu_{F_Y} + z_{F_Y}^*\sigma_{F_Y} = \mu_{F_Y}(1 + z_{F_Y}^* V_{F_Y}) = 50(1 + (-1.644)(0.125)) = 39.73$$

$$z^* = \mu_Z + z_Z^*\sigma_Z = \mu_Z(1 + z_Z^* V_Z) = \mu_Z(1 + (-0.9713)(0.05)) = 0.9514\mu_Z$$

8. Determine the value of the remaining random variable from $g = 0$.

$$g(f_y^*, z^*, p^*) = 0 \quad \Rightarrow \quad p^* = f_y^* z^* \frac{4}{L} = 0.6300\mu_Z$$

We also need to obtain an improved estimate of μ_P for use in the next iteration. Using Equation 8.21,

$$\mu_P = \frac{p^*}{(1 + \alpha_P \beta_{\text{target}} V_P)} = \frac{0.6300\mu_Z}{1{+}(0.7171)(2.5)(0.15)} = 0.4965\mu_Z$$

We now need to iterate until {α} converges. A summary of subsequent cycles is given in Table 8.2.

After four cycles, {α} has converged (for practical purposes); hence, no more iterations are needed. The design factors (based on mean values) are obtained using Equation 8.11.

$$\gamma_F = \frac{f_y^*}{\mu_{F_y}} = \frac{37.48}{50} = 0.7496$$

$$\gamma_Z = \frac{z^*}{\mu_Z} = \frac{0.9474\mu_Z}{\mu_Z} = 0.9474$$

$$\gamma_P = \frac{p^*}{\mu_P} = \frac{0.5918\mu_Z}{0.4901\mu_Z} = 1.208$$

Table 8.2 Iteration results for Example 8.3

	Iteration			
	1	*2*	*3*	*4*
f_Y^*(start)	50	39.73	37.53	37.48
z^*(start)	μ_Z	$0.9514\mu_Z$	$0.9462\mu_Z$	$0.9474\mu_Z$
p^*(start)	$0.8333\mu_Z$	$0.6300\mu_Z$	$0.5919\mu_Z$	$0.5919\mu_Z$
μ_P(start)	$0.8333\mu_Z$	$0.4965\mu_Z$	$0.4901\mu_Z$	$0.4901\mu_Z$
α_F	−0.6574	−0.7981	−0.8010	−0.8014
α_Z	−0.3885	−0.4302	−0.4210	−0.4207
α_P	0.7171	0.5535	0.5535	0.5532
f_Y^*(end)	39.73	37.53	37.48	37.48
z^*(end)	$0.9514\mu_Z$	$0.9462\mu_Z$	$0.9474\mu_Z$	$0.9474\mu_Z$
p^*(end)	$0.6300\mu_Z$	$0.5919\mu_Z$	$0.5919\mu_Z$	$0.5918\mu_Z$
μ_P(end)	$0.4965\mu_Z$	$0.4901\mu_Z$	$0.4901\mu_Z$	$0.4901\mu_Z$

Therefore, the design equation in terms of mean values can be written as

$$(0.7496)\mu_{F_Y}(0.9474)\mu_Z \geq (1.208)\mu_P\left(\frac{L}{4}\right)$$

Combining coefficients, and keeping only two significant digits, the final form of the design equation is

$$(0.71)\mu_{F_Y}\mu_Z \geq (1.2)\mu_P\left(\frac{L}{4}\right)$$

Note that this result is strictly valid for the assumed conditions only (i.e., $F_Y = 50$ ksi and $L = 20$ ft). The factors may change for other values of these variables, and they will change if the coefficients of variation or the correlation coefficients are different.

Example 8.4

Consider the following limit state equation involving resistance (R), dead load effect (D), and live load effect (L):

$$g(R,D,L) = R - (D + L) = R - D - L$$

A possible design equation corresponding to this limit state equation, in LRFD format, is

$$\phi R_n \geq \gamma_D D_n + \gamma_L L_n$$

where R_n, D_n, and L_n are nominal values of the loading. Given the following parameters, determine the partial safety factors to achieve $\beta = 3.0$:

R is lognormal	$V_R = 13\%$	$\lambda_R = 1.10$
D is normal	$V_D = 10\%$	$\lambda_D = 1.05$
L is Type I extreme distribution	$V_L = 25\%$	$\lambda_L = 1.0$

There are three unknown mean values, and only two are allowed in the analysis. Since we are interested in determining the required resistance given the loading, it is logical to keep the mean resistance as one unknown. For this example, the mean value of the dead load will be used as the second unknown, and we will consider a specific relationship between live load and dead load. For this example, assume $\mu_L/\mu_D = 3$.

Again, we will follow the step-by-step procedure for the first iteration and summarize the results of subsequent iterations.

1. Formulate limit state function. This has been done.
2. Obtain an initial design point. Assume $d^* = \mu_D$ and $l^* = \mu_L = 3\mu_D$ (since the live load-to-dead load ratio to be considered is 3). Then, from $g = 0$, $r^* = 4\mu_D$. Also, we need to express μ_R in terms of μ_D for the first iteration. Using $g = 0$ evaluated at the mean values, $\mu_R = 4\mu_D$.
3. Determine the equivalent normal parameters for R and L.

For lognormal variables, closed-form solutions for the equivalent normal parameters were derived in Example 5.8. When the COV is less than about 0.2, these expressions simplify to (see Example 5.10)

$$\sigma_R^e \approx r^* V_R$$

$$\mu_R^e \approx r^*[1 - \ln(r^*) + \ln(\mu_R)]$$

Hence, plugging in the appropriate information for the first iteration, we get

$$\sigma_R^e = r^* V_R = 4\mu_D(0.13) = 0.52\mu_D$$

$$\mu_R^e = r^*[1 - \ln(r^*) + \ln(\mu_R)]$$

$$= r^*\left[1 - \ln\left(\frac{r^*}{\mu_R}\right)\right]$$

$$= 4\mu_D\left[1 - \ln\left(\frac{4\mu_D}{4\mu_D}\right)\right]$$

$$= 4\mu_D$$

For Type I variables, we can follow the approach used in Chapter 5. First, we must determine the distribution parameters u and a:

$$a = \sqrt{\frac{\pi^2}{6\sigma_L^2}} = \frac{\pi}{\sqrt{6}(V_L\mu_L)} = \frac{5.130}{\mu_L} = \frac{1.710}{\mu_D}$$

$$u = \mu_L - \frac{0.5772}{a} = 3\mu_D - \frac{0.5772}{\left(1.710/\mu_D\right)} = 2.662\mu_D$$

Then,

$$F_L(l^*) = \exp[-\exp(-a(l^* - u))]$$

$$f_L(l^*) = a\{\exp(-a(l^* - u))\}\exp[-\exp(-a(l - u))]$$

Substituting $l^* = 3\mu_D$ for the first iteration, we get

$$F_L(l^*) = 0.5706; \quad f_L(l^*) = \frac{0.5474}{\mu_D}$$

Thus, the equivalent normal parameters for Q are

$$\sigma_L^e = 0.7173\mu_D; \quad \mu_L^e = 2.872\mu_D$$

4. Determine $\{G\}$:

$$G_1 = -\frac{\partial g}{\partial R}\sigma_R^e\Big|_{\text{design point}} = -\sigma_R^e$$

$$G_2 = -\frac{\partial g}{\partial D}\sigma_D\bigg|_{\text{design point}} = +\sigma_D = V_D\mu_D$$

$$G_3 = -\frac{\partial g}{\partial L}\sigma_L^e\bigg|_{\text{design point}} = +\sigma_L^e$$

Thus, for the first iteration,

$$G_1 = -0.52\mu_D$$

$$G_2 = 0.10\mu_D$$

$$G_3 = 0.7173\mu_D$$

5. Calculate $\{\alpha\}$. Since no correlation information is available, we will assume that the variables are uncorrelated.

$$\{\alpha\} = \frac{[\rho]\{G\}}{\sqrt{\{G\}^T[\rho]\{G\}}}$$

$$= \frac{\begin{bmatrix} 1 & 0 & 0 \\ 0 & 1 & 0 \\ 0 & 0 & 1 \end{bmatrix}\begin{Bmatrix} -0.52\mu_D \\ -0.10\mu_D \\ 0.7173\mu_D \end{Bmatrix}}{\sqrt{\begin{Bmatrix} -0.52\mu_D \\ -0.10\mu_D \\ 0.7173\mu_D \end{Bmatrix}^T \begin{bmatrix} 1 & 0 & 0 \\ 0 & 1 & 0 \\ 0 & 0 & 1 \end{bmatrix}\begin{Bmatrix} -0.52\mu_D \\ -0.10\mu_D \\ 0.7173\mu_D \end{Bmatrix}}} = \begin{Bmatrix} -0.5832 \\ 0.1122 \\ 0.8045 \end{Bmatrix}$$

6. Determine the design point in reduced variates for $n - 1 = 2$ variables. Choosing the variables for D and L, we get

$$z_D^* = \alpha_D\beta = 0.1122(3.0) = 0.3366$$

$$z_L^* = \alpha_L\beta = 0.8045(3.0) = 2.414$$

7. Determine the corresponding design points in original coordinates.

$$d^* = \mu_D + z_D^*\sigma_D = \mu_D(1 + z_D^* V_D) = 1.034\mu_D$$

$$l^* = \mu_L^e + z_L^* \sigma_L^e = 2.872\mu_D + 2.414(0.7173\mu_D) = 4.604\mu_D$$

Note the use of the equivalent normal parameters in determining l^*.

8. Determine the remaining design point from $g = 0$. This gives $r^* = d^* + l^* = 5.637\mu_D$. We also need to get an updated measure of μ_R for use in the next iteration. Using Equation 8.21,

$$\mu_R = \frac{r^*}{1+\alpha_R \beta V_R} = \frac{5.638\mu_D}{1+(-0.5832)(3.0)(0.13)} = 7.298\mu_D$$

The subsequent iterations are summarized in Table 8.3.

After five cycles, $\{\alpha\}$ appears to have converged for practical purposes. Therefore, the design factors (based on nominal values of the variables) are

$$\phi = \frac{r^*}{R_n} = \frac{r^*}{\mu_R / \lambda_R} = \lambda_R \frac{r^*}{\mu_R} = (1.10)\frac{6.921\mu_D}{8.429\mu_D} = 0.9032$$

$$\gamma_D = \lambda_D \frac{d^*}{\mu_D} = (1.05)\frac{1.015\mu_D}{\mu_D} = 1.066$$

Table 8.3 Iteration results for Example 8.4

	Iteration				
	1	*2*	*3*	*4*	*5*
r^*(start)	$4\mu_D$	$5.638\mu_D$	$6.683\mu_D$	$6.908\mu_D$	$6.921\mu_D$
d^*(start)	μ_D	$1.034\mu_D$	$1.020\mu_D$	$1.016\mu_D$	$1.015\mu_D$
l^*(start)	$3\mu_D$	$4.604\mu_D$	$5.663\mu_D$	$5.892\mu_D$	$5.906\mu_D$
μ_R(start)	$4\mu_D$	$7.298\mu_D$	$8.249\mu_D$	$8.425\mu_D$	$8.429\mu_D$
α_R	−0.5832	−0.4868	−0.4618	−0.4588	−0.4586
α_D	0.1122	0.0664	0.0532	0.0511	0.0510
α_L	0.8045	0.8710	0.8854	0.8871	0.8872
r^*(end)	$5.638\mu_D$	$6.683\mu_D$	$6.908\mu_D$	$6.921\mu_D$	$6.921\mu_D$
d^*(end)	$1.034\mu_D$	$1.020\mu_D$	$1.016\mu_D$	$1.015\mu_D$	$1.015\mu_D$
l^*(end)	$4.604\mu_D$	$5.663\mu_D$	$5.892\mu_D$	$5.906\mu_D$	$5.906\mu_D$
μ_R(end)	$7.298\mu_D$	$8.249\mu_D$	$8.425\mu_D$	$8.429\mu_D$	$8.429\mu_D$

$$\gamma_L = \lambda_L \frac{l^*}{\mu_L} = (1.0)\frac{5.906\mu_D}{3\mu_D} = 1.969$$

Note that the true mean values, not the equivalent mean values, are used in this final step because the bias factors are defined in terms of true mean values. The final form of the design equation (keeping only two significant digits) is therefore

$$(0.90)R_n \geq (1.1)D_n + (2.0)L_n$$

8.6 DEVELOPMENT OF A BRIDGE DESIGN CODE

The previous sections described various issues associated with developing reliability-based design codes and a general approach for addressing these issues. In this section, the presented approach is demonstrated on the development of an LRFD bridge design code (Nowak, 1995). The work involved the development of load models, resistance models, limit states, and acceptance criteria. This description deals only with those aspects related to structural reliability.

8.6.1 Scope

It was assumed that the code provided design criteria for the following highway bridge structures:

- Steel girder bridges (composite and non-composite)
- Reinforced concrete bridges (T-beams)
- Prestressed concrete bridges (AASHTO girders)

The design provisions were developed for the ULSs of flexural capacity (bending moment) and shear. Calculations were performed for spans ranging from 30 to 200 ft (9 to 60 m) and girder spacings from 4 to 12 ft (1.2 to 3.6 m).

8.6.2 Objectives

Code provisions must ensure that the safety level of the designed structures is adequate. In the LRFD code, safety was provided through the selection of conservative load and resistance factors. Load and resistance factors were determined for each limit state considered. Safety was measured in terms of the reliability index. Therefore, the acceptance criteria were based on closeness to the preselected target value of β. The code objective was to minimize the discrepancy between the reliability index of designed structures and the target index β_T.

In this study, the Rackwitz–Fiessler procedure was used to calculate values of β. Load was treated as a normal random variable and resistance was treated as a lognormal random variable. The derivation of the statistical parameters of load and resistance (bias factors and coefficients of variation) is described by Nowak (1993), Nowak and Hong (1991), Hwang and Nowak (1991), and Tabsh and Nowak (1991). Dead load was considered as a sum of three components: factory-made components, cast-in-place concrete, and asphalt. Factory-made components (structural steel, precast concrete) typically have the lowest degree of variation. Cast-in-place concrete has a higher COV. Asphalt wearing surface was considered separately. The model for live load included a static and dynamic component. Dynamic load is usually considered as a fraction of the static live load. The nominal live load was calculated according to AASHTO (2002). The results are summarized in Table 8.4 for loads and Table 8.5 for resistance (Nowak, 1993). The relatively large degree of variation in the live load model is due to inaccurate girder distribution factors specified by AASHTO (2002). For more information on load and resistance parameters for bridges, see Chapters 6 and 7.

Table 8.4 Statistical parameters for load components considered in study

Load component	*Bias factor*[a]	*COV*
Dead load		
Factory-made components	1.03	0.08
Cast-in-place components	1.05	0.10
Asphalt wearing surface[b]	1.00	0.25
Live load and dynamic load	1.0–1.8	0.18

[a] The bias factor is defined as the ratio of the mean value to the nominal value.
[b] The mean thickness of the asphalt surface was assumed to be 3.5 in.

Table 8.5 Statistical parameters for resistance considered in study

Material	*Limit state*	*Bias factor*	*COV*
Steel	Moment	1.12	0.10
	Shear	1.14	0.105
Reinforced concrete	Moment	1.14	0.13
	Shear	1.20	0.155
Prestressed concrete	Moment	1.05	0.075
	Shear	1.15	0.14

8.6.3 Frequency of demand

An inventory of the national bridges indicated that most of the slab-on-girder structures were simply supported with spans under 100 ft (30 m) and girder spacing from 5 to 8 ft (1.5 to 2.4 m). The most common structural types were steel girders, reinforced concrete T-beams, and prestressed concrete AASHTO-type girders.

In this study, the calculations were performed for all three types of material (steel, reinforced concrete, and prestressed concrete). The spans covered the range from 30 to 200 ft (9 to 60 m). Girder spacing considered in the study included 4, 6, 8, 10, and 12 ft (1.2, 1.8, 2.4, 3.0, and 3.6 m). Spans of 30 ft to 100 ft and girder spacings of 6–8 ft were considered the most common configurations encountered; all other spans and spacings were considered less common.

8.6.4 Target reliability level

For various design situations covered by the code, the acceptable safety levels must be established. These levels, conveniently expressed in terms of target reliability indices, serve as a basis for the development of design criteria (load and resistance factors). Selection of target indices is a multidisciplinary task. It involves structural safety analysis, economic analysis, and even the consideration of political decisions.

Reliability indices below the target value, β_T, are generally not acceptable. However, some lower values may be justified in some special cases (for example, to maintain the simplicity of the format). On the other hand, reliability indices higher than β_T are practically inevitable. This is especially the case with design conditions that do not govern (e.g., a beam designed for flexure may have an index β for shear much larger than β_T). Reliability indices higher than the target value are often justified to keep the code format simple.

In the development of a new code, it is convenient to compare the new provisions to the old (existing) code. Therefore, in this study, reliability indices were calculated for girder bridges designed using AASHTO (2002). The basic design formula is

$$1.3D + 2.17(L + I) < \phi R_n \tag{8.22}$$

where D is the dead load (or load effect), L is the live load (or load effect), I is the dynamic load (or load effect), R_n is the nominal load-carrying capacity, and ϕ is the resistance factor. Resistance factors specified by AASHTO (2002) are shown in Table 8.6.

The reliability analysis was performed for steel girders, reinforced concrete T-beams, and prestressed concrete girders. The results are shown in

Table 8.6 Resistance factors corresponding to AASHTO (2002)

Material	*Moment*	*Shear*
Steel	1.00	1.00
Reinforced concrete	0.90	0.85
Prestressed concrete	1.00	0.90

Figures 8.7 through 8.9 for moments and Figures 8.10 through 8.12 for shears. There is a considerable variation in the β values. This variation is caused by variations in the statistical parameters of load and resistance.

Selection of the target reliability indices can be based on the indices for current codes, evaluation of performance of existing structures, and

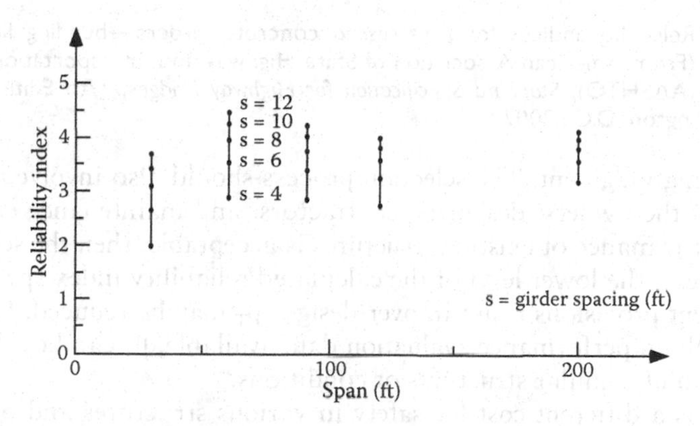

Figure 8.7 Reliability indices for steel girders—bending limit state. (From American Association of State Highway and Transportation Officials (AASHTO), *Standard Specification for Highway Bridges*, 17th Edition, Washington, D.C., 2002.)

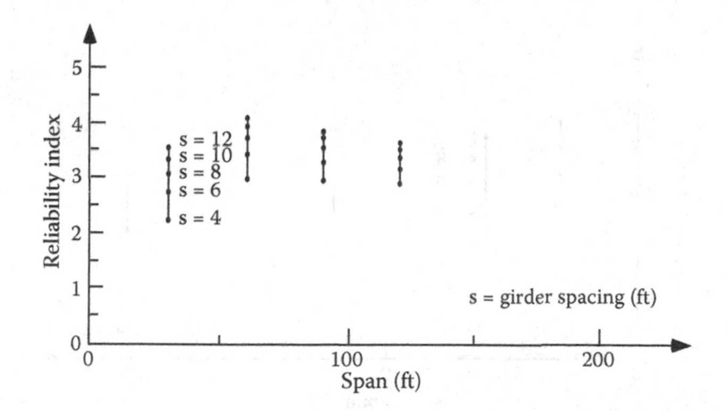

Figure 8.8 Reliability indices for reinforced concrete T-beams—bending limit state. (From American Association of State Highway and Transportation Officials (AASHTO), *Standard Specification for Highway Bridges*, 17th Edition, Washington, D.C., 2002.)

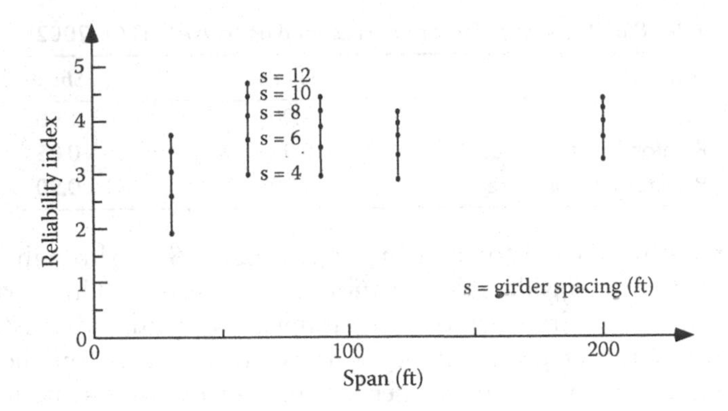

Figure 8.9 Reliability indices for prestressed concrete girders—bending limit state. (From American Association of State Highway and Transportation Officials (AASHTO), *Standard Specification for Highway Bridges*, 17th Edition, Washington, D.C., 2002.)

engineering judgment. The selection process should also involve representatives of the owners, designers, contractors, and maintenance engineers. If the performance of existing structures is acceptable, then the selected β_T should be at the lower level of the calculated reliability index spectrum. If the current provisions result in over-design, β_T may be reduced. For structures with no performance evaluation data available, β_T can be selected by comparison to similar structures or conditions.

There is a different cost for safety in various structures and parts. For example, increasing the safety level in a beam connection costs less than increasing the safety in a beam itself. A connection also generally fails in a brittle, catastrophic manner. Therefore, target reliability indices for

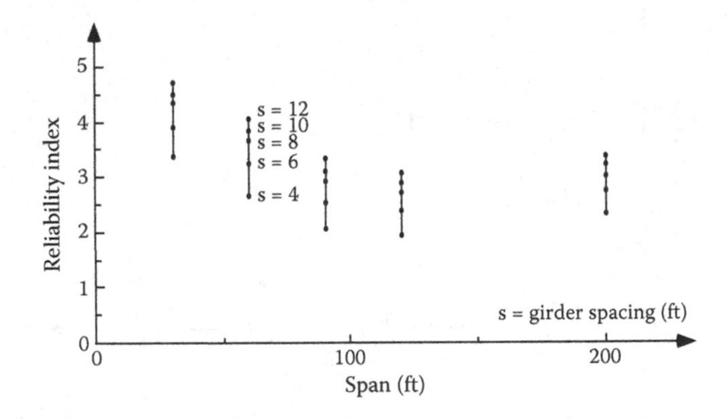

Figure 8.10 Reliability indices for steel girders—shear limit state. (From American Association of State Highway and Transportation Officials (AASHTO), *Standard Specification for Highway Bridges*, 17th Edition, Washington, D.C., 2002.)

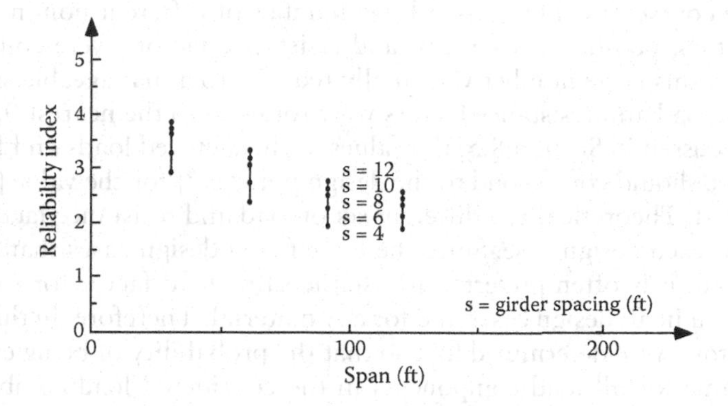

Figure 8.11 Reliability indices for reinforced concrete T-beams—shear limit state. (From American Association of State Highway and Transportation Officials (AASHTO), *Standard Specification for Highway Bridges*, Washington, D.C., 17th Edition, 2002.)

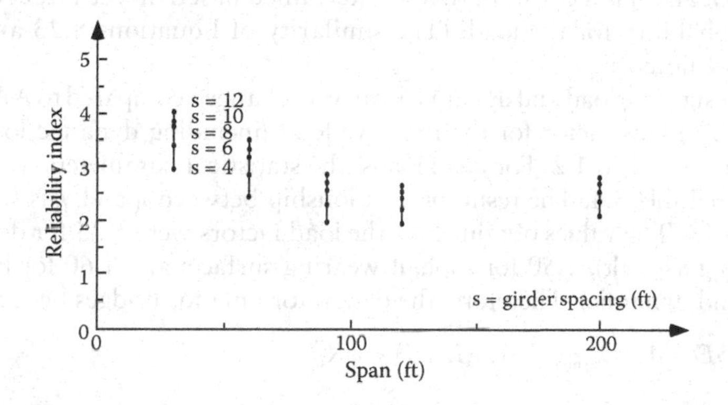

Figure 8.12 Reliability indices for prestressed concrete girders—shear limit state. (From American Association of State Highway and Transportation Officials (AASHTO), *Standard Specification for Highway Bridges*, 17th Edition, Washington, D.C., 2002.)

connections should be relatively higher. Such considerations must be taken into account in the selection of β_T values. In the present study, the target reliability index was selected to be 3.5 for moments and shears.

8.6.5 Load and resistance factors

The design formula chosen for the study was the same general equation shown in Equation 8.22. However, load and resistance factors were calculated so that reliability indices for bridges designed using the new code would be close to the target value $\beta_T = 3.5$.

In the course of calibration, a large number of different nominal loads, load factors, nominal resistances, and resistance factors were considered. However, this large number was finally reduced to a manageable size. The values of load and resistance factors were rounded to the nearest 0.05.

As discussed in Section 8.5, the values of the factored loads and factored resistance should correspond to the design point $\{x^*\}$ for the value β_T being considered. Theoretically, a different set of load and resistance factors may result for each design case. Since there are many design cases that must be considered, it is often preferred to use the same load factor for each load component in all design cases and for any material. Therefore, in this study, load factors were determined first so that the probability of being exceeded is the same for all load components in the considered load combination. The load factor for each load can be calculated using

$$\gamma_i = \lambda_i(1 + \eta V_i) \tag{8.23}$$

where λ_i is the mean-to-nominal ratio (bias factor) for load i, V_i is the COV for load i, and η is a constant that is determined based on the target exceedance probability for the load. (The similarity of Equations 8.23 and 8.20 should be noted.)

The design live load and dynamic load were changed compared to AASHTO (2002). The bias factor for the new live load (including dynamic load) was chosen to be 1.1 to 1.2. For dead loads, the statistical parameters were those shown in Table 8.4. The resulting relationship between η and γ_i is shown in Figure 8.13. The values obtained for the load factors were 1.25 for dead load (excluding asphalt), 1.50 for asphalt wearing surface, and 1.60 for live load (static and dynamic). Therefore, the design formula for bridges became

$$1.25D + 1.5D_{\text{asphalt}} + 1.6(L + I) < \phi R_n \tag{8.24}$$

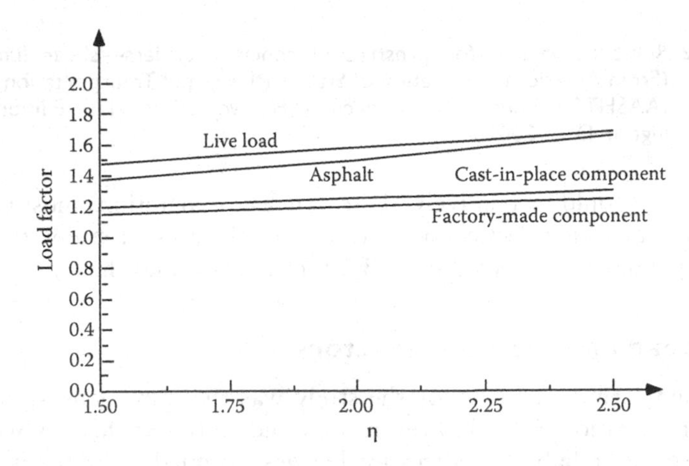

Figure 8.13 Relationship between load factors (γ_i) and η.

Then, values of the resistance factor corresponding to β_T were determined by trial and error. The number of different ϕ factors for consideration was limited to two or three values for each material and limit state because they were rounded to the nearest 0.05. For each considered value of ϕ, the reliability indices were calculated for bridges designed using the three considered types of material. The results are shown in Figures 8.14 through 8.16 for moments and Figures 8.17 through 8.19 for shears. Two values of the live load factor were considered: γ = 1.60 and γ = 1.70. The best fit to the target reliability level, β_T = 3.5, was obtained for the live load factor γ = 1.60 and the resistance factors presented in Table 8.7.

For comparison, the reliability indices for the AASHTO LRFD (2012) code are also plotted versus β corresponding to the AASHTO Standard (2002) in Figure 8.20 for moments and Figure 8.21 for shears. It is clear that the calibrated code results in a considerably reduced scatter of β values. In the first edition of AASHTO LRFD (2012), the live load factor was increased to 1.70 to account for a projected higher traffic volume in the future.

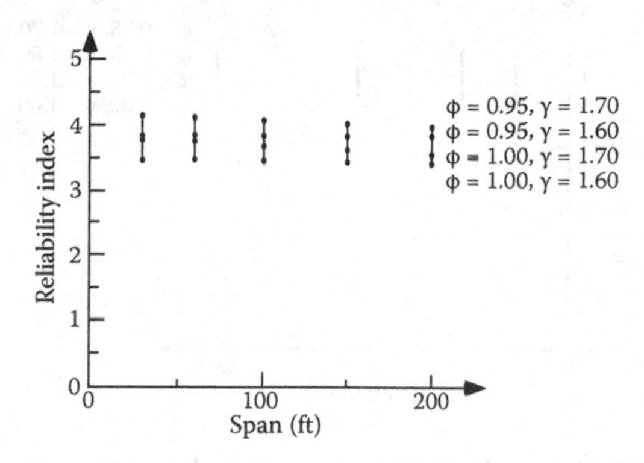

Figure 8.14 **Reliability indices for steel girders—bending limit state (new code).**

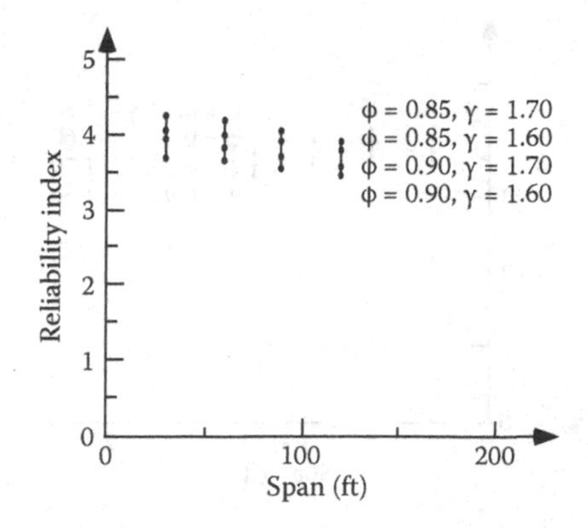

Figure 8.15 **Reliability indices for reinforced concrete T beams—bending limit state (new code).**

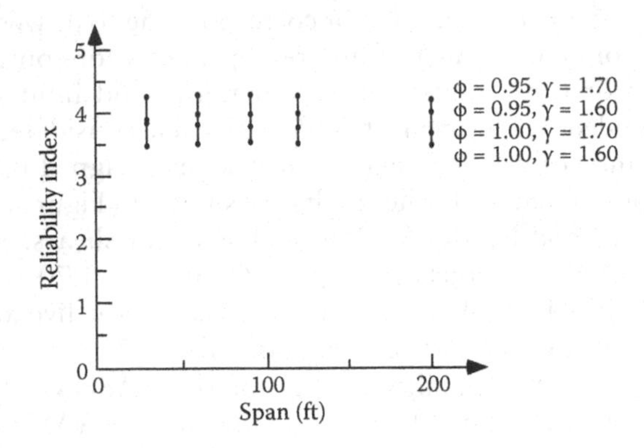

Figure 8.16 Reliability indices for prestressed concrete girders—bending limit state (new code).

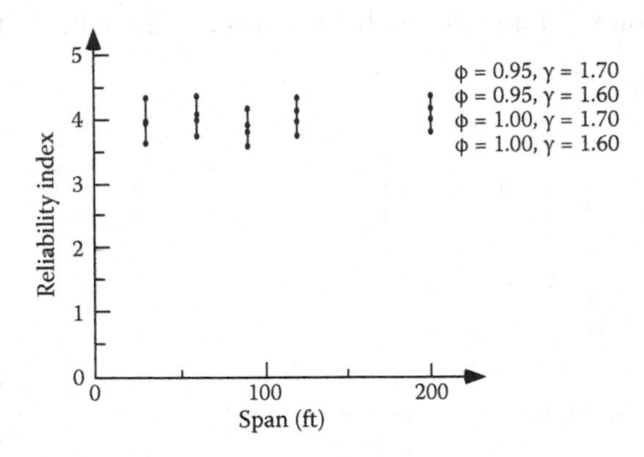

Figure 8.17 Reliability indices for steel girders—shear limit state (new code).

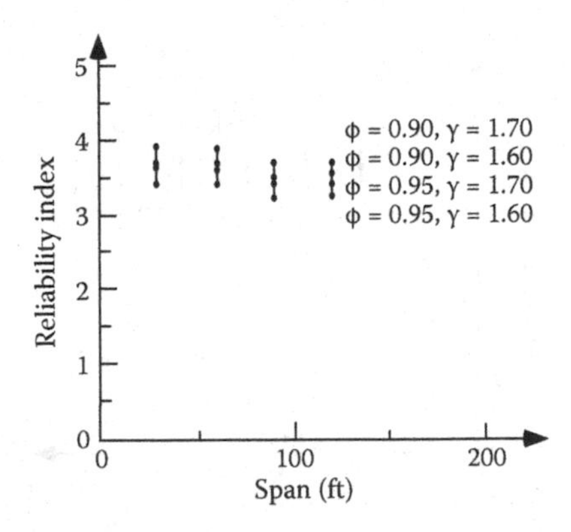

Figure 8.18 Reliability indices for reinforced concrete T-beams—shear limit state (new code).

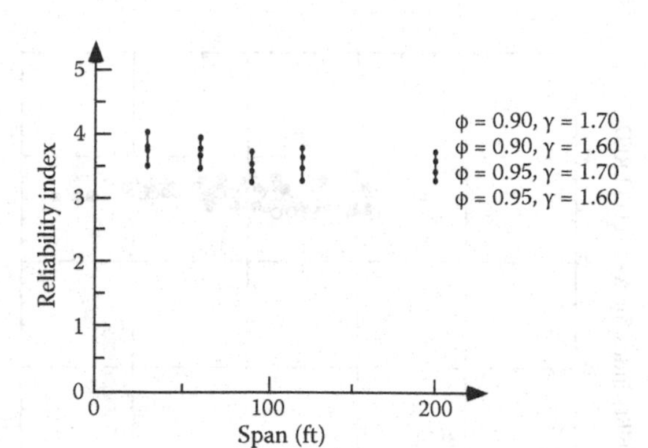

Figure 8.19 Reliability indices for prestressed concrete girders—shear limit state (new code).

Table 8.7 Calculated resistance factors

Material	*Moment*	*Shear*
Steel	1.00	1.00
Reinforced concrete	0.90	0.90
Prestressed concrete	1.00	0.90

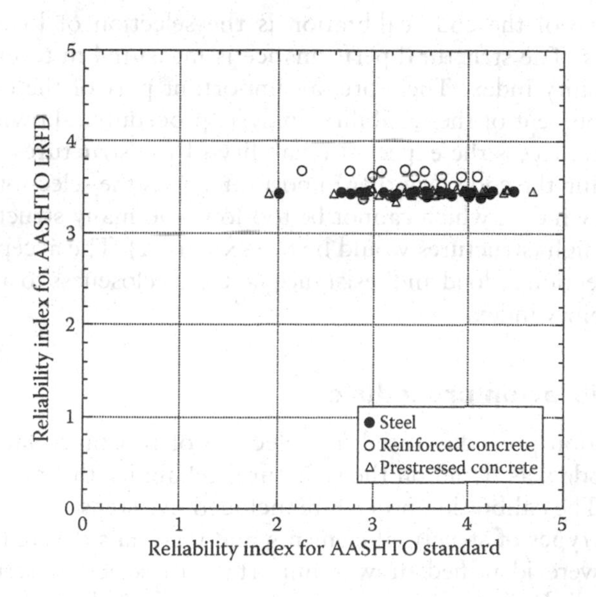

Figure 8.20 Comparison of reliability indices for AASHTO LRFD (2012) and AASHTO Standard (2002)—bending limit state.

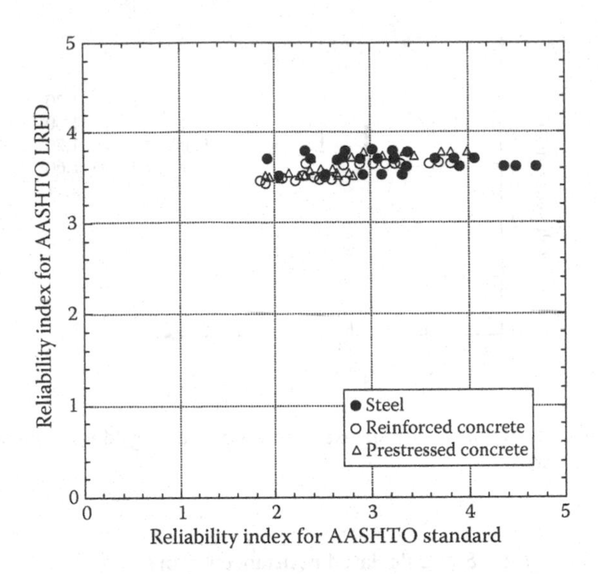

Figure 8.21 Comparison of reliability indices for AASHTO LRFD (2012) and AASHTO Standard (2002)—shear limit state.

8.7 EXAMPLE OF THE CODE CALIBRATION—ACI 318

8.7.1 Scope

The objective of the code calibration is the selection of load and resistance factors. The structural performance is measured in terms of the so-called reliability index. Therefore, an important part of the code writing is the development of the reliability analysis procedure, allowing the code committee to access the expected reliability of the structures that will be designed using the code. Another important step is the selection of the target reliability index, which cannot be too low (too many structures would fail) or too high (structures would be too expensive). The acceptability criterion in selection of load and resistance factors is closeness to a preselected target reliability index.

8.7.2 Calibration procedure

The calibration procedure used for selection of resistance factors for the ACI 318 Code was based on the structural reliability theory according to Chapter 5. The calibration procedure included five steps.

First, the types of structural elements and materials covered by the ACI 318 Code were identified. It was important to select the representative dimensions and reinforcement ratios for structural elements. Depending on the structural types, the typical load component ratios, $D/(D + L)$, were determined, where D = dead load effect and L = live load effect.

The second step was to select the statistical models for load components that are needed for reliability analysis. For each time-varying load, two load components were considered: an arbitrary point-in-time value and the maximum lifetime value. The models included the mean value and COV. It was assumed that the database for loads available in the literature was sufficient for the purpose of calibration. For load combinations, Turkstra's rule was applied (Turkstra, 1970; Turkstra and Madsen, 1980; Section 6.7.3).

The third step, selection of the resistance models, was one of the most important tasks in the calibration process. The available database for material properties was reviewed. The quality of materials (concrete and steel) had improved over the years and it was not reflected in the design formulas, particularly in the resistance factors.

The next step in the calibration process was focused on the reliability analysis procedures. Various available procedures were presented in Chapter 5. In this study, safety was measured in terms of the reliability index.

The developed procedure was applied to calculate the reliability indices for the considered structural types and materials, for various ratios of load components, designed exactly according to ACI 318-08 (no under-design or over-design). The analysis was performed for two sets of statistical models for resistance: (a) based on the material test data from the 1970s and early 1980s, and (b) based on new material test data provided by the industry.

The next important step in the calibration process was the selection of the target reliability index for all design cases, including all of the considered structural types and materials. The reliability indices corresponding to the current practice, as calculated in the previous step, were used as an important reference in the selection process. The target reliability level depends on consequences of failure and cost of increasing/decreasing the safety margin by a unit (marginal cost of safety).

The final step in this calibration was the calculation and selection of resistance factors for the considered limit states and the design cases corresponding to the load factors specified in the ASCE-7/98 Standard. The acceptance criterion was closeness to the target reliability index. The resistance factors were rounded to the nearest 0.05. To check the consistency of the results, reliability indices were calculated using the proposed resistance factors and the results were compared with the target values.

8.7.3 Reliability analysis

Load and resistance parameters are random variables; therefore, it is convenient to measure the structural performance in terms of the reliability index, β. Various procedures for calculation of β are presented in Chapter 5.

The reliability index, β, can be considered as a function of the probability of failure, P_F (Equation 5.15),

$$\beta = -\Phi^{-1}(P_F)$$

where Φ^{-1} is the inverse standard normal distribution function.

The reliability analysis procedure used in this calibration includes the following steps:

- Prepare input data:
  Structural type and limit state
  Nominal values of load components: D, L, S, W, and E
  Load and resistance factors: γ_D, γ_L, γ_S, γ_W, γ_E, ϕ
- Calculate load parameters: the mean total load, corresponding COV, and standard deviation.
- Calculate the nominal resistance.
- Determine the statistical parameters of R.
- Calculate the reliability index, β (Equation 5.14)

$$\beta = \frac{\mu_R - \mu_Q}{\sqrt{\sigma_R^2 + \sigma_Q^2}}$$

where μ_R is the mean value of resistance, μ_Q is the mean value of the total load effect, σ_R is the standard deviation of resistance, and σ_Q is the standard deviation of the total load effect.

8.7.4 Target reliability index

The optimum value of the target reliability index, β_T, can be determined based on two parameters: consequences of failure and incremental cost of safety (Madsen et al., 1986; Melchers, 1987; Section 8.4). In general, the larger is the expected cost of failure, the larger is β_T. However, β_T also depends on the financial considerations. If additional reliability can be achieved at a low cost, then β_T can be larger than otherwise acceptable minimum level, and if it is very costly to increase β_T, then even β_T that is lower than the otherwise required value can be acceptable. However, the selection of the optimum β_T requires a considerable database.

The code provisions in ACI 318 apply to various types of structural components, and it is assumed that β_T is to be selected for primary members (important components), and failure of a component can cause failure of other components. For secondary members, β_T can be reduced. However, there is practically no basis available to determine the consequences of failure and cost of safety for the structural components considered in this study. Because of insufficient input data, in this study, the target reliability levels were selected on the basis of the existing code (ACI 318-08).

The relationship between the element reliability index, β_e, and system reliability index, β_s, depends on the type of the system (parallel, series, or mixed) and the degree of correlation between the elements. In general, statically determinate structures can be treated as series systems, $\beta_s < \beta_e$, and parallel systems, $\beta_s > \beta_e$. The difference between β_s and β_e depends on the coefficient of correlation, ρ; β_s increases for increased ρ for series systems, and β_s decreases for increased ρ for parallel systems.

It was assumed that the reliability indices for components designed using the ACI 318-08 Code are acceptable. The Code provisions have been used for over 30 years. Therefore, the corresponding values of β were calculated using the "old material data." These reliability indices were assumed to be a lower limit for acceptable values of β_T. For each type of component, a large variation of β's as a function of load ratio is an indication that the load factors are not properly selected.

The target β is 3.5 for most of the components except for columns. Special consideration is required for slabs. In cast-in-place slabs, there is a considerable degree of load sharing, and the system reliability is much larger than β for a segment of 1 ft (1 m). This justifies a reduced value of $\beta_T = 2.5$. For precast slab panels, the degree of load sharing can be similar to beams.

8.7.5 Results of reliability analysis

The reliability indices corresponding to various categories of structural types and materials were reviewed and compared to the target values. For each case (structural type, material, and limit state), the resulting reliability indices were calculated; as an example, some results for $f_c' = 4000$ psi are presented in graphs on Figures 8.22 through 8.24.

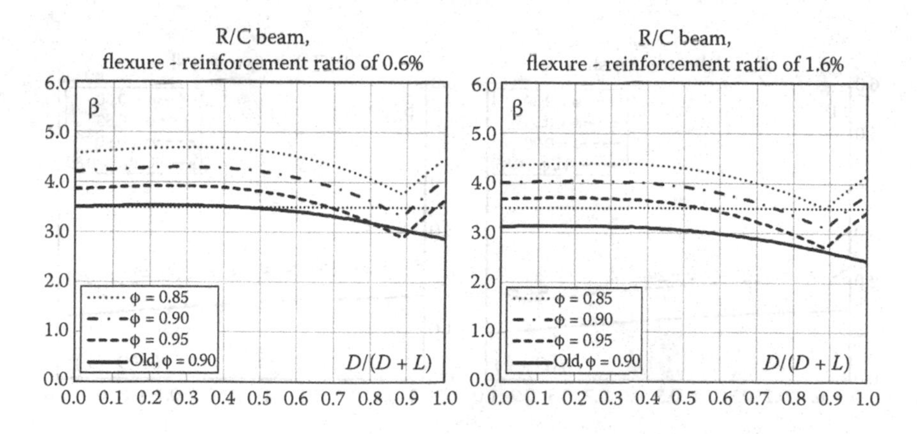

Figure 8.22 Reliability index vs. load ratio for R/C beam in flexure.

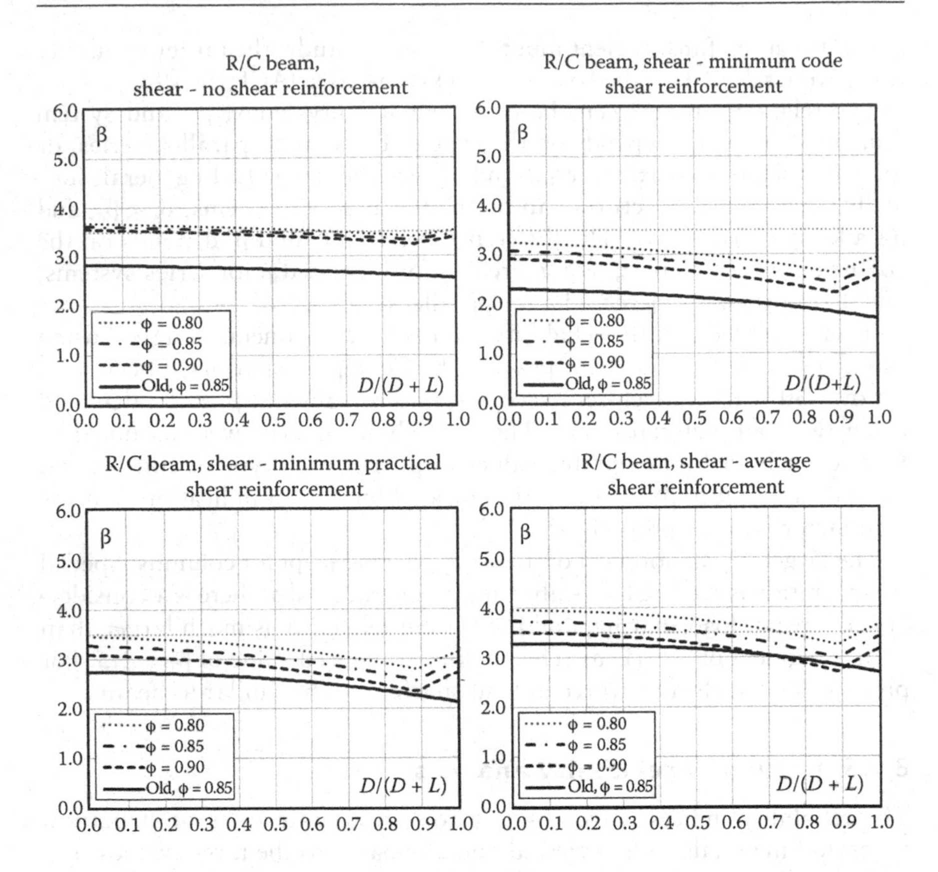

Figure 8.23 Reliability index vs. load ratio for R/C beam in shear.

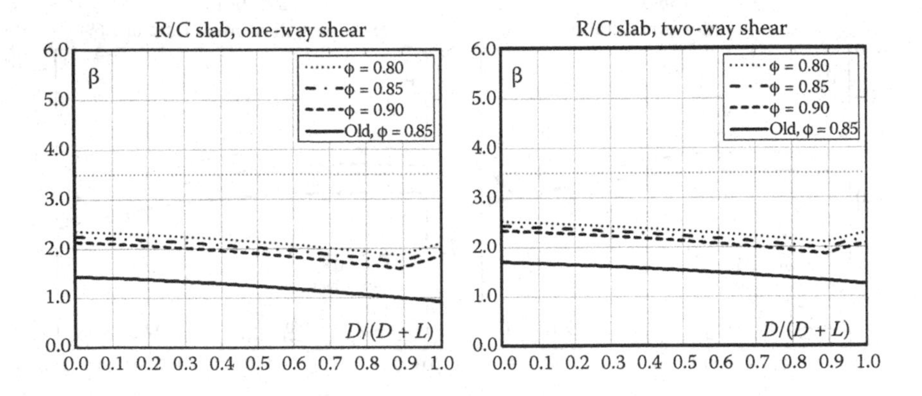

Figure 8.24 Reliability index vs. load ratio for R/C slab.

For the design cases using ACI 318-08 and using the proposed design formula, the calculations were carried out for several values of resistance factor, ϕ, to select one that provides the best fit to the target reliability index. The considered ϕ factors were rounded to 0.05. The resulting reliability indices are plotted for the recommended resistance factor, ϕ, and in addition for two other values of resistance factor, $\phi + 0.05$ and $\phi - 0.05$.

The reliability-based calibration performed for structural types and materials covered by the ACI 318-08 allows for increase of a resistance factor by about 10%.

8.8 CONCLUDING REMARKS

The theory of probability and statistics provides a convenient tool for the development of rational design and evaluation criteria. The major steps in the development of a code include the definition of the scope, code objective, frequency of demand, desired closeness to the target, and code format. The scope of the code should be clearly identified to avoid ambiguities and improper use. In current codes, the objective is to provide criteria that will result in designs with a reliability index as close to the target index as feasible. An analysis of the frequency of demand allows for the identification of the most important safety checks. The concept of a space metric (β-metric) serves as a measure of closeness to the target safety level. Usually, under-design is penalized more than over-design. Selection of the code format is very important from the user's point of view.

The basic approach has been demonstrated using examples of recently developed/updated codes for the design of highway bridges (AASHTO LRFD, 2012) and building structures (ACI 318). The new provisions are derived using the available statistical data on loads and resistance parameters. Load and resistance factors are calculated so that the reliability indices for bridges designed by the new code are close to the preselected target value. The probability-based approach results in a uniform safety level for various spans and materials.

PROBLEMS

8.1 Redo Example 8.3 assuming there is no correlation, and compare your answer to the results of Example 8.3.

8.2 Redo Example 8.3 for a target β of 3.

8.3 Redo Example 8.4 for a ratio of $\mu_L/\mu_D = 2$.

8.4 Redo Example 8.4 for a target β of 2.5.

8.5 Verify that the design equation derived in Example 8.3 does lead to a β value of 2.5. To do this, assume that the mean value of the design load is 10 kips. Solve the design equation for the required mean resistance. Then, use the methods presented in Chapter 5 to show that the resulting reliability index, β, is 2.5. Try other values of the design load if you wish.

8.6 Verify that the design equation derived in Example 8.4 does lead to a β value of 3. To do this, assume that the nominal dead load is 10.5 kips. Solve the design equation for the required nominal resistance noting that we assumed $\mu_L/\mu_D = 3$. Then, use the methods presented in Chapter 5 to show that the resulting reliability index, β, is 3.

8.7 Consider a design code for beams made of a new material. The test results show that the mean-to-nominal ratio for resistance is 1.10, and the COV of resistance is 12.5%. There are four load components: dead, live, snow, and wind. The parameters for these load components are given in Table P8.7a. The typical load proportions and frequencies are given in Table P8.7b. If the design formula is

$$1.2D + 1.6(L + S + W) < \phi R_n$$

and the target reliability index is 3.25, select the optimum value for ϕ.

Table P8.7a Parameters to consider for Problem 8.7

	Arbitrary point in time		*50-Year extreme*	
Load	*Bias*	*COV*	*Bias*	*COV*
D (dead)	1.05	0.1	1.05	0.1
L (live)	0.25	0.55	1.0	0.25
S (snow)	0.2	0.73	0.82	0.26
W (wind)	0.33	0.59	0.78	0.37

Table P8.7b Load proportions and frequencies for Problem 8.7

	% of total load			
Frequency	*Dead*	*Live*	*Snow*	*Wind*
Sometimes	30	50	10	10
Often	30	30	30	10
Often	30	30	10	30
Often	40	40	10	10
Sometimes	40	20	30	10
Sometimes	40	20	10	30
Seldom	50	50	0	0
Seldom	30	0	10	60

Hints: 1) Consider the following weighting system. An “often” event is weighted twice as much as a “sometimes” event. A “sometimes” event is weighted twice as much as “seldom” event; 2) Use Turkstra’s rule to determine the parameters for the combined load effect; and 3) The optimum value of ϕ is equal to $\Sigma\, w_i\varphi_i$, where w_i is the weight factor and ϕ_i is the resistance factor for case i.

Chapter 9

System reliability

In the preceding chapters, we have focused on evaluating the reliability of individual structural components. However, most structural systems consist of many interconnected structural components. Therefore, it is important to distinguish between the reliability of each component and the reliability of the entire system. This chapter provides an introduction to the topic of system reliability.

9.1 ELEMENTS AND SYSTEMS

In the preceding sections, we have focused almost exclusively on the calculation of the reliability (i.e., reliability index) of individual components or members of structures. However, most buildings and bridges consist of a system of interconnected components and members. Using the techniques presented earlier, we can design or analyze individual members in the context of structural reliability. However, we have not talked about how the system as a whole performs or how to calculate the reliability of the structure as a whole. When considering system reliability, it is important to recognize that the failure of a single component may or may not mean failure of the structure. (In the present context, the term "failure of a single component" implies the occurrence of a load effect that exceeds the capacity of the considered component.) Consequently, the reliability of an individual member may or may not be representative of the reliability of the entire structural system. The objective of this chapter is to present some techniques to determine (or at least estimate or bound) the probability of failure for a whole structure (system).

There are two extreme types of structural elements (components) that are commonly considered in system reliability analyses. These extreme types are *brittle* members and *ductile* members. A member is classified as brittle if the member becomes completely ineffective after it fails. Figure 9.1a shows the load–displacement curve of a brittle member. Examples of members that can be classified as brittle include unreinforced concrete

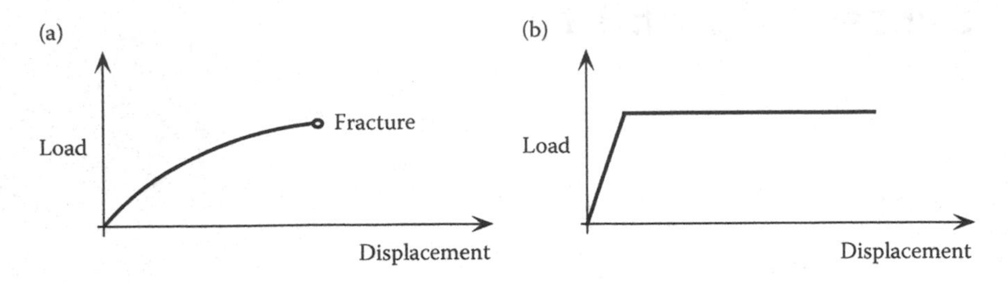

Figure 9.1 Load–displacement curves for brittle and ductile members. (a) Brittle member. (b) Ductile member.

Figure 9.2 Symbols used to distinguish brittle and ductile elements.

members in tension or a timber beam. Contrary to this, a ductile element is able to maintain its load-carrying capacity after it "fails." Figure 9.1b shows the load–displacement curve for a ductile material. In this case, the member "fails" once the yield point (kink) is reached, but the member can still sustain load for larger displacements. An example of a member with ductile behavior would be a member made of low-carbon steel.

In conducting reliability analyses, it is convenient to distinguish brittle and ductile members by using different symbols. The symbols shown in Figure 9.2 will be used in this chapter for this purpose.

9.2 SERIES AND PARALLEL SYSTEMS

To begin our study of system reliability, we will consider two idealized types of systems. In a *series* system, the failure of one member leads to immediate failure of the entire system. Figure 9.3a shows a length of chain as an example of a series system. When one link fails, the chain is useless in carrying load. In contrast, for a *parallel* system, all of the members must fail before the system fails. Figure 9.3b shows a cable supporting a weight. The cable is composed of several strands of wire, and failure of the cable requires failure of all the strands.

In reality, most structures cannot be classified as either series or parallel. They may not fail when a single member fails, but they can fail before all members fail.

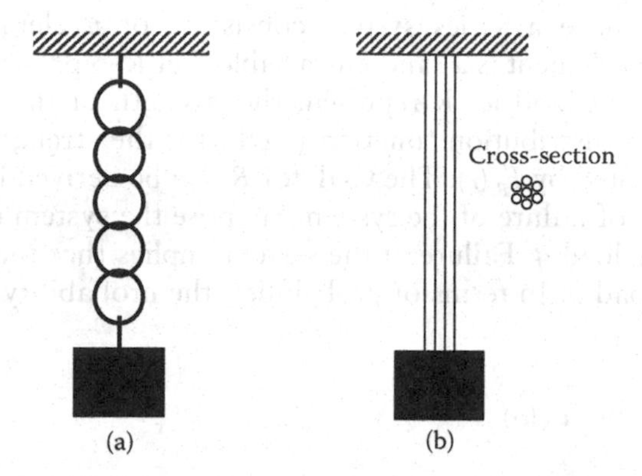

Figure 9.3 Two examples of structural systems. (a) Series system. (b) Parallel system.

9.2.1 Series systems

A series system is sometimes referred to as a weakest-link system because failure of the system corresponds to failure of the weakest element in the system. Figure 9.4 shows a statically determinate truss system that can be classified as a series system. If any one of the truss members fails, the truss will be unable to carry its load without collapsing.

Figure 9.5 shows some examples of series systems using the symbols for brittle and ductile materials discussed earlier. In the top case, each element of the system is a brittle element, and the failure of any one of these will lead to failure. In the bottom case, each element is a ductile element. If the elements follow the material behavior shown in Figure 9.1b, then the chain will undergo excessive deformation as soon as one of the elements yields. This would constitute failure of the system.

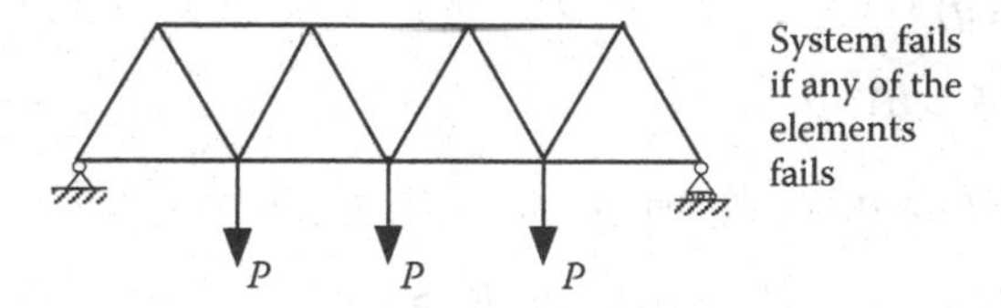

Figure 9.4 A statically determinate truss. This is an example of a series system.

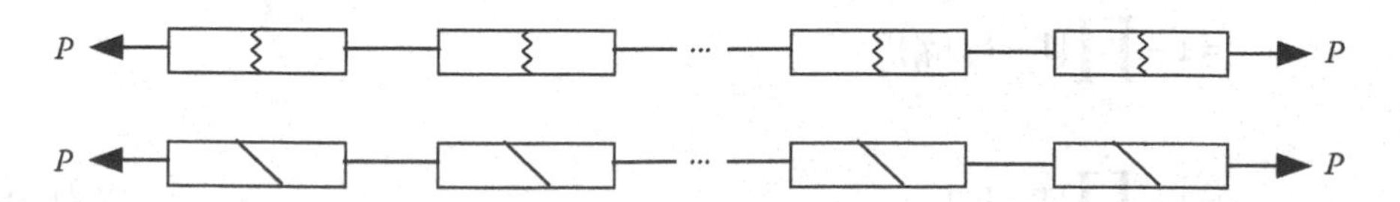

Figure 9.5 Examples of series systems using the symbols shown in Figure 9.2.

Suppose we have a series system consisting of n elements, and the strength of each element is a random variable. Let R_i represent the strength of the ith element, and let R represent the strength of the entire system. The cumulative distribution function (CDF) for the strength of each element is represented by $F_{R_i}(r)$. The CDF for R can be derived by considering the probability of failure of the system. Suppose the system is subjected to a *deterministic* load q. Failure of the system implies that the strength R is less than the load q. In terms of probability, the probability of failure, P_f, would be

$$P_f = P(R \le q) = F_R(q) \tag{9.1}$$

(Note that the use of ≤ instead of < is acceptable because we are assuming that R is a continuous random variable.)

Now, when the load q is applied to the system, the load q_i in each of the constituent members of the system depends on the geometry and layout of the system. For example, in the simple chain system shown in Figure 9.3a, the applied load $q = P$ would be the load applied to the system, and each element of the chain would be subjected to the same load $q_i = P$. In the case of the truss shown in Figure 9.4, the magnitude of the forces P could be considered as the load applied to the system. However, the load in each of the members would depend on the geometry of the truss and therefore would be different from member to member.

With this interpretation, and assuming that the strengths of the elements are all *statistically independent*, we can calculate the probability of failure in Equation 9.1 as follows:

$$\begin{aligned}
P_f &= F_R(q) \\
&= P(R \le q) \\
&= 1 - P(R > q) \\
&= 1 - P[(R_1 > q_1) \cap (R_2 > q_2) \cap \ldots \cap (R_n > q_n)] \\
&= 1 - P(R_1 > q_1)\, P(R_2 > q_2) \ldots\ldots P(R_n > q_n) \\
&= 1 - [1 - P(R_1 \le q_1)][1 - P(R_2 \le q_2)] \ldots\ldots [1 - P(R_n \le q_n)] \\
&= 1 - \prod_{i=1}^{n} [1 - F_{R_i}(q_i)] \\
&= 1 - \prod_{i=1}^{n} [1 - P_{f_i}]
\end{aligned} \tag{9.2}$$

In Equation 9.2, P_{f_i} is the probability of failure of the ith element. The multiplication of probabilities shown in the fifth line of the derivation above is permissible because we assumed that the strengths are independent; hence, the events $R_1 > q_1$, $R_2 > q_2$, etc. are also statistically independent for the series system. (Statistical independence is discussed in Chapter 2.)

Example 9.1

Consider the series system shown in Figure 9.6. It is a series system because the structure will collapse if either the beam or the cable fails.

Member AB is a steel beam (W10 × 54) with a yield stress $F_y = 36$ ksi and a plastic section modulus $Z = 66.6$ in^3. Member BC is a steel bar with diameter $D = 1$ in and an ultimate strength of $F_u = 58$ ksi. Assume that the force P is deterministic, and the dead load (self-weight) of the members can be neglected. Calculate the reliability of the system. The following statistical parameters are assumed for the resistance R of the members:

For R_{AB}: $\lambda_{AB} = 1.07$ $V_{AB} = 0.13$ R_{AB} is lognormal

For R_{BC}: $\lambda_{BC} = 1.14$ $V_{BC} = 0.14$ R_{BC} is normal

First, we need to determine the mean and standard deviation of the two resistances. The resistance of the beam is based on its nominal plastic moment capacity:

$$R_n \text{ for AB} = ZF_y = (66.6)(36) = 2398 \text{ kip-in}$$

Therefore, the mean value and standard deviation of the resistance of AB are

$$\mu_{R_{AB}} = \lambda_{AB} R_n = (1.07)(2398) = 2566 \text{ kip-in}$$

$$\sigma_{R_{AB}} = \mu_{R_{AB}} V_{AB} = (2566)(0.13) = 334 \text{ kip-in}$$

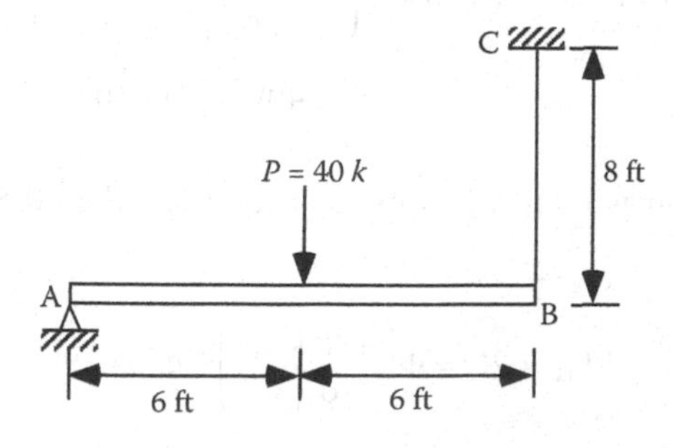

Figure 9.6 Two-element series system considered in Example 9.1.

Since R_{AB} follows a lognormal distribution, the relevant distribution parameters are $\mu_{\ln R}$ and $\sigma_{\ln R}$. Using Equations 2.48 and 2.49,

$$\sigma_{\ln R} = \sqrt{\ln\left(1+V_{AB}^2\right)} = 0.129$$

$$\mu_{\ln R} = \ln(\mu_{R_{AB}}) - 0.5\sigma_{\ln R}^2 = 7.84$$

For the steel bar, the nominal resistance is the cross-section area multiplied by the ultimate stress:

$$R_n \text{ for BC} = AF_u = \frac{\pi}{4}D^2F_u = \frac{\pi}{4}(1)^2(58) = 45.6 \text{ kips}$$

Therefore, the mean value and standard deviation of the resistance of BC are

$$\mu_{R_{BC}} = \lambda_{BC}R_n = (1.14)(45.6) = 52.0 \text{ kips}$$

$$\sigma_{R_{BC}} = \mu_{R_{BC}}V_{BC} = (52.0)(0.14) = 7.28 \text{ kips}$$

Now, we must evaluate how to define failure of the elements and the overall system. Based on a static analysis of the system, the vertical reaction at A and the force in the cable are both $0.5P$ = 20 kips. The beam AB will fail when the moment capacity is less than the maximum moment, which is

$$M_{max} = \frac{PL}{4} = \frac{(40 \text{ kips})(12 \text{ ft})}{4} = 120 \text{ kip-ft} = 1440 \text{ kip-in}$$

Therefore, the probability of failure of member AB is

$$P_f \text{ for AB} = P(R_{AB} < 1440) = \Phi\left(\frac{\ln(1440) - \mu_{\ln R}}{\sigma_{\ln R}}\right)$$

$$= \Phi(-4.40) = 5.41\times10^{-6}$$

The steel bar will fail when its capacity is less than $0.5P$ = 20 kips, i.e.,

$$P_f \text{ for BC} = P(R_{BC} < 20) = \Phi\left(\frac{20 - \mu_{R_{BC}}}{\sigma_{R_{BC}}}\right) = \Phi(-4.39)$$

$$= 5.67\times10^{-6}$$

The overall system will fail if either the beam or the bar fails. If we assume that the capacities of the beam and bar are independent, we can use Equation 9.2 to determine the probability of failure of the system.

$$P_f = 1 - \prod_{i=1}^{2}[1 - P_{f_i}]$$

$$= 1 - [1 - (5.41 \times 10^{-6})][1 - (5.67 \times 10^{-6})]$$

$$= 1.11 \times 10^{-5}$$

Observe that the probability of failure is larger than the probability of failure of either component.

In reliability studies, it is often more convenient to convert probabilities of failure to reliability indices using Equation 5.15 and then to compare the indices.

$$\beta_{AB} = -\Phi^{-1}(P_f \text{ for AB}) = -\Phi^{-1}(5.41 \times 10^{-6}) = 4.40$$

$$\beta_{BC} = -\Phi^{-1}(P_f \text{ for BC}) = -\Phi^{-1}(5.67 \times 10^{-6}) = 4.39$$

$$\beta_{system} = -\Phi^{-1}(P_f \text{ for system}) = -\Phi^{-1}(1.11 \times 10^{-5}) = 4.24$$

Observe that β_{AB} and β_{BC} are larger than the reliability index for the entire system. This is a common characteristic of a series system. The lower value of β_{system} tells us that the system is "less reliable" than either of its components.

Example 9.2

Consider a series system with an arbitrary number of elements. Let n denote the number of elements in the system. If the probability of failure for each element is 0.05, then the probability of failure for the system, P_f, is

$$P_f = 1 - \prod_{i=1}^{n}[1 - P_{f_i}]$$

$$= 1 - [1 - 0.05]^n$$

$$= 1 - [0.95]^n$$

The probabilities of failure for various values of n are given in Table 9.1. Also shown are some β values that are back-calculated from the

Table 9.1 Probabilities of failure for the series system in Example 9.2

n	P_f	$\beta = -\Phi^{-1}(P_f)$
1	0.05	1.64
2	0.0975	1.295
3	0.1426	1.07
5	0.2262	0.75
10	0.4013	0.25

probabilities of failure using Equation 5.15. Observe that the probability of failure increases (β decreases) as the number of elements increases.

9.2.2 Parallel systems

A parallel system can consist of ductile or brittle elements. Some examples of parallel systems are shown in Figure 9.7. These systems qualify as parallel systems because failure of all components is required for the overall system to fail.

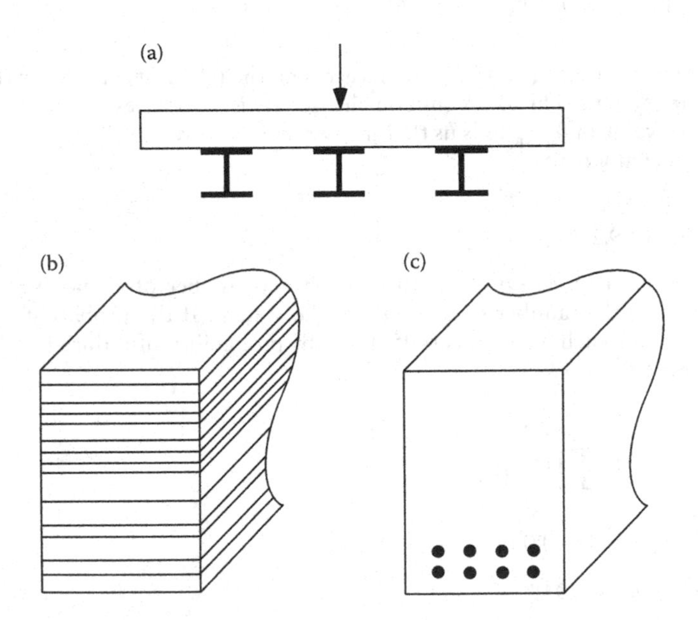

Figure 9.7 Some examples of parallel systems. (a) Three closely spaced beams supporting a slab. (b) A timber beam made of noncomposite boards stacked together. (c) Rebars in a reinforced concrete beam.

9.2.2.1 *Parallel systems with perfectly ductile elements*

A parallel system with *n perfectly ductile* elements is in a state of failure when all of its elements fail (i.e., yield). Let R_i represent the strength of the *i*th element in such a system. The system strength will be the sum of all the strengths of the elements, or

$$R = \sum_{i=1}^{n} R_i \tag{9.3}$$

If the strengths of the individual elements are all *uncorrelated* normal random variables, then the system strength is also normal with distribution parameters μ_R and σ_R^2 given by

$$\mu_R = \sum_{i=1}^{n} \mu_{R_i} \qquad \sigma_R^2 = \sum_{i=1}^{n} \sigma_{R_i}^2 \tag{9.4}$$

According to the central limit theorem, it is usually reasonable to assume R to be normally distributed even in cases where the R_i variables are non-normal provided that n is not too small.

The probability of failure of the system can be determined as follows. Assume a *deterministic* load q is acting on the system, and let q_i be the portion of the total load taken by the *i*th element. Then, failure of this element corresponds to the event $R_i < q_i$, and the probability of failure of the element is given by Equation 9.1. With this information, the probability of failure of the system can be determined as follows:

$$\begin{aligned} P_f &= P[(R_1 < q_1) \cap (R_2 < q_2) \cap \ldots \cap (R_n < q_n)] \\ &= P(R_1 < q_1)\, P(R_2 < q_2) \ldots\ldots P(R_n < q_n) \\ &= F_{R_1}(q_1)\, F_{R_2}(q_2) \ldots\ldots F_{R_n}(q_n) \\ &= \prod_{i=1}^{n} F_{R_i}(q_i) \\ &= \prod_{i=1}^{n} P_{f_i} \end{aligned} \tag{9.5}$$

The multiplication of probabilities is allowed because we have assumed that all strengths are independent *and* because we have neglected any possibility of load redistribution in the system once one or more elements start to yield. Thus, the failure events ($R_i < q_i$) are all independent.

Now, consider a special case of this same system of uncorrelated, perfectly ductile elements. If the R_i variables are identically distributed [meaning that they have the same probability density function (PDF) and CDF], then the mean and variance given in Equation 9.4 become

$$\mu_R = n\mu_{R_i} \qquad \sigma_R^2 = n\sigma_{R_i}^2 \tag{9.6}$$

The corresponding coefficient of variation (COV) would be

$$V_R = \frac{\sigma_R}{\mu_R} = \frac{\sqrt{n\sigma_{R_i}^2}}{n\mu_{R_i}} = \frac{1}{\sqrt{n}}\frac{\sigma_{R_i}}{\mu_{R_i}} = \frac{1}{\sqrt{n}}V_{R_i} \tag{9.7}$$

Thus, the COV for a system of n parallel, identically distributed elements is smaller than the COV of each element.

Example 9.3

Consider a parallel system consisting of two perfectly ductile elements as shown in Figure 9.8. The strengths of both elements are normally distributed with distribution parameters

$$\mu_{R_1} = \mu_{R_2} = 5\ \text{kN} \qquad V_{R_1} = V_{R_2} = 0.20$$

Determine the mean, standard deviation, and COV of the strength (resistance) of the combined system.

The standard deviation of the strength of each element is simply

$$\sigma_{R_1} = \mu_{R_1} V_{R_1} = 1\ \text{kN} = \sigma_{R_2}$$

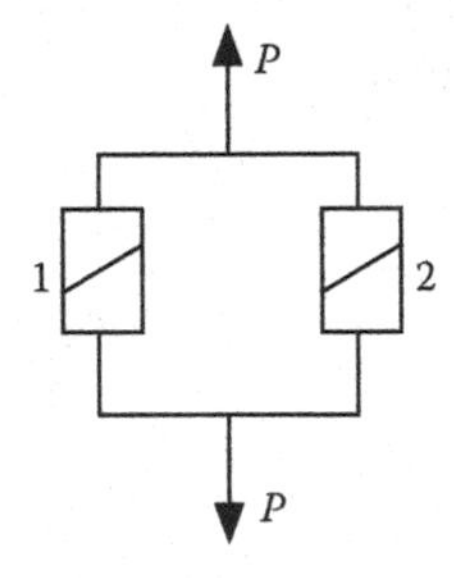

Figure 9.8 Parallel system considered in Example 9.3.

Since both elements are identically distributed, we can apply Equations 9.6 and 9.7 with $n = 2$. The result is

$$\mu_R = n\mu_{R_i} = 10 \text{ kN}$$

$$\sigma_R^2 = n\sigma_{R_i}^2 = 2 \text{ kN} \Rightarrow \sigma_R = \sqrt{2} \text{ kN}$$

$$V_R = \frac{1}{\sqrt{n}} V_{R_i} = \frac{1}{\sqrt{2}}(0.20) = 0.14$$

Also, we can conclude that the strength of the system is normally distributed since the total strength is the sum of two independent normal random variables.

9.2.2.2 Parallel systems with brittle elements

Consider a parallel system consisting of *n perfectly brittle* elements. Such a system is illustrated in Figure 9.9. For this type of system, if one of the brittle elements fails, then it loses its capacity to carry load. The load must be redistributed to the remaining $(n - 1)$ elements. If, after the load is redistributed, the system does not fail, the load can be increased until the next element fails. Then, the load is redistributed among the remaining $(n - 2)$ elements. This process of failure and load redistribution continues until overall failure of the system.

Let $R_1, R_2, \ldots, R_n$ represent the strengths of elements 1, 2, ..., *n*. Furthermore, assume that the strengths are ordered such that $R_1 < R_2 < \ldots < R_n$. With these assumptions, the strength of the system, *R*, is

$$R = \max\,[nR_1, (n-1)R_2, (n-2)R_3, \ldots, 2R_{n-1}, R_n] \tag{9.8}$$

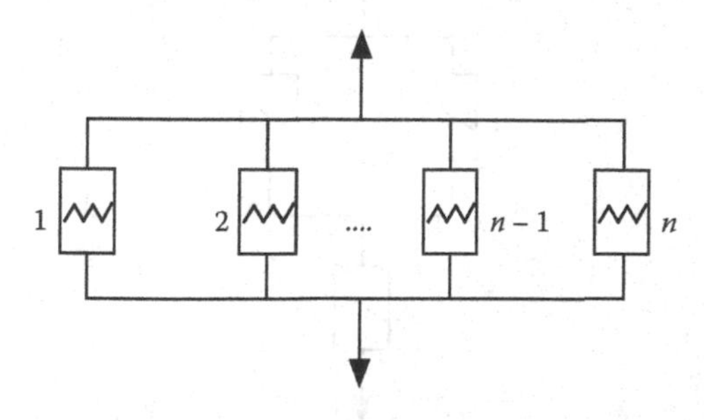

Figure 9.9 Parallel system consisting of brittle elements.

9.2.3 Hybrid (combined) systems

Many actual structures can be considered as a combination of series and parallel systems. Such systems are referred to as hybrid or combined systems. A schematic example of a hybrid system is shown in Figure 9.10. In that example system, elements 1 and 2 are in parallel, and the combination of 1 and 2 is in series with element 3. Many simple hybrid systems can be analyzed using the techniques discussed in Sections 9.2.1 and 9.2.2. This is demonstrated in the following example.

Example 9.4

Consider the steel frame structure shown in Figure 9.11. The column supports are fixed, and the beam-column joints are modeled as pinned connections. The applied loading consists of a vertical force S (snow) and a horizontal force W (wind). To simplify the problem, assume that the compression capacity of columns AB and CD is considerably larger than the loads.

Assume that the loads are deterministic, but the member resistances are random variables. The statistical parameters of the resistances are given below in Table 9.2.

Determine the required design (nominal) value of resistance for elements AB, BC, and CD so that the system reliability is $\beta_s = 4.5$.

For a system reliability analysis, the system can be represented as a hybrid system as shown in Figure 9.11b. The overall system will fail if the moment-carrying capacities of both columns are exceeded or if the moment-carrying capacity of the beam is exceeded. Thus, the two columns behave as a parallel system, and this system is in series with the horizontal beam.

Let P_{f_s} represent the probability of failure of the overall system, and let $P_{f_{AB}}$, $P_{f_{BC}}$, and $P_{f_{CD}}$ represent the probabilities of failure of elements AB, BC, and CD. For combined systems such as this one, it

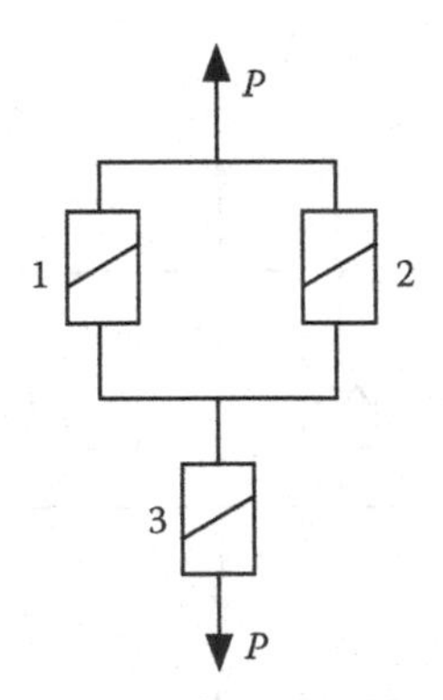

Figure 9.10 An example of a hybrid (combined) system.

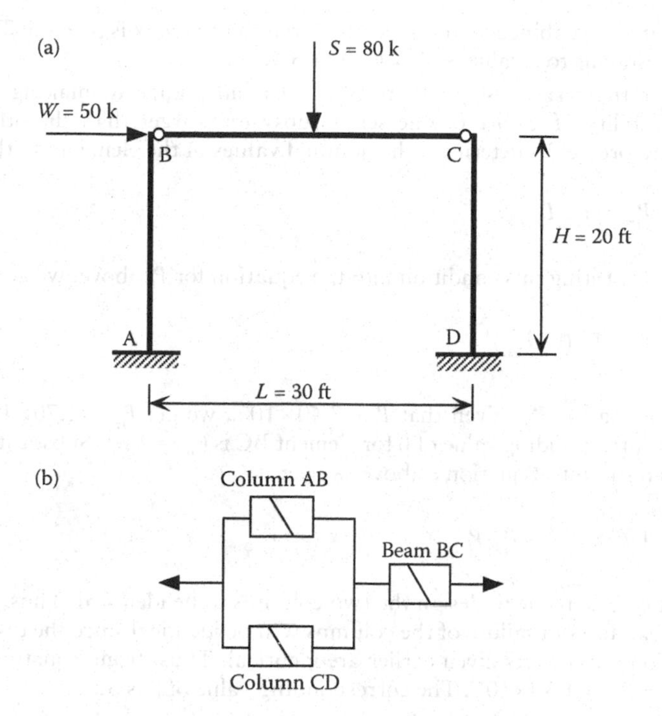

Figure 9.11 Frame structure considered in Example 9.4. (a) Actual frame with loading. (b) System reliability analysis model.

is convenient to break up the problem into subsystems and then to analyze the subsystems. Columns AB and CD behave as a parallel subsystem; hence, the probability of failure of the subsystem is, according to Equation 9.5,

$$P_{f_{AB}} P_{f_{CD}} \qquad \text{(a)}$$

That subsystem forms a series system with element BC; hence, the probability of failure of the entire system is, according to Equation 9.2,

$$P_{f_s} = 1 - (1 - P_{f_{BC}})(1 - P_{f_{AB}} P_{f_{CD}}) \qquad \text{(b)}$$

Table 9.2 Statistical parameters for members in the frame shown in Figure 9.11

Element	*Bias factor*	*COV*
AB	1.10	0.125
BC	1.08	0.14
CD	1.10	0.125

For the combined system, the target reliability index is $\beta_s = 4.5$. This corresponds to a value of $P_{f_s} = \Phi(-\beta) = 3.40\times10^{-6}$.

For the series system, there is no real advantage to making the probability of failure of one series subsystem larger than the other. Therefore, let us determine the nominal values of the elements so that

$$P_{f\mathrm{BC}} = P_{f\mathrm{AB}} P_{f\mathrm{CD}} \tag{c}$$

Substituting this condition into the equation for P_{f_s} above, we get

$$P_{f_s} = 1-(1-P_{f\mathrm{BC}})^2 \tag{d}$$

Solving for $P_{f\mathrm{BC}}$ given that $P_{f_s} = 3.40\times10^{-6}$, we get $P_{f\mathrm{BC}} = 1.70\times10^{-6}$. The corresponding value of β for element BC is $\beta_{\mathrm{BC}} \approx 4.65$. Substituting this result into Equation c above, we get

$$1.70\times10^{-6} = P_{f\mathrm{AB}} P_{f\mathrm{CD}} \tag{e}$$

It is reasonable to design the two columns to be identical. Thus, the probabilities of failure of the columns will be identical since the distribution parameters given earlier are identical. Thus, from Equation e, $P_{f\mathrm{AB}} = P_{f\mathrm{CD}} = 1.30\times10^{-3}$. The corresponding value of β is 3.01.

The above calculations have given us the target β values for each member. Now, we must determine the corresponding nominal design values needed to achieve those reliability indices. The first step is to determine the limit state function for each element. For each column, the limit state function is

$$g(R) = R - \left(\frac{W}{2}H\right) = R - 500 \text{ kip-ft} \tag{f}$$

where R is the moment capacity of either AB or CD and $WH/2$ is the moment "demand" at the base of each column. For element BC, the limit state function is

$$g(R_{\mathrm{BC}}) = R_{\mathrm{BC}} - \left(\frac{SL}{4}\right) = R_{\mathrm{BC}} - 600 \text{ kip-ft} \tag{g}$$

Observe that both limit state functions are linear equations. Therefore, we can use Equation 5.18 to determine the reliability index for each element. For column AB,

$$\beta_{\mathrm{AB}} = \frac{-500+\mu_{R_{\mathrm{AB}}}}{\sqrt{\sigma^2_{R_{\mathrm{AB}}}}} = \frac{-500+\lambda_{R_{\mathrm{AB}}}R_n}{\sqrt{(V_{R_{\mathrm{AB}}}\lambda_{R_{\mathrm{AB}}}R_n)^2}} = \frac{-500+\lambda_{R_{\mathrm{AB}}}R_n}{V_{R_{\mathrm{AB}}}\lambda_{R_{\mathrm{AB}}}R_n} = 3.01 \tag{h}$$

where R_n is the nominal resistance of column AB. Substituting the given values of the bias factor and COV and solving Equation h for R_n gives

$$R_n \text{ for AB} = 729 \text{ kip-ft} = R_n \text{ for CD} \tag{i}$$

Similarly, for element BC,

$$\beta_{BC} = \frac{-600 + \mu_{R_{BC}}}{\sqrt{\sigma^2_{R_{BC}}}} = \frac{-600 + \lambda_{R_{BC}} R_n}{V_{R_{BC}} \lambda_{R_{BC}} R_n} = 4.65 \tag{j}$$

which leads to

$$R_n \text{ for BC} = 1592 \text{ kip-ft} \tag{k}$$

9.3 RELIABILITY BOUNDS FOR STRUCTURAL SYSTEMS

9.3.1 Boolean variables

In system reliability, it is often convenient to work with variables known as Boolean variables. For example, consider a system with n elements. In our analysis, it is assumed that each element can exist in only one of two states:

- failure
- non-failure (safe)

To describe these two states, we can use two Boolean variables, S_i and F_i, which are defined as follows for the ith element:

$$S_i = \begin{cases} 1 & \text{if the element is in a non-failure state} \\ 0 & \text{if the element is in a failure state} \end{cases} \tag{9.9}$$

$$F_i = 1 - S_i = \begin{cases} 0 & \text{if the element is in a non-failure state} \\ 1 & \text{if the element is in a failure state} \end{cases} \tag{9.10}$$

Based on the states of the elements, we can define the state of the system using Boolean variables also. For convenience, let $\bar{S}$ and $\bar{F}$ be vectors containing the values of S_i and F_i, respectively, for all elements in the system. In other words,

$$\bar{S} = \{S_1, S_2, \ldots, S_n\} \tag{9.11a}$$

$$\bar{F} = \{F_1, F_2, \ldots, F_n\} \tag{9.11b}$$

We can define a system function, $S_S(\bar{S})$ (or $F_S(\bar{F})$), which indicates the state of the entire system. The function is defined as

$$S_S(\bar{S}) = \begin{cases} 1 & \text{if the } \textit{system} \text{ is in a non-failure state} \\ 0 & \text{if the } \textit{system} \text{ is in a failure state} \end{cases} \tag{9.12a}$$

or

$$F_S(\bar{F}) = 1 - S_S(\bar{S}) = \begin{cases} 0 & \text{if the } \textit{system} \text{ is in a non-failure state} \\ 1 & \text{if the } \textit{system} \text{ is in a failure state} \end{cases} \tag{9.12b}$$

Now, let us see how systems can be described by these Boolean variables. For a series system, failure of one element means system failure. Therefore, the system function can be expressed as

$$S_S(\bar{S}) = S_1 S_2 \ldots S_n = \prod_{i=1}^{n} S_i \tag{9.13}$$

If the ith element is in a failure state, then $S_i = 0$ and hence the product in Equation 9.13 is equal to zero. If none of the elements has failed, then all of the S_i values are equal to 1 and hence the product is equal to 1.

Now, consider a parallel system with perfectly ductile elements. If at least one element is in a non-failure state, then the system is in a non-failure state. Therefore, the corresponding system function can be written as

$$S_S(\bar{S}) = 1 - \prod_{i=1}^{n} (1 - S_i) \tag{9.14}$$

Hence, if one element is in a non-failure state, then $(1 - S_i) = 0$ for that element and the product in Equation 9.13 will be equal to zero. The resulting function will be equal to 1, which it should be since the system is in a non-failure state.

Using Boolean variables, we can also calculate expected values and probabilities. Consider the ith element in a system. If the state of the element is random, then the Boolean variable S_i is a discrete random variable. There are only two possible values (1 and 0), and there is a probability associated with each value. Thus, the expected value of S_i can be calculated using the formula presented in Chapter 2 for a discrete random variable, i.e.,

$$E(S_i) = (1)[P(S_i = 1)] + (0)\ [P(S_i = 0)] = P(S_i = 1) \tag{9.15}$$

Note that $P(S_i = 1)$ is the same as $P(F_i = 0)$. Also, note that $P(F_i = 0) + P(F_i = 1) = 1$ because the sum of all the probabilities over the entire range of possible discrete values must be equal to 1. Substituting these results into Equation 9.15, we get

$$E(S_i) = P(S_i = 1) = 1 - P(F_i = 1) \tag{9.16}$$

Similarly, we can determine the expected value of F_i as follows:

$$E(F_i) = (1)[P(F_i = 1)] + (0)[P(F_i = 0)] = P(F_i = 1) \tag{9.17}$$

Finally, the probability of failure of the entire system, P_f, can be found using the same approach.

$$\begin{aligned} E[F_S(\bar{F})] &= (1)P[F_S(\bar{F}) = 1] + (0)P[F_S(\bar{F}) = 0] \\ &= P[F_S(\bar{F}) = 1] \\ &= P_f \end{aligned} \tag{9.18}$$

9.3.2 Series systems with positive correlation

In the earlier discussion of series systems, it was assumed that the elements were all independent and hence uncorrelated. When the elements are uncorrelated, the analysis of the system reliability is relatively straightforward. If correlation exists among some or all of the elements, the exact calculation of the probability of failure of the system is typically quite difficult if not impossible. However, simple bounds can be derived for a series system with *positive* correlation between pairs of elements. In other words, these bounds apply when the correlation coefficient ρ_{ij} is greater than or equal to zero.

For series systems with positive correlation, the probability of failure must satisfy

$$\max_i \{P[F_i = 1]\} \leq P_f \leq 1 - \prod_{i=1}^{n} (1 - P[F_i = 1]) \tag{9.19}$$

The lower bound is the probability of failure when all elements are fully correlated ($\rho_{ij} = 1$). If they are fully correlated, then all elements will tend to fail when one fails; hence, the probability of failure will correspond to the largest probability of failure among the constituent elements. The upper bound is the probability of failure when all elements are uncorrelated. The derivation of these bounds can be found in Ang and Tang (1984) and Thoft-Christensen and Baker (1982).

Example 9.5

Consider two perfectly ductile elements of a series system as shown in Figure 9.12. Assume that the reliability index for each element (1 and 2) is $\beta_e = 3.5$. The strengths of both elements are normally distributed. Determine the upper and lower bounds on the failure probability of the system.

The lower bound is the maximum failure probability of the individual elements. Since the two elements are identically distributed, the lower bound is simply

$$\max_i\{P[F_i = 1]\} = P(F_1 = 1) = P(F_2 = 1) = \Phi(-\beta_e)$$

$$= \Phi(-3.5) = 0.000233$$

For the upper bound, we use the value of $P(F_i = 1)$ just computed and substitute into Equation 9.19.

$$1 - \prod_{i=1}^{n}(1 - P[F_i = 1]) = 1 - (1 - 0.000233)^2 = 0.000466$$

The exact probability of failure of the system is between these bounds.

$$0.000233 \le P_f \le 0.000466$$

The exact value of P_f depends on the coefficient of correlation, ρ.

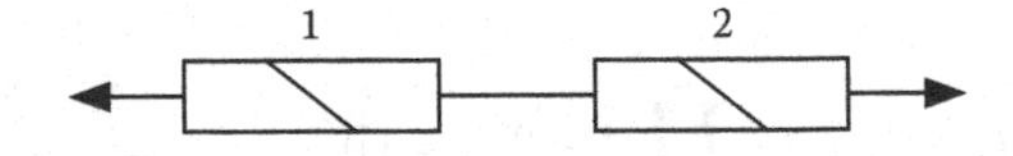

Figure 9.12 Series system considered in Example 9.5.

9.3.3 Parallel systems with positive correlation

Analogous to the case of a series system with positive correlation, we can determine bounds on the failure probability of a parallel system by considering the extreme cases of perfect correlation and no correlation. Consider a parallel system with n ductile elements such as the one shown in Figure 9.13. The bounds on the probability of failure for a parallel system with positive correlation are as follows:

$$\prod_{i=1}^{n} P[F_i = 1] \;\le\; P_f \;\le\; \min_i\{P[F_i = 1]\} \tag{9.20}$$

The lower bound represents the case where the elements are all uncorrelated and the system fails only if all the elements fail. On the other hand, the upper bound represents the case in which all elements are perfectly correlated. In this case, the safest element determines the reliability of the system.

Example 9.6

Consider the structural system shown in Figure 9.14. Let the reliability indices for bars AB and CD be $\beta_{AB} = 3.00$ and $\beta_{CD} = 3.25$. The system fails (i.e., cannot carry any additional load) if both members yield. Therefore, the system can be considered a parallel system. The probability of failure can be approximated by calculating the lower and upper bounds using Equation 9.20.

The upper bound is based on the smallest failure probability among all the elements. In this case, member CD has the largest reliability index and hence the smallest failure probability.

$$\min_i\{P[F_i = 1]\} = P(F_{CD} = 1) = \Phi(-3.25) = 0.000577$$

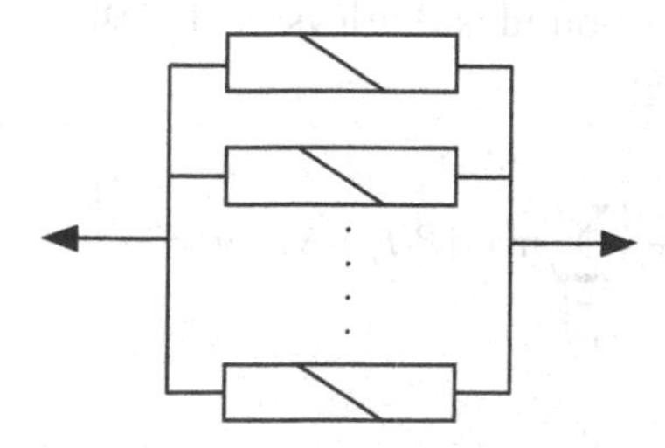

Figure 9.13 Parallel system with n ductile elements.

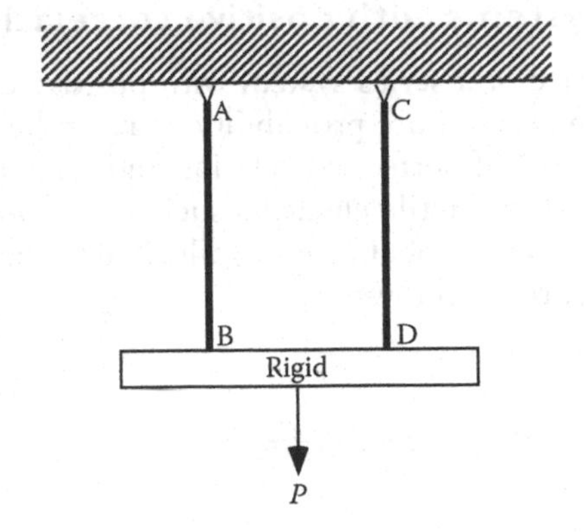

Figure 9.14 System considered in Example 9.6.

The lower bound is based on the product of the failure probabilities, or

$$\prod_{i=1}^{n} P[F_i = 1] = \Phi(-3.00)\Phi(-3.25) = (0.00135)(0.000577)$$

$$= 7.79 \times 10^{-7}$$

For $0 \le \rho \le 1$, the exact probability of failure must be between these bounds.

$$7.79 \times 10^{-7} \le P_f \le 5.77 \times 10^{-4}$$

9.3.4 Ditlevsen bounds for a series system

For many practical cases, the bounds derived earlier for a series system are too far apart to be really useful. An alternative set of bounds, known as the Ditlevsen bounds, tend to be more narrow.

The Ditlevsen upper bound is (Ditlevsen, 1979)

$$P_f \le \sum_{i=1}^{n} P(F_i = 1) - \left\{ \sum_{\substack{i=2 \\ j<i}}^{n} \max[P(F_i = 1 \cap F_j = 1)] \right\} \tag{9.21}$$

and the Ditlevsen lower bound is

$$P_f \geq P(F_1 = 1) + \max\left\langle \sum_{i=2}^{n} \left\{ P(F_i = 1) - \sum_{j=1}^{i-1} P(F_i = 1 \cap F_j = 1) \right\},\ 0 \right\rangle \tag{9.22}$$

Two comments are in order here. First, the numbering scheme of the elements can actually affect the Ditlevsen bounds. Therefore, one needs to consider many different numbering schemes to obtain the best (narrowest) bounds. Second, these bounds utilize joint probabilities that may be difficult to calculate.

9.4 SYSTEMS WITH EQUALLY CORRELATED ELEMENTS

In the previous section, we discussed how to calculate bounds for the reliability of systems using the extreme cases of uncorrelated ($\rho = 0$) and fully correlated ($\rho = 1$) elements. There is another special case that can be considered involving systems in which the correlation coefficient is the same for all pairs of elements and its value is somewhere between 0 and 1. This section discusses how the probability of failure of such systems can be calculated.

Consider a system with n elements. The strength of the ith element ($i = 1, 2, \ldots, n$) will be denoted by R_i. To be able to calculate the failure probability exactly, we must make the following assumptions:

- The strengths of the elements are all normally distributed.
- The strengths are all equally correlated, and the correlation coefficient is ρ.
- All applied loads are deterministic and constant in time.
- All elements are designed so that they have the same reliability index β_e.

9.4.1 Series systems with equally correlated elements

For a series system composed of n elements, Stuart (1958) derived the following formula for the probability of failure:

$$P_f = 1 - \int_{-\infty}^{\infty} \left\{ \Phi\left(\frac{\beta_e + t\sqrt{\rho}}{\sqrt{1-\rho}} \right) \right\}^n \phi(t)\, dt \tag{9.23}$$

where β_e is the reliability index for each element, $\Phi()$ and $\phi()$ are the standard normal CDF and PDF, respectively, and ρ is the correlation coefficient that is common among all pairs of elements. Values of P_f based on Equation 9.23 for various values of n, ρ, and β_e are tabulated in Tables 9.3 through 9.6. (These values were generated by evaluating Equation 9.23 numerically using Simpson's rule with 10,000 intervals between −500 and 500.) For $\beta_e = 3.00$, the probability of failure is plotted in Figure 9.15 as a function of the value of the coefficient of correlation, ρ, and the number of elements, n.

Once the probability of failure is known, the corresponding reliability index for the system can be determined using Equation 5.15. Figure 9.16 provides a comparison of how β for the system varies with the correlation coefficient ρ and the number of elements. Two values of β_e are considered ($\beta_e = 2$ and $\beta_e = 3$).

Example 9.7

Calculate the probability of failure for the truss shown in Figure 9.17. Assume $\beta_e = 3.0$ for all elements and $\rho = 0.85$ between elements.

The truss is statically determinate. Therefore, failure of one member will lead to failure of the structure. This allows us to represent the truss as a series system with equally correlated elements. Therefore, using Equation 9.23, the probability of failure is

$$P_f = 1 - \int_{-\infty}^{\infty} \left\{ \Phi\left(\frac{\beta_e + t\sqrt{\rho}}{\sqrt{1-\rho}} \right) \right\}^3 \phi(t)\, dt$$

where the exponent is equal to 3 since there are three members. Using Table 9.4, we find

$$P_f = 29.05 \times 10^{-4}$$

What is the probability of failure for each element? It is

$$(p_f)_{\text{element}} = \Phi(-\beta_e) = \Phi(-3) = 13.50 \times 10^{-4}$$

Table 9.3 Values of P_f (after multiplying by 10^{-3}) based on Equation 9.23 for various values of n and ρ when $\beta_e = 2$

ρ	n								
	2	3	4	5	6	7	8	9	10
0.05	44.82	66.24	87.03	107.2	126.8	145.9	164.4	182.3	199.8
0.1	44.63	65.69	85.97	105.5	124.4	142.6	160.1	177.1	193.5
0.15	44.40	65.04	84.74	103.6	121.6	138.9	155.5	171.5	186.8
0.2	44.13	64.29	83.36	101.4	118.6	135.0	150.6	165.5	179.8
0.25	43.82	63.44	81.81	99.07	115.4	130.8	145.4	159.3	172.5
0.3	43.46	62.49	80.10	96.50	111.8	126.3	139.9	152.7	165.0
0.35	43.05	61.42	78.22	93.72	108.1	121.5	134.1	146.0	157.2
0.4	42.58	60.23	76.17	90.73	104.1	116.6	128.2	139.1	149.3
0.45	42.05	58.91	73.95	87.54	99.97	111.4	122.1	132.0	141.3
0.5	41.45	57.47	71.55	84.15	95.58	106.0	115.7	124.7	133.1
0.55	40.77	55.87	68.96	80.55	90.97	100.5	109.2	117.2	124.7
0.6	40.00	54.12	66.17	76.72	86.13	94.64	102.4	109.6	116.2
0.65	39.13	52.19	63.15	72.64	81.04	88.58	95.43	101.7	107.5
0.7	38.14	50.06	59.89	68.30	75.67	82.24	88.19	93.62	98.62
0.75	37.00	47.69	56.33	63.63	69.97	75.58	80.63	85.22	89.43
0.8	35.68	45.02	52.41	58.57	63.86	68.51	72.67	76.43	79.86
0.85	34.10	41.95	48.02	52.99	57.22	60.89	64.16	67.09	69.76
0.9	32.14	38.30	42.92	46.64	49.75	52.43	54.79	56.90	58.80
0.95	29.48	33.58	36.54	38.86	40.76	42.38	43.79	45.03	46.14

Table 9.4 Values of P_f (after multiplying by 10^{-4}) based on Equation 9.23 for various values of n and ρ when $\beta_e = 3$

ρ	n								
	2	3	4	5	6	7	8	9	10
0.05	26.97	40.41	53.81	67.19	80.54	93.86	107.1	120.4	133.6
0.1	26.95	40.35	53.70	67.01	80.27	93.48	106.6	119.8	132.8
0.15	26.92	40.27	53.54	66.75	79.87	92.93	105.9	118.8	131.7
0.2	26.88	40.16	53.32	66.38	79.33	92.19	104.9	117.6	130.2
0.25	26.83	40.00	53.02	65.89	78.61	91.20	103.7	116.0	128.2
0.3	26.76	39.80	52.62	65.24	77.68	89.93	102.0	113.9	125.7
0.35	26.66	39.53	52.10	64.42	76.49	88.34	99.96	111.4	122.6
0.4	26.54	39.18	51.45	63.39	75.03	86.39	97.49	108.3	119.0
0.45	26.38	38.74	50.64	62.13	73.26	84.05	94.55	104.8	114.7
0.5	26.18	38.19	49.64	60.61	71.14	81.30	91.12	100.6	109.9
0.55	25.92	37.52	48.44	58.79	68.66	78.11	87.19	95.93	104.4
0.6	25.60	36.69	46.99	56.66	65.78	74.45	82.72	90.65	98.25
0.65	25.20	35.69	45.28	54.16	62.47	70.30	77.71	84.76	91.50
0.7	24.70	34.48	43.26	51.28	58.69	65.61	72.11	78.25	84.09
0.75	24.07	33.01	40.87	47.94	54.39	60.34	65.89	71.10	76.02
0.8	23.28	31.24	38.05	44.07	49.49	54.44	59.00	63.25	67.23
0.85	22.26	29.05	34.69	39.56	43.88	47.77	51.32	54.59	57.63
0.9	20.89	26.27	30.58	34.19	37.33	40.11	42.62	44.90	47.00
0.95	18.91	22.49	25.20	27.40	29.26	30.87	32.29	33.56	34.72

Table 9.5 Values of P_f (after multiplying by 10^{-6}) based on Equation 9.23 for various values of n and ρ when $\beta_e = 4$

	n								
ρ	2	3	4	5	6	7	8	9	10
0.05	63.34	95.01	126.7	158.3	190.0	221.6	253.3	285.0	316.6
0.1	63.34	95.00	126.7	158.3	190.0	221.6	253.2	284.9	316.5
0.15	63.33	94.98	126.6	158.3	189.9	221.5	253.1	284.7	316.2
0.2	63.32	94.95	126.6	158.2	189.7	221.3	252.8	284.3	315.8
0.25	63.30	94.90	126.5	158.0	189.5	220.9	252.3	283.7	315.0
0.3	63.27	94.81	126.3	157.7	189.0	220.3	251.5	282.7	313.8
0.35	63.23	94.67	126.0	157.2	188.3	219.4	250.3	281.1	311.8
0.4	63.15	94.44	125.6	156.5	187.3	217.9	248.4	278.7	308.8
0.45	63.03	94.10	124.9	155.4	185.7	215.8	245.6	275.2	304.6
0.5	62.86	93.59	123.9	153.9	183.5	212.7	241.7	270.3	298.6
0.55	62.59	92.86	122.5	151.7	180.3	208.5	236.3	263.7	290.7
0.6	62.22	91.81	120.6	148.6	176.0	202.9	229.1	254.9	280.2
0.65	61.67	90.35	117.9	144.6	170.4	195.4	219.8	243.6	266.9
0.7	60.90	88.33	114.3	139.1	162.9	185.9	208.0	229.5	250.3
0.75	59.82	85.57	109.5	132.1	153.4	173.8	193.3	212.0	230.1
0.8	58.27	81.82	103.2	122.9	141.4	158.7	175.2	190.9	205.8
0.85	56.07	76.68	94.83	111.2	126.2	140.1	153.1	165.4	177.0
0.9	52.80	69.51	83.62	95.95	107.0	117.0	126.2	134.8	142.8
0.95	47.54	58.85	67.80	75.28	81.74	87.46	92.59	97.27	101.6

Table 9.6 Values of P_f (after multiplying by 10^{-8}) based on Equation 9.23 for various values of n and ρ when $\beta_e = 5$

	n								
ρ	2	3	4	5	6	7	8	9	10
0.05	57.33	86.00	114.7	143.3	172.0	200.7	229.3	258.0	286.7
0.1	57.33	86.00	114.7	143.3	172.0	200.7	229.3	258.0	286.6
0.15	57.33	85.99	114.7	143.3	172.0	200.7	229.3	258.0	286.6
0.2	57.33	85.99	114.7	143.3	172.0	200.6	229.3	258.0	286.6
0.25	57.33	85.99	114.6	143.3	172.0	200.6	229.3	257.9	286.6
0.3	57.33	85.98	114.6	143.3	171.9	200.6	229.2	257.8	286.5
0.35	57.32	85.97	114.6	143.2	171.8	200.4	229.0	257.6	286.2
0.4	57.31	85.93	114.5	143.1	171.7	200.2	228.7	257.2	285.7
0.45	57.29	85.87	114.4	142.9	171.4	199.8	228.2	256.5	284.8
0.5	57.25	85.75	114.2	142.5	170.8	199.0	227.2	255.2	283.2
0.55	57.18	85.54	113.8	141.9	169.8	197.7	225.4	253.0	280.5
0.6	57.05	85.18	113.1	140.7	168.2	195.5	222.5	249.4	276.2
0.65	56.84	84.57	111.9	138.9	165.6	191.9	218.0	243.8	269.3
0.7	56.48	83.58	110.1	136.0	161.5	186.5	211.1	235.3	259.2
0.75	55.88	81.98	107.2	131.5	155.2	178.3	200.9	223.0	244.6
0.8	54.91	79.45	102.7	124.8	146.1	166.6	186.4	205.6	224.2
0.85	53.31	75.48	95.88	114.9	132.9	149.9	166.1	181.7	196.7
0.9	50.61	69.17	85.54	100.3	113.9	126.6	138.4	149.6	160.2
0.95	45.68	58.59	69.22	78.36	86.43	93.69	100.3	106.4	112.1

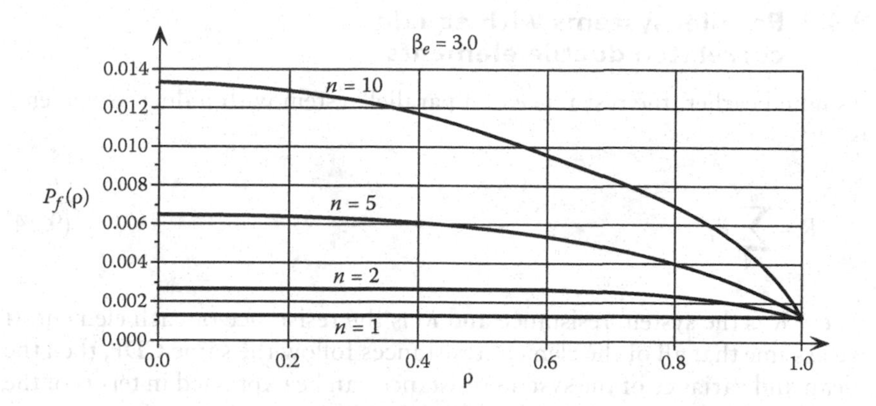

Figure 9.15 Probability of failure for series systems with equally correlated elements based on Equation 9.23. $\beta_e = 3.0$.

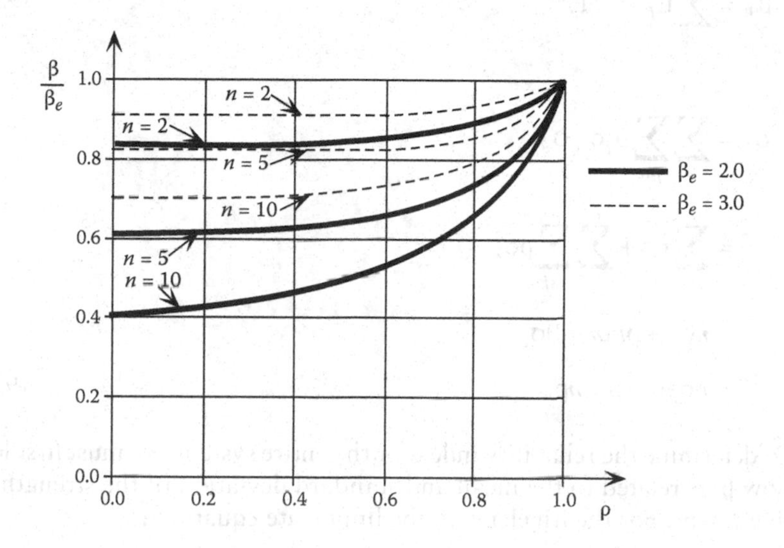

Figure 9.16 Reliability indices for a series system with equally correlated elements.

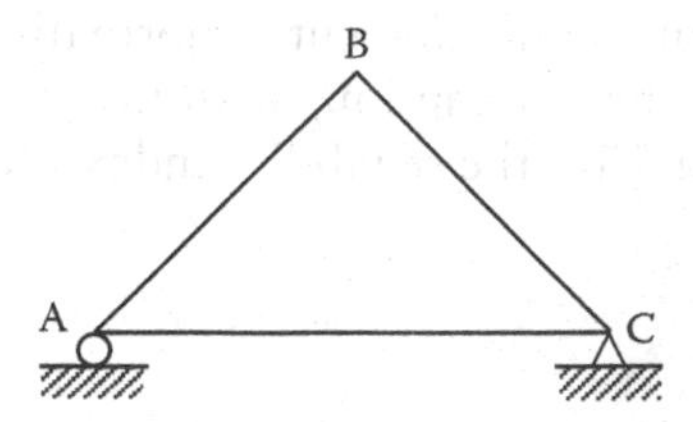

Figure 9.17 Truss considered in Example 9.7.

9.4.2 Parallel systems with equally correlated ductile elements

As noted earlier, the resistance of a parallel system with n ductile elements is

$$R = \sum_{i=1}^{n} R_i \tag{9.24}$$

where R is the system resistance and R_i is the resistance of each element. If we assume that all of the element resistances follow the same CDF, then the mean and variance of the system resistance can be expressed in terms of the element parameters μ_e and σ_e^2 as follows:

$$\mu_R = \sum_{i=1}^{n} \mu_{R_i} = n\mu_e \tag{9.25}$$

$$\begin{aligned} \sigma_R^2 &= \sum_{i=1}^{n}\sum_{j=1}^{n} \rho_{ij}\sigma_{R_i}\sigma_{R_j} \\ &= \sum_{i=1}^{n} \sigma_e^2 + \sum^{n}\sum^{n}_{i \neq j} \rho\sigma_e^2 \\ &= n\sigma_e^2 + \rho n(n-1)\sigma_e^2 \\ &= n\sigma_e^2(1-\rho+n\rho) \end{aligned} \tag{9.26}$$

To determine the reliability index for the entire system, we must first look at how β_e is related to the mean and standard deviation of the strength for each element. For the ith element, the limit state equation is

$$g(R_i) = R_i - q_i \tag{9.27}$$

where q_i is the load effect in the ith member for the loading condition considered. The strength of the ith element is normally distributed (as are all elements according to our assumption) with mean μ_e and standard deviation σ_e. Using Equation 5.18, the reliability index for the element is

$$\beta_e = \frac{\mu_e - q_i}{\sigma_e} \tag{9.28}$$

If we solve Equation 9.28 for q_i, we get

$$q_i = \mu_e - \beta_e\sigma_e \tag{9.29}$$

Since μ_e, β_e, and σ_e are the same for all elements, Equation 9.29 requires q_i to be the same for all elements. Thus, $q_{tot} = nq_i$. Using Equation 9.29, this becomes

$$q_{tot} = n\mu_e - n\beta_e\sigma_e \tag{9.30}$$

Now, the limit state equation for the entire system is

$$g(R) = R - q_{tot} \tag{9.31}$$

and the reliability index for the system is

$$\beta_{system} = \frac{\mu_R - q_{tot}}{\sigma_R} \tag{9.32}$$

Substituting Equations 9.25, 9.26, and 9.30 into Equation 9.32, we get the following relationship between β_{system}, β_e, and ρ for a parallel system with equally correlated ductile elements:

$$\begin{aligned}
\beta_{system} &= \frac{n\mu_e - (n\mu_e - n\beta_e\sigma_e)}{\sqrt{n\sigma_e^2(1-\rho+n\rho)}} \\
&= \frac{n\beta_e\sigma_e}{\sigma_e\sqrt{n(1-\rho+n\rho)}} \\
&= \beta_e\frac{\sqrt{n^2}}{\sqrt{n(1-\rho+n\rho)}} \\
&= \beta_e\sqrt{\frac{n}{(1-\rho+n\rho)}}
\end{aligned} \tag{9.33}$$

Figures 9.18 and 9.19 illustrate how the probability of failure of the system varies with the value of the correlation coefficient and the number of elements. In Figure 9.18, the probabilities of failure are calculated assuming an element reliability index $\beta_e = 3.0$.

Example 9.8

Consider the structure shown in Figure 9.20a. Each component (member AB and member BC) is built up of two angles as shown in Figure 9.20b. This structure can be modeled as a combination of series and parallel subsystems. Each component can be considered as a parallel subsystem with two elements each. The two components constitute a series system. The combined hybrid system model is shown in Figure 9.21.

It is assumed that the reliability indices for components AB and BC are the same; the value of the index will be referred to as β_c. Furthermore, it is assumed that the reliability indices for the angles AB_1, AB_2, BC_1, and BC_2 are all equal to β_e. The strengths of the angles

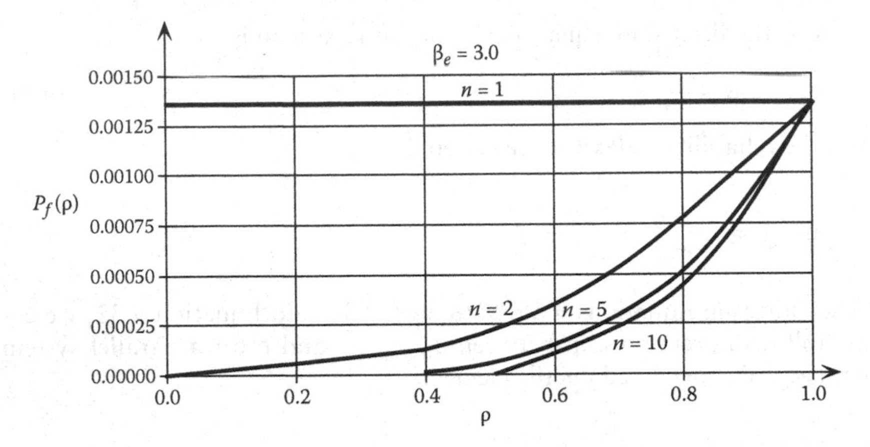

Figure 9.18 Probability of failure for a parallel system with equally correlated elements (β_e = 3.0).

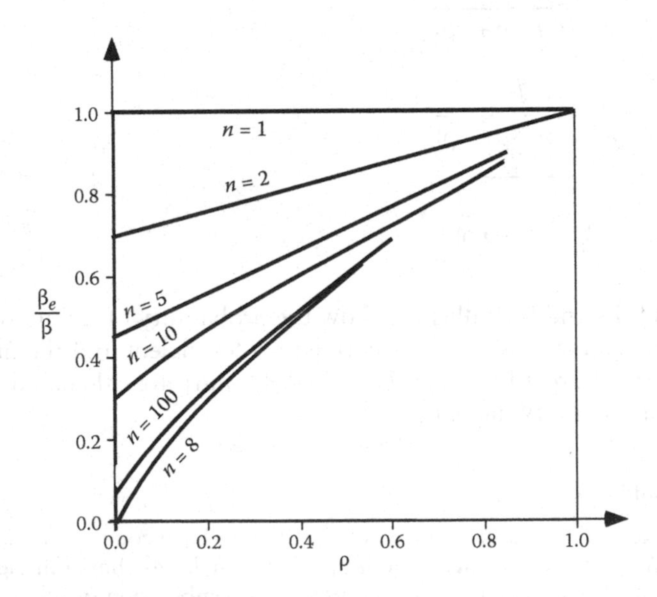

Figure 9.19 Reliability indices for parallel system with equally correlated elements.

are assumed to be equally correlated with the correlation coefficient equal to ρ_e. Determine the probability of failure of the system.

A single component (AB or BC) is considered first. Since the two angles in each component form a parallel system with equally correlated elements, the reliability index for the component (β_c) is calculated using Equation 9.33 with $n = 2$:

$$\beta_c = \beta_e \sqrt{\frac{n}{(1-\rho_e + n\rho_e)}} = \beta_e \sqrt{\frac{2}{1+\rho_e}} \tag{a}$$

The probability of failure of the whole truss, P_f, can be calculated using Equation 9.23 for $n = 2$. However, note that the "element" index β_e in Equation 9.23 corresponds to the component index β_c for this problem. Furthermore, the correlation coefficient used must be the correlation coefficient for the components, ρ_c. With these substitutions, Equation 9.23 becomes

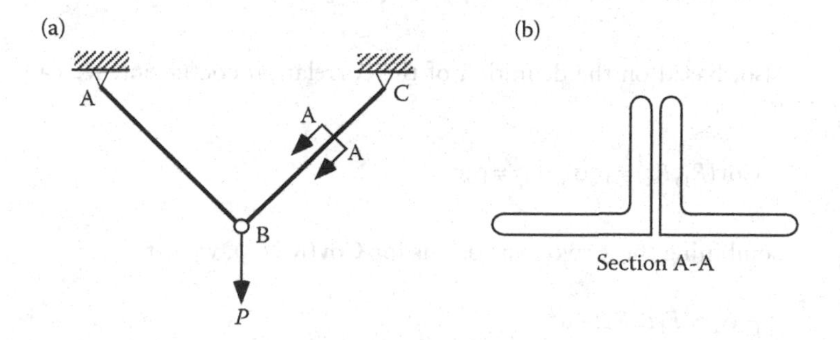

Figure 9.20 Structure considered in Example 9.8. (a) Overall structural system. (b) Cross-section of members.

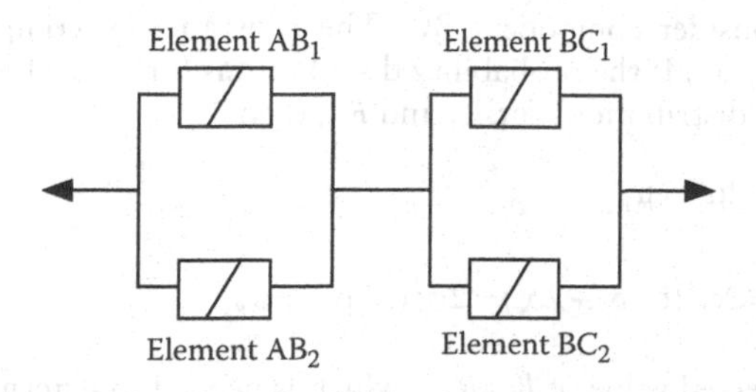

Figure 9.21 Reliability model for the system considered in Example 9.8.

$$P_f = 1 - \int_{-\infty}^{\infty} \left\{ \Phi\left(\frac{\beta_c + t\sqrt{\rho_c}}{\sqrt{1-\rho_c}} \right) \right\}^2 \phi(t)\, dt \tag{b}$$

To evaluate this integral, we need to determine ρ_c. As we will show next, it is NOT equal to ρ_e.

Let R_1 and R_2 represent the strengths of the angles in component AB. The strength of component AB, R_{AB}, is equal to $R_1 + R_2$. Based on Equations 9.25 and 9.26, the mean and variance of R_{AB} are

$$\mu_{R_{AB}} = 2\mu_e$$

$$\sigma^2_{R_{AB}} = 2\sigma^2_e(1 - \rho_e + 2\rho_e) = 2\sigma^2_e(1 + \rho_e)$$

Now, we need to do some creative mathematics. First, using the fundamental definition of covariance discussed in Chapter 2, we can write

$$\text{Cov}(R_1, R_2) = E[R_1 R_2] - \mu_{R_1}\mu_{R_2} = E[R_1 R_2] - \mu^2_e$$

Also, based on the definition of the correlation coefficient, we can write

$$\text{Cov}(R_1, R_2) = \rho_e \sigma_{R_1} \sigma_{R_2} = \rho_e \sigma^2_e$$

Combining these two expressions for $\text{Cov}(R_1,R_2)$, we get

$$\rho_e \sigma^2_e = E[R_1 R_2] - \mu^2_e$$

Rearranging this expression, we get

$$E[R_1 R_2] = \mu^2_e + \rho_e \sigma^2_e$$

Now, consider component BC. The resistance of component BC, R_{BC}, is $R_3 + R_4$. If the probability distributions for R_3 and R_4 are identical to the distributions for R_1 and R_2, then

$$\mu_{R_{BC}} = 2\mu_e = \mu_{R_{AB}}$$

$$\sigma^2_{R_{BC}} = 2\sigma^2_e(1 - \rho_e + 2\rho_e) = 2\sigma^2_e(1 + \rho_e) = \sigma^2_{R_{AB}}$$

The expected value of $R_{AB}R_{BC}$, which is needed to determine ρ_c, is

$$E[R_{AB}R_{BC}] = E[(R_1 + R_2)(R_3 + R_4)] = 4E[R_1R_2]$$

since all resistances are assumed to be identically distributed. Substituting our earlier result for $E[R_1R_2]$, we get

$$E[R_{AB}R_{BC}] = 4\left[\mu_e^2 + \rho_e\sigma_e^2\right]$$

Now, we can determine ρ_c by applying the basic definition:

$$\rho_c = \frac{\mathrm{Cov}(R_{AB}, R_{BC})}{\sigma_{R_{AB}}\sigma_{R_{BC}}}$$

$$= \frac{E[R_{AB}R_{BC}] - \mu_{R_{AB}}\mu_{R_{BC}}}{\sigma_{R_{AB}}\sigma_{R_{BC}}}$$

$$= \frac{4\left[\mu_e^2 + \rho_e\sigma_e^2\right] - (2\mu_e)^2}{\left\{\sqrt{2\sigma_e^2(1+\rho_e)}\right\}^2}$$

$$= \frac{2\rho_e\sigma_e^2}{\left[\sigma_e^2(1+\rho_e)\right]}$$

$$= \frac{2\rho_e}{1+\rho_e} \qquad \text{(c)}$$

The probability of failure of the system can be determined by substituting Equations a and c into Equation b above. Given specific information on the element reliability index and the element correlation coefficient, we can evaluate this integral. For example, let $\rho_e = 0.8$ and $\beta_e = 2.85$. Then,

$$\beta_c = \beta_e\sqrt{\frac{2}{1+\rho_e}} = 2.85\sqrt{\frac{2}{1+0.8}} = 3.00$$

$$\rho_c = \frac{2\rho_e}{1+\rho_e} = \frac{2(0.8)}{1+0.8} = 0.889$$

Entering Table 9.4 (since $\beta_c = 3$) with $n = 2$ and $\rho = \rho_c = 0.889$, we find (by interpolation)

$$P_f = 21.19 \times 10^{-4}$$

and the reliability index is

$$\beta_{\text{system}} = -\Phi^{-1}(P_f) = 2.86$$

9.5 SYSTEMS WITH UNEQUALLY CORRELATED ELEMENTS

In the previous section, we considered the special case of a system in which all elements were equally correlated. In real structures, this may or may not be the case. In general, we may need to consider a system in which the elements are unequally correlated. This is the subject of this section.

We will make the following assumptions:

- The strengths of the elements are all *normally distributed* with identical distribution parameters μ_e and σ_e.
- All applied loads are deterministic and constant in time.
- All elements are designed so that they have the same reliability index β_e.

9.5.1 Parallel system with ductile elements

Consider a parallel system with n perfectly ductile elements. The strength of the ith element ($i = 1, 2, \ldots, n$) will be denoted by R_i. The strength of the entire system is R. The correlation matrix describing the correlations between elements is

$$[\rho] = \begin{bmatrix} 1 & \rho_{12} & \cdots & \rho_{1n} \\ \rho_{21} & 1 & & \rho_{2n} \\ \vdots & & \ddots & \\ \rho_{n1} & \rho_{n2} & \cdots & 1 \end{bmatrix} \tag{9.34}$$

The reliability index for the system, β_{system}, is

$$\beta_{\text{system}} = \frac{\mu_R - q_{\text{tot}}}{\sigma_R} \tag{9.35}$$

where

$$\mu_R = n\mu_e \tag{9.36}$$

and

$$\sigma_R^2 = \sum_{i=1}^{n}\sum_{j=1}^{n}\rho_{ij}\sigma_{R_i}\sigma_{R_j}$$

$$= \sum_{i=1}^{n}\sigma_e^2 + \sum^{n}\sum_{i\neq j}^{n}\rho_{ij}\sigma_e^2$$

$$= n\sigma_e^2 + \sigma_e^2\left(\sum^{n}\sum_{i\neq j}^{n}\rho_{ij}\right)$$

$$= \sigma_e^2\left[n + \sum^{n}\sum_{i\neq j}^{n}\rho_{ij}\right] \tag{9.37}$$

Substituting these results into Equation 9.35, and also using Equation 9.30, which assumes that all elements are subjected to the same load, the system reliability can be found to be

$$\beta_{\text{system}} = \frac{n\mu_e - (n\mu_e - n\beta_e\sigma_e)}{\sqrt{\sigma_e^2\left[n + \sum^{n}\sum_{i\neq j}^{n}\rho_{ij}\right]}}$$

$$= \frac{n\beta_e\sigma_e}{\sqrt{\sigma_e^2\left[n + \sum^{n}\sum_{i\neq j}^{n}\rho_{ij}\right]}}$$

$$= \beta_e\sqrt{\frac{n^2}{n + \sum^{n}\sum_{i\neq j}^{n}\rho_{ij}}}$$

$$= \beta_e\sqrt{\frac{n}{1 + \frac{1}{n}\sum^{n}\sum_{i\neq j}^{n}\rho_{ij}}} \tag{9.38}$$

Now, define an "average correlation" coefficient, $\bar{\rho}$, as

$$\bar{\rho} = \frac{1}{n(n-1)} \sum^{n} \sum^{n}_{i \neq j} \rho_{ij} \tag{9.39}$$

Combining Equation 9.39 with Equation 9.38, we get

$$\beta_{\text{system}} = \beta_e \sqrt{\frac{n}{1+(n-1)\bar{\rho}}} \tag{9.40}$$

Observe that Equation 9.40 has the same form as Equation 9.33 except for the different correlation coefficient.

Example 9.9

Calculate the system reliability index for the parallel system shown in Figure 9.22. The element reliability index for all elements is $\beta_e = 4.0$. The correlation matrix is given as

$$[\rho] = \begin{bmatrix} 1 & 0.4 & 0.2 & 0.1 \\ 0.4 & 1 & 0.2 & 0.3 \\ 0.2 & 0.2 & 1 & 0.5 \\ 0.1 & 0.3 & 0.5 & 1 \end{bmatrix}$$

First, we must calculate the average correlation coefficient $\bar{\rho}$:

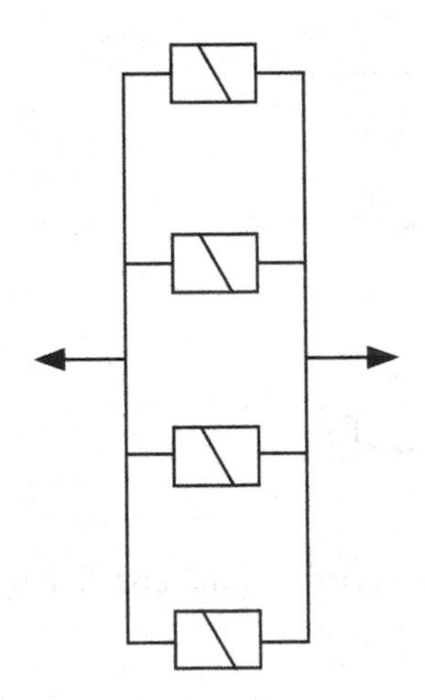

Figure 9.22 Parallel system considered in Example 9.9.

$$\bar{\rho} = \frac{1}{n(n-1)} \sum_{i \neq j}^{n} \sum^{n} \rho_{ij}$$

$$= \frac{1}{4(3)} [0.4 + 0.2 + 0.1 + 0.4 + 0.2 + 0.3 + 0.2 +$$

$$0.2 + 0.5 + 0.1 + 0.3 + 0.5]$$

$$= \frac{3.4}{12}$$

$$= 0.283$$

Substituting into Equation 9.40, we get

$$\beta_{\text{system}} = \beta_e \sqrt{\frac{n}{1 + (n-1)\bar{\rho}}}$$

$$= 4.0 \sqrt{\frac{4}{1 + (3)(0.283)}}$$

$$= 5.88$$

The corresponding probability of failure is 2.05×10^{-9}.

Before leaving this example, let us consider two extreme cases. First, assume that all ρ_{ij} are zero. This would correspond to a parallel system composed of elements with uncorrelated strengths. Then, $\bar{\rho} = 0$ and hence

$$\beta_{\text{system}} = \beta_e \sqrt{\frac{n}{1 + (n-1)\bar{\rho}}}$$

$$= 4.0 \sqrt{\frac{4}{1 + (3)(0)}}$$

$$= 8.0$$

The corresponding probability of failure is $\Phi(-8) = 6.22 \times 10^{-16}$. Now, consider the case in which the strengths of all elements are perfectly correlated ($\rho_{ij} = 1$). Then, $\bar{\rho} = 1$ and hence

$$\beta_{\text{system}} = \beta_e \sqrt{\frac{n}{1 + (n-1)\bar{\rho}}}$$

$$= 4.0 \sqrt{\frac{4}{1 + (3)(1)}}$$

$$= 4.0$$

The corresponding probability of failure is $\Phi(-4) = 3.17 \times 10^{-5}$.

In Section 9.3.3, we determined that bounds for parallel systems with positive correlation are

$$\prod_{i=1}^{n} P[F_i = 1] \;\leq\; P_f \;\leq\; \min_i\{P[F_i = 1]\}$$

In the context of this example, the lower bound is

$$\begin{aligned}\prod_{i=1}^{4} P[F_i = 1] &= \prod_{i=1}^{4} P(\text{element failure})\\ &= [\Phi(-4)]^4\\ &= 1.01\times10^{-18}\end{aligned}$$

The upper bound is

$$\min_i\{P[F_i = 1]\} = \Phi(-4) = 3.17\times10^{-5}$$

Note that there is a discrepancy between the lower bound estimate of failure probability (for independent failure events) and the probability calculated using Equation 9.40 with $\bar{\rho}=0$. The reason for the difference is the fact that Equation 9.30 assumes that the total load is equally distributed to all elements. In reality, the total load will be distributed (at ultimate) in proportion to the load-carrying capacity of each element. Thus, the elements are actually partially correlated through the applied load, and this is implicitly accounted for to some degree in Equation 9.40. This leads to the slightly larger estimate of the failure probability of the system.

9.5.2 Series system

Consider a series system with n elements. If all $\rho_{ij} = \rho$, then the probability of failure can be calculated using Equation 9.23. If the ρ_{ij} are different for various combinations of elements, then the probability of failure can be approximated by using (Thoft-Christensen and Baker, 1982)

$$P_f = 1 - \int_{-\infty}^{\infty}\left\{\Phi\left(\frac{\beta_e + t\sqrt{\bar{\rho}}}{\sqrt{1-\bar{\rho}}}\right)\right\}^n \phi(t)\,dt \tag{9.41}$$

where $\bar{\rho}$ is the average correlation coefficient defined by Equation 9.39. This approximation is typically very good for small values of n (number of elements). The approximation can be improved for larger n by using the following formula:

$$P_f = P_f(n,\bar{\rho}) - [P_f(2,\bar{\rho}) - P_f(2,\rho_{\max})] \tag{9.42}$$

where $P_f(n,\bar{\rho})$ is calculated using Equation 9.41 using the actual value of n and the average correlation coefficient, $P_f(2,\bar{\rho})$ is calculated using Equation 9.41 with $n = 2$ and the average correlation coefficient, and $P_f(2,\rho_{\max})$ is calculated using Equation 9.41 with $n = 2$ and the *maximum* correlation coefficient among all ρ_{ij}.

Example 9.10

Calculate the probability of failure for the series system shown in Figure 9.23.

The reliability index of each element is $\beta_e = 3.00$. The correlation matrix is

$$[\rho] = \begin{bmatrix} 1 & 0.4 & 0.2 & 0.1 \\ 0.4 & 1 & 0.2 & 0.3 \\ 0.2 & 0.2 & 1 & 0.5 \\ 0.1 & 0.3 & 0.5 & 1 \end{bmatrix}$$

This correlation matrix is the same as the one used in Example 9.9. Referring to that example, we found the average coefficient, $\bar{\rho}$, to be 0.283.

First, we will estimate the probability of failure using Equation 9.41 and Table 9.4. With $n = 4$ and $\bar{\rho} = 0.283$, by interpolation,

$$P_f = 52.76 \times 10^{-4}$$

To improve the estimate, we will use Equation 9.42. For this problem, $\rho_{\max} = 0.5$. Thus,

$$P_f = P_f(4,\bar{\rho}) - [P_f(2,\bar{\rho}) - P_f(2,0.5)]$$

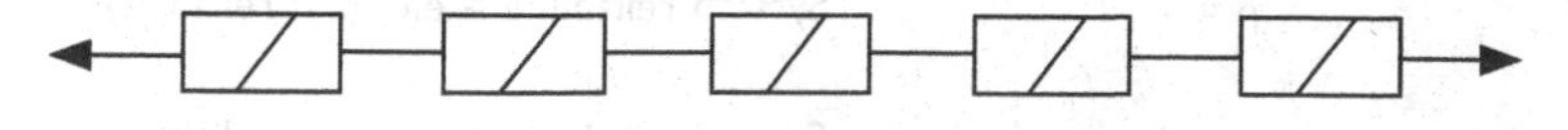

Figure 9.23 Series system considered in Example 9.10.

Using Table 9.4,

$$P_f(4, \bar{\rho}) = 52.76 \times 10^{-4}$$

$$P_f(2, \bar{\rho}) = 26.78 \times 10^{-4}$$

$$P_f(2, 0.5) = 26.18 \times 10^{-4}$$

$$P_f = \{52.76 - (26.78 - 26.18)\} \times 10^{-4} = 52.16 \times 10^{-4}$$

9.6 SUMMARY

The field of system reliability is an area in which there is a great deal of ongoing research. In this chapter, we have covered only the basic concepts of system reliability for relatively simple, idealized systems.

As the examples have illustrated, the effect of correlation on the overall reliability of a system is generally very important. In Figure 9.24 below, this effect is shown schematically on a continuous line. A word description is provided in Table 9.7. Remember that perfectly correlated implies "identical"; if one element is weak, so are all others. Real systems have partially correlated elements.

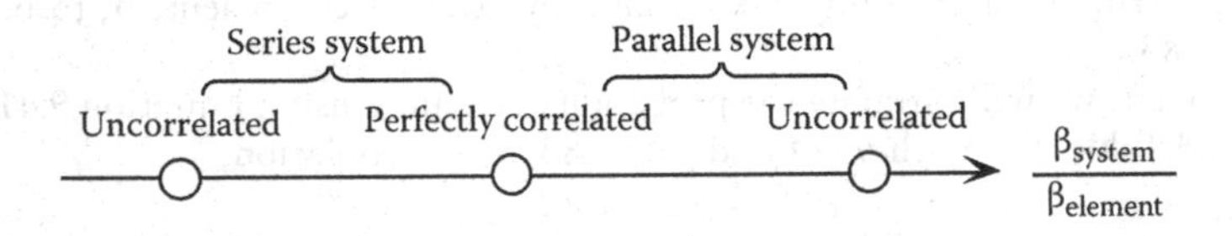

Figure 9.24 Ratio of system reliability to element reliability as a function of correlation.

Table 9.7 Summary of series and parallel systems

System	*Correlation*	*Description*
Series	$\rho = 0$	System reliability < element reliability
	$\rho = 1$	System reliability = element reliability
Parallel	$\rho = 0$	System reliability > element reliability
	$\rho = 1$	System reliability = element reliability

PROBLEMS

9.1 Consider spot welds in cold-formed channels. The shear capacity of the spot welds determines the capacity of the cantilever shown in Figure P9.1. The spot welds can be considered as a parallel system, since all of them must fail before the system fails.
The shear force per weld is denoted by V_i and is a constant. The weld strength for each weld is R_i. The shear strength is determined by a stress–strain curve, as shown in Figure P9.2.
Two extreme cases can be considered:

1. Perfectly correlated welds (if one is bad, all others are equally bad)
2. Uncorrelated welds (if one is bad, the others may be bad, good, or average)

Determine the COV for the two extreme cases, and discuss how they would relate to the COV for a general case of partially correlated welds.

9.2 Consider the three-bar structural system shown below in Figure P9.3a. The system model used for reliability analysis is shown in Figure P9.3b. Let R_i be the strength of the *i*th element. Assume that R_1 and R_2 are normally distributed with mean values of 10 kN and standard deviations of 2 kN. Assume R_3 is lognormally distributed with a mean value of 20 kN and a standard deviation of 3.5 kN. The strengths are all independent.

A. What is the probability of failure of the system when it is loaded with a single force Q = 15 kN?
B. Repeat step A for a load of 20 kN.

9.3 Repeat Problem 9.2, part A, above assuming that all strengths are perfectly correlated.

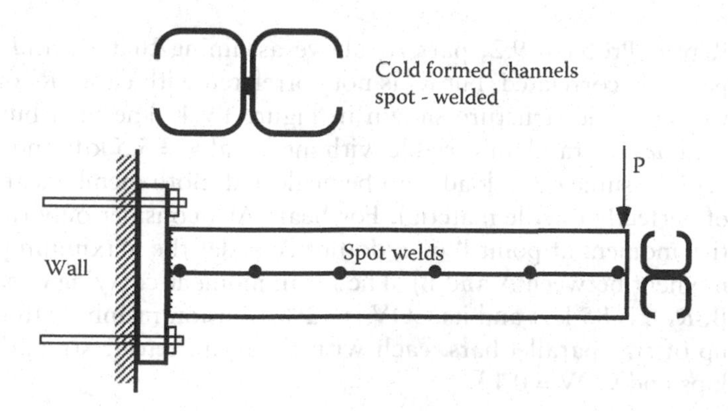

Figure P9.1 Spot weld in a cantilever constructed of cold-formed steel.

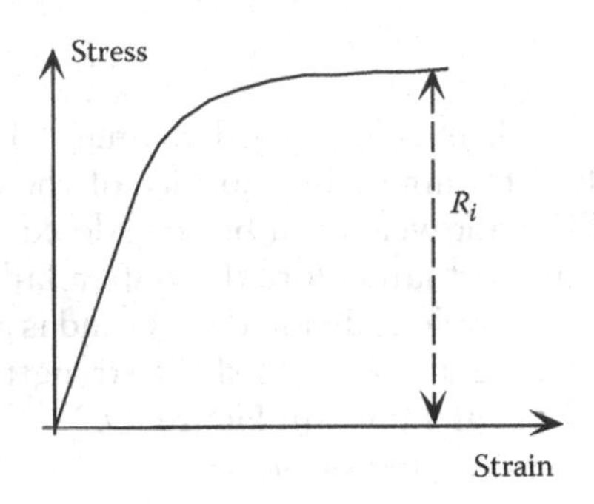

Figure P9.2 Stress–strain relationship for a spot weld.

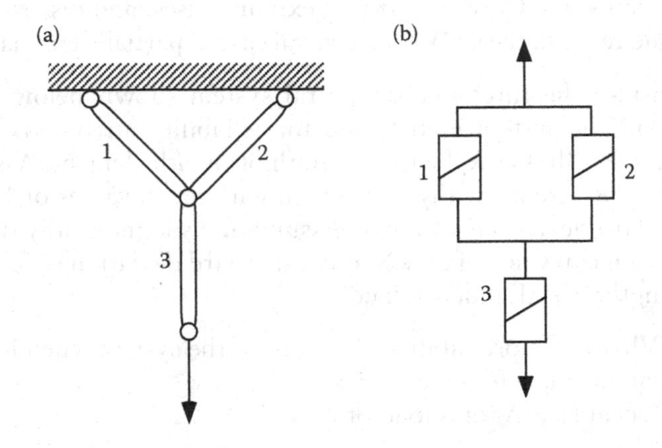

Figure P9.3 Information for Problem 9.2. (a) Three-bar system. (b) System analysis model.

9.4 Repeat Problem 9.2, part A, above assuming that R_1 and R_2 are perfectly correlated, but R_3 is not correlated with either R_1 or R_2.

9.5 Consider the structure shown in Figure P9.4. The distributed live load, w, is a random variable with mean value = 5.5 k/ft and COV = 0.15. Assume other loads can be neglected. Both members are made of perfectly ductile material. For beam AC, consider only the negative moment at point B (i.e., do not consider the maximum positive moment between A and B). The mean moment-carrying capacity at B, R_{AC}, is 85 k-ft and its COV is 12%. Tension member DB is made up of two parallel bars, each with the mean tensile strength of 35 kips and COV = 0.13.

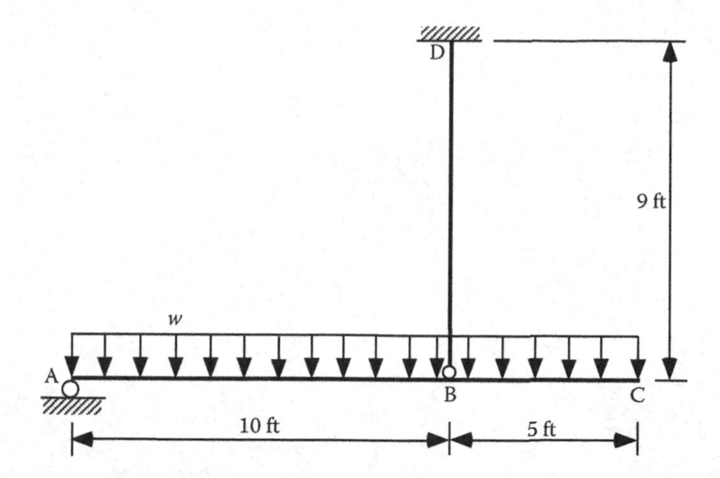

Figure P9.4 Structure for Problem 9.5.

A. Calculate the system reliability index assuming that all members are uncorrelated.
B. Repeat A for the case in which all members are perfectly correlated.
C. Repeat A for the case in which the two bars in DB are perfectly correlated but member AC is uncorrelated with DB.

[illegible]

[illegible]

Chapter 10

Uncertainties in the building process

10.1 INTRODUCTION

10.1.1 Human error

Human error is defined as a departure from acceptable practice. Errors are an inevitable part of all human activities. They add a considerable degree of uncertainty to design and construction activities. In fact, surveys indicate that it is a dominant cause of structural failures in buildings and bridges. Errors can be categorized according to causes and consequences. Structural reliability is determined by error control. The problem of error control can be approached from two directions: 1) reduce error frequency or 2) minimize consequences. Checking calculations and job inspections are used to control the error quantity. Sensitivity analyses can be performed to identify the severity of consequences.

Example 10.1

Consider the following example of human error. Suppose, according to design calculations, that two 3/8-inch-diameter steel bolts are needed in a connection. By mistake, four 3/16-inch-diameter bolts are actually used. The result is a 50% reduction of bolt cross-section area and hence a 50% reduction in capacity.

Example 10.2

An example of human error that resulted in a tragic collapse was the walkway of the Hyatt Regency Hotel in Kansas City. There were two levels of walkways suspended by steel hanger bars as shown in Figure 10.1. In the original design, Figure 10.1a, the weight of each walkway was transferred to the hanger bar through a nut. However, the design was changed, as shown in Figure 10.1b, without recalculating the loads. The nut in the upper walkway was forced to support the weight from both walkways. The overloaded nut failed and the walkways collapsed killing more than 100 people.

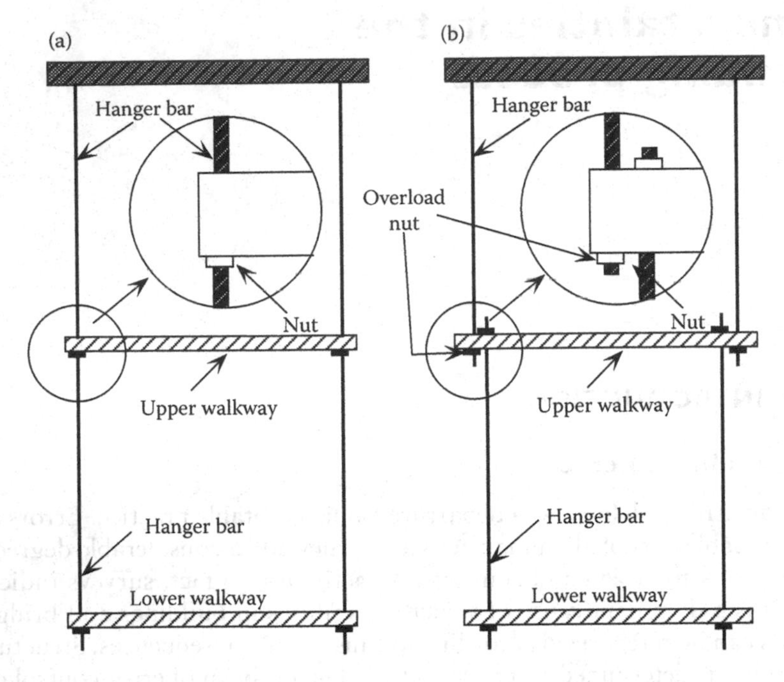

Figure 10.1 Details of walkway hanger bar connection considered in Example 10.2. (a) Original design. (b) As-built configuration.

10.1.2 Categories of uncertainty

The uncertainties involved in the building process can be put into categories, depending on the source, as shown in Figure 10.2. The two major sources of uncertainty are natural hazards and man-made hazards.

Natural hazards are due to wind, earthquake, temperature differentials, snow load, or ice accretion. Also included are natural variations of structural material properties [such as strength, modulus of elasticity (MOE), dimensions] and loads (for example, weight of people, furniture, or trucks on bridges).

Man-made hazards can be subdivided into two groups: from within the building process and from outside the building process. The latter includes uncertainties due to fires, gas explosions, collisions, and similar causes. The former contains uncertainties due to acceptable practice (including innovations, unique structures, use of new materials, and new types of structures) and uncertainties due to departures from acceptable practice. Practice is acceptable if no significant number of the most knowledgeable engineers find it unacceptable. Common practice is not necessarily acceptable. Acceptable practice is not necessarily common. Departures from acceptable practice are human errors.

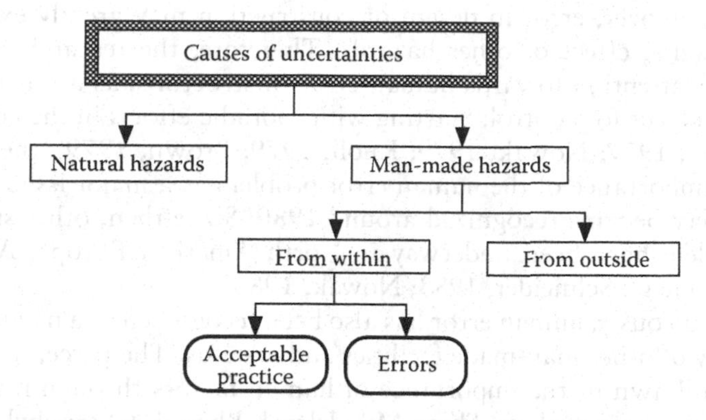

Figure 10.2 Classification of uncertainties in the building process.

10.1.3 Theoretical and actual failure rates

Structural reliability theory has advanced considerably in the last 30 years and has found many applications to structural design standards and procedures. Through improved understanding of the risk of structural failure, it has become possible to optimize investment in that part of structural safety that is effectively controlled by traditional safety factors. However, a considerable discrepancy has been observed between the theoretical and actual failure rates.

The calculated probabilities of failure for buildings and bridges are typically between 10^{-6} and 10^{-8} and the observed values are at least an order of magnitude higher. For example, bridges in the United States have a failure probability rate of 10^{-3} to 10^{-5} annually. The failure rates are much higher for very large and unique structures. Because of a small population, even one collapse drastically increases the overall rate of failure. In North America, there are about 40 long-span suspension bridges and about 70 large cantilever bridges. The Quebec Bridge, a cantilever with a span of 550 m, collapsed twice during construction, in 1907 and then again in 1914, before it was completed in 1917. In 1940, the Tacoma Narrows Bridge, with a suspended span of 850 m, disintegrated under gusty winds of about 40 mph blowing over the Puget Sound.

The discrepancy between the theoretical and actual failure rates is due to an incomplete theoretical model. Most of the failures occur due to human errors that are not included in the analysis. The reliability is usually calculated with the assumption that the design parameters vary randomly due to causes within acceptable practice.

Surveys of structural failures of buildings and bridges indicate that human error is the major cause (Fraczek, 1979; Allen, 1977; Matousek,

1977). Moreover, error in design or construction may greatly exacerbate the damaging effect of other hazards. Therefore, the research front has turned its attention towards human error in structures as an increasingly effective object for control. Starting with sporadic efforts in the late 1970s (Rackwitz, 1977; Nowak, 1979; Knoll, 1979; Brown, 1979), the breadth and the importance of the human error problem as a major issue in structural safety became recognized around 1980. Since then, other studies of the problem have been underway in North America, Europe, Australia, and Japan (e.g., Schneider, 1983; Nowak, 1986).

Simultaneously, human error has also been recognized as a major issue in the safety of other man-made facilities and systems. The perception of risk has been drawn to the importance of human factors through major incidents or disasters such as Three Mile Island, Bhopal, Chernobyl, and the oil spill in Alaska. Risk assessment research in the nuclear power industry has shown that the safety level is determined by human factors and man–machine interactions/interfaces (Swain and Guttmann, 1983). In aviation, about 70% of all accidents have been attributed to crew error, and similar figures are thought to apply to the shipping and offshore industry. For effective control of the safety of any man-made system, an understanding of human error is now essential. The concurrent increased emphasis on product liability has increased the demands of flawless performance of engineered systems.

Therefore, the control of errors is an important part of the strategy to improve the reliability of structures. It involves the reduction of causes (frequency of occurrence) and consequences. Frequency of errors can be reduced by inspections and checking. Consequences of errors can be controlled through identification of the consequential errors using sensitivity analysis.

10.1.4 Previous research

Research into human error in structures has approached the subject from two opposite main directions: fundamental studies and frameworks for application. Fundamental studies seek to deepen the understanding through disaggregated error statistics, phenomenological models, or heuristic models. They range from experimental studies of the commission of errors to simplified mathematical models or even stochastic process models. Fundamental studies proceed empirically from observation of human behavior to a quantitative description that ultimately will serve to account for the occurrence of human error in structures. They are within the purview of natural and social science of technology. Much of the knowledge required for the ultimate understanding of the problem of human error in structures must await developments within the social sciences and psychology.

On the other hand, frameworks for application to structural engineering or construction belong in the realms of engineering science, operations research, or management science. The emphasis is pragmatic, concerned less with detailed understanding than with effective and efficient control. These studies have been developed simultaneously, in parallel but independently, with the fundamental studies. In the pioneering study by Pugsley (1973), the notion of error-prone structures was introduced, and a framework was sketched to identify such structures by a combination of attributes. This work was refined by other researchers who have also explored the use of fuzzy set concepts in the subject area (Brown, 1979). Lind (1983) gave a framework for determining the optimal allocation of engineering time to design modeling, materials testing, and inspection of construction in relation to a given target level of design reliability. In several studies, Nowak (1979) and Nowak and Carr (1985a, 1985b) have given a systematic framework that relates human errors to their effect on structural reliability via sensitivity coefficients.

During the past century, structural engineering theory has been dominated by the needs of quantitative analysis: determine the loads and find the stresses. Excellent special tools are now available to facilitate this task. Control of human error in structures requires a broadening of the approach and a change in emphasis. Much research remains to be done; research has until the present proceeded from individuals' initiatives, and some aspects that are clearly of fundamental importance remain unexplored. One example is the sociology of the construction organization. It has frequently been pointed out that good human relations and communications with clear definitions of responsibilities is important in error control, but this aspect has not been explored formally in the structural context.

Summing up, our knowledge of human error in structures constitutes a major part of our knowledge of structural failure. This knowledge of the profession is broad, but incomplete, disorganized, largely inaccessible, perishable, and mostly intuitive—as well as partly scientific. Effective means to organize this knowledge to design structures or to predict structural behavior has been lacking until recently. It has largely prevented progress in research directed towards improved frameworks for practical control of the hazard of structural human error.

10.2 CLASSIFICATION OF ERRORS

Classification of errors may be useful in the selection of efficient control measures. Errors can be put into categories with regard to causes and consequences. The analysis of causes may allow for identification of the occurrence mechanisms and lead to a reduction in the frequency. The consequential errors can be prevented by additional control measures and by special design methods.

Errors can be considered with regard to person involved (i.e., designer, architect, draftsman, contractor, construction worker, manufacturer, user, or owner), phase of the building process (i.e., planning, design, fabrication, transportation, construction, operation, use, or demolition), place (e.g., office, jobsite, or factory), reason (such as ignorance, negligence, or carelessness), frequency of occurrence, or mechanism of occurrence.

With regard to the mechanism of occurrence, there are three fundamental types of errors: errors of concept, errors of execution, and errors of intention (Nowak and Carr, 1985b). *Conceptual error* is an unintentional departure from the accepted practice due to insufficient knowledge. *Error of execution* is an unintentional departure from what one believed to be accepted practice. *Error of intention* is an intentional departure from the acceptable practice. The alternative paths with regard to acceptable practice are shown in Figure 10.3.

Examples of conceptual errors include cases when the designer 1) was not aware of the most applicable method, model, tool, or information; 2) did not know which method was most applicable; 3) did not know how to use the method, model, tool, or information; 4) did not do something because

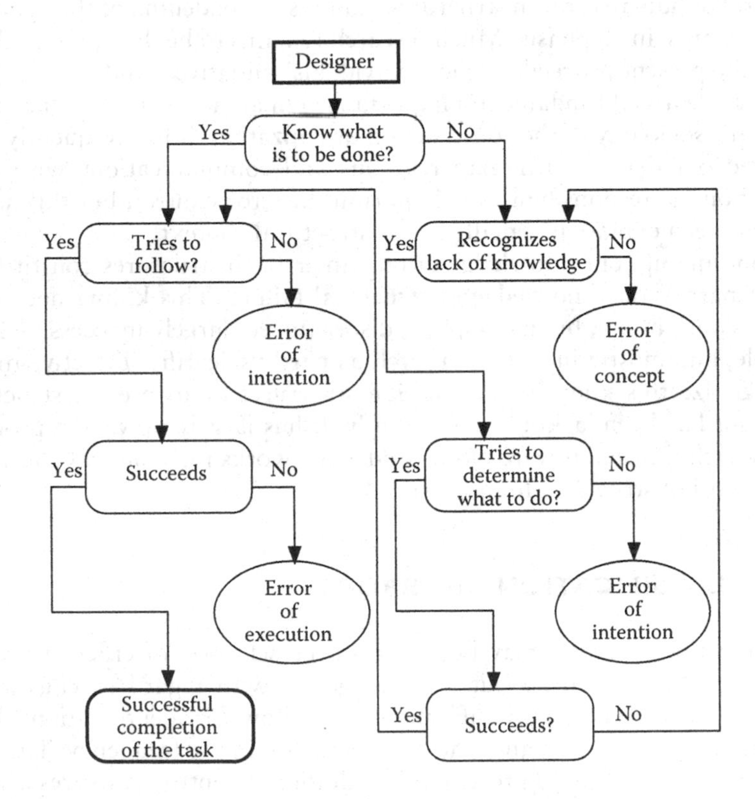

Figure 10.3 Alternative paths with regard to acceptable practice.

he or she did not know it should be done; 5) did not know the acceptable level of effort or care; 6) did not know the possible consequences; 7) did not understand assumptions or limitations; or 8) used simplifying assumptions that were incorrect.

Errors of execution include cases when the designer misread, miswrote, misdrew, misheard, misspoke, misoperated, forgot, lost, misplaced, left out, did not think of, did not hear, or did not see something. These errors also occur when an individual hears and sees the proper information but does not recognize it.

Examples of intentional errors include cases when the designer performed poorly: 1) because of expediency; 2) to save time, money, energy, or bother; 3) to avoid responsibility or liability; 4) to avoid embarrassing someone else; 5) because he or she was requested or required by supervisor, contract, or regulations to complete current work or obtain future work; 6) because he or she did not have the capability to do the work according to accepted practice; 7) because it is impossible for anyone to do the work under accepted practice; 8) because the individual accepted risk that was recognized as unacceptable; or 9) because the individual chose to depart from common practice without acceptable reason.

Error may result in a reduced or increased reliability (hazardous or opulent errors). The consequences of errors may vary from minor serviceability problems to an overall collapse. The structural performance may be affected directly or indirectly (e.g., by increasing the probability of other errors). The resulting failure may be localized or it may trigger failure of other components leading to progressive collapse.

An error changes the analytical model of the structure. It may affect loads, load-carrying capacity, or the overall interpretation of behavior. Gross errors cause drastic departures from the assumed theoretical model (for example, placing the beam upside down or forgetting about the reinforcing steel).

In practice, there are an infinite number of possible errors. However, for each structure, errors can be grouped depending on the affected part(s) of the theoretical model. *Parametric errors* cause changes in the statistical distribution functions of parameters of load and load-carrying capacity. Examples include use of the wrong grade of material and under- or overestimation of load. *Modal errors* result in changes in the mode of structural behavior. Examples of modal errors include omission of a relevant mode of failure from the analysis or incorrect interpretation of the structural behavior.

In most structural engineering tasks, errors in design and construction can be characterized by two parameters: frequency and consequences. The frequency of errors can be lowered by employing a control scheme. Moreover, the relationship between consequences of errors and structural safety can be defined by sensitivity functions. The sensitivity functions often become a basis for an error control strategy.

Nowak and Carr (1985b) noted that a chain of errors, causes, and consequences can be identified as follows: a human failing causes a human error, which causes a structural error, which causes a service error, which may cause failure and consequently causes losses. An example of the argument stated above is the ignorance of an inexperienced engineer designing a reinforced concrete slab; that ignorance may lead to an incorrect number of steel reinforcing bars, which may lead to poor strength and consequently failure causing damage to the structure and injury to the users.

In real life, the number of possible errors and consequences in a structure is infinite. Nevertheless, a moderate number of possible consequences can be established for each structure by determining the controlling parameters in the design and construction. The major consequential relationship is between structural error and service error. Going back to the example of the reinforced concrete slab, the possible structural errors would include insufficient area of steel, effective depth, dimensions, or design loads. Of course, these are consequential errors because they produce service errors, such as insufficient bending, shear, and deflection resistance. Sensitivity analysis can be used to determine the consequential relationship between structural errors and the resulting service errors. For each structural error, the sensitivity analysis can determine the impact of such an error on the load or load-carrying capacity of the structure and consequently on its serviceability and safety. This procedure can be applied for a structural element or for a whole structure.

10.3 ERROR SURVEYS

Even though the actual failure rate of structures is higher than the theoretical value (calculated without considering human errors), it is still very small. Therefore, the failure database is limited, making it difficult to develop structural reliability models that account for errors. The available data sources include failure surveys in North America (Fraczek, 1979; Allen, 1977) and Europe (Matousek, 1977).

Fraczek (1979) and Allen (1977) reported on a large survey conducted in an attempt to develop a profile of the practices, activities, and circumstances making up the design and construction of concrete structures that lead to errors. The data showed that only about 10% of failures were traceable to stochastic variability in loads and capacities; the remaining 90% were due to other causes, including mostly design and construction errors. About half of the errors occurred in design, and the other half occurred in construction.

Most of the design errors resulting in failure were due to misconception or lack of consideration of structural behavior. These errors were detected during the service life of structures and most resulted in serviceability problems. Design errors were far more prevalent than construction errors in elements that required close attention to detail such as connections and joints.

Most of the collapses occurred during construction, mainly as a result of inadequate formwork or temporary bracing. Construction errors resulting in structural failures were due to incorrect procedures such as improper bracing, omissions, misplacements, or overloading. The scope of the survey included errors that were caught before construction. Although a large number of these kinds of errors occurred, only a few were reported, presumably because records of such errors were usually not made. Therefore, the survey did not provide representative information on errors that were corrected in time.

The European information documented by Matousek (1977) was drawn from structural failure dossiers of insurance companies, books, journals, newspapers, and direct personal information from engineering firms and contractors. They observed that most of the damage occurred during construction of the bearing structure. This fact is not surprising, because utilization of a structure is thoroughly envisaged during analysis and design, while different stages of erection often are poorly taken into account or even neglected. It was found that errors in planning led to a larger amount of building and equipment damage, while consequences of errors in construction were more severe with respect to injury or death of people.

The cause of failures was always found in human unreliability. About 75% of the instances of damage and 90% of the costs of damage were due to human error. About 45% of failures were due to defects in design, 49% were due to construction, and 6% occurred because of use and inadequate maintenance. A large portion of the mistakes leading to failure could have been detected, without any additional control, by adequate checking by the person next involved in the building process. Most of the mistakes could be detected in time by additional control. About 15% of the total cases would escape in spite of this additional control. Additional inspection and checking during planning and design were found to be the most efficient method of error control. Additional control during the construction process would help to avoid most cases of failure in which persons are injured or killed.

Errors that resulted in a collapse are just the "tip of the iceberg" compared to the total population of errors, as shown in Figure 10.4. Many errors are detected, most of them by the person who committed the mistake, others by various levels of checkers and inspectors. The error control systems are often biased to opulent errors (those resulting in conservative over-design). Of the errors that slip through the control system, only very few result in failure.

To increase the database for study of errors, a survey was conducted by a research team at the University of Michigan. The survey concentrated on detected errors and near-failure cases. Interviews were arranged with engineers and managers working for design offices and construction companies, with projects including office buildings, parking structures, nuclear

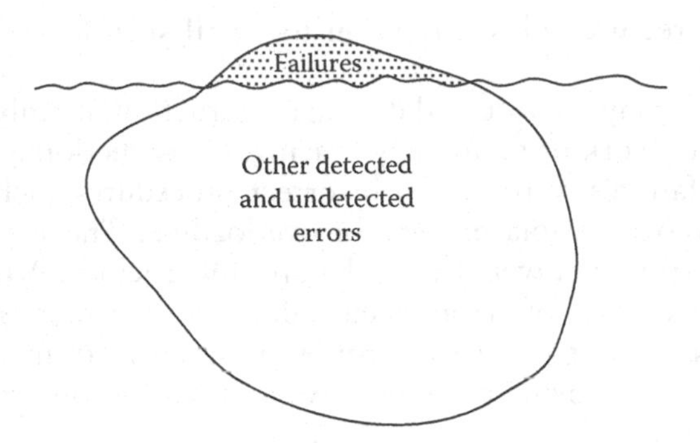

Figure 10.4 Proportion of errors leading to failure relative to the total population of possible errors.

power plants, bridges, and others. The natural aversion to admit commission of errors was an important barrier that had to be overcome to get unbiased results. Therefore, the interviews were conducted with the agreement that the identities of projects and companies would not be revealed.

The survey identified many causes of errors. The most common causes included an incomplete understanding of structural behavior (design criteria, assumptions, and boundary conditions), departure from typical loading cases (e.g., thermal effects or tornado), loads during construction stages being poorly considered in the analysis, time pressure, specification ambiguity, discontinuity in design process, lack of experience, communication problems, lack of coordination, and undefined goals of structural use.

Errors in the design process depend on the method of design and experience of the designer and checker. There are three major ways to perform design calculations: by hand, by computer, and by using standard tables and charts. In hand calculations, the typical errors include calculation errors, omissions of critical loading conditions (shortcuts or rules of thumb are not always safe), use of wrong units, and incorrect interpretation of the code. The survey of errors in design by computer indicated that errors in computer outputs are usually difficult to detect. Furthermore, it was noted that 1) it is easy to add errors when updating a program, 2) assumptions of the design program are often unknown to the user, 3) programs are sometimes not suited for some components, and 4) complicated programs increase the probability of errors. In design using standard tables and charts, errors may result due to misuse of tables or charts (design component may not fit the standard, or elements may be selected without knowing the design forces or loading conditions).

The results of the surveys mentioned above indicate that errors occur due to a very wide variety of causes. Any structural part or any phase of the building process can be affected. However, an efficient error control strategy must concentrate on the most consequential errors. Identification of the consequential errors is accomplished through sensitivity analysis.

Example 10.3

Some examples of the common causes of errors observed in error surveys include the following:

- Incomplete understanding of the behavior of the structure.
- Poor judgment and overlooking the problem.
- Calculation errors are not detected.
- Change of use is a frequent error, e.g., loading the structure with loads for which it was not designed.
- Contractor interprets the design and drawings to his own advantage.
- Organizational problems; lack of continuity.
- Trying to fit numbers in wrong formulas.
- Errors often occur when information is copied from different sources without understanding.
- Specification ambiguity.
- Errors caused by inexperienced engineers, designers, and inspectors.
- Poor inspection or no regulations to provide good inspectors.
- Lack of coordination between field engineers.
- Communication problems.
- Undefined goals so that a change of use may be expected.
- Little attention given to the boundary conditions and supports.
- Incomplete design and ignorance of some important forces such as torsion or buckling.
- Time pressures, especially for inexperienced engineers.
- Lack of clear and well-understood design criteria.
- Complex set of load conditions (different for various structural parts).
- Abnormal loading.
- Use of load combination for the wrong building.
- Departure from typical causes (unusual loading such as thermal effects, tornado, missiles).

10.4 APPROACH TO ERRORS

As reflected in Figure 10.5, the probability of failure depends mostly on the control of causes and consequences of errors. The causes cover frequencies of occurrence and reasons. Frequency of errors can be reduced by inspections, calculation checking, improving the work environment, or use of special design and construction techniques. Motivation has been identified

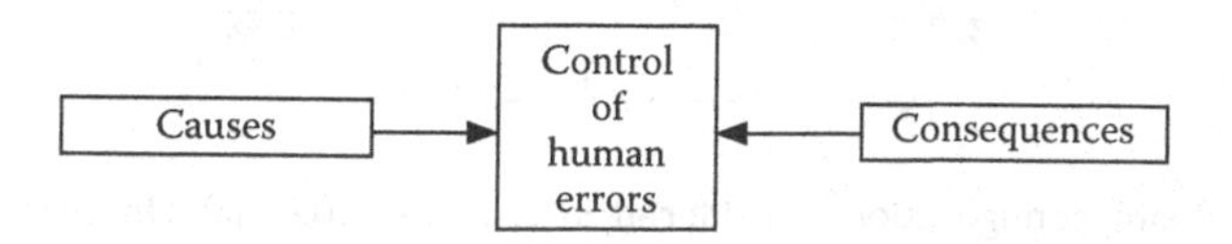

Figure 10.5 An approach to controlling human error.

as the most important factor affecting human performance in the building process (Schneider, 1983). Other factors affecting performance are knowledge, experience, and physiological conditions.

Error frequency can be reduced by eliminating or reducing the opportunity. A foolproof approach is based on the use of design and/or construction procedures that are easy to understand and follow. For example, to avoid use of a wrong bolt size on the jobsite, only one bolt size can be specified, thus eliminating access to any other sizes.

Consequences of errors can be controlled through the identification of the consequential errors using sensitivity analysis. The objective of the sensitivity analysis is to relate error magnitude and structural reliability. Human error may affect parameters or modes of structural behavior. For each considered parameter, a reliability analysis is performed to determine the reliability corresponding to various errors. Only error consequences are considered (rather than causes). For example, the effective depth of steel reinforcement, d, is a critical parameter in the design of reinforced concrete beams. To develop a sensitivity function for d, the reliability is calculated for various possible values of d. Multiple calculations reduce errors as well as optimize the design.

Example 10.4

The concept of sensitivity analysis is demonstrated on a simple beam design. The beam is designed to resist a uniformly distributed load as shown in Figure 10.6a. The calculated reliability index is $\beta = 3.5$.

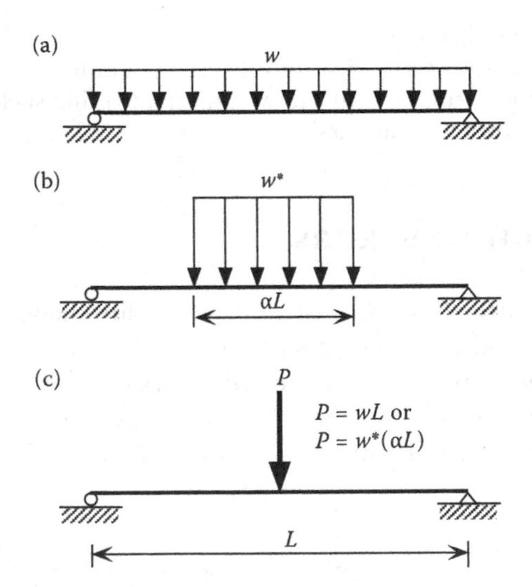

Figure 10.6 Beam configuration considered in Example 10.4. (a) Uniform load acting over the entire span. (b) Uniform load acting over a central portion of the span. (c) Uniform load with a concentrated load at midspan.

As is often the case during construction or use, the load can be piled in the central portion of the beam, as shown in Figure 10.6b and c, rather than spread over the whole length. The loaded portion of the span is denoted by αL with $0 < \alpha < 1$, where L is the beam length. The reliability index, β, for the beam is calculated for various values of α. For $\alpha = 1$, $\beta = 3.5$. Three types of material are considered: structural steel, prestressed concrete, and wood, with coefficients of variation of resistance of 0.105, 0.065, and 0.225, respectively. The sensitivity functions are plotted in Figure 10.7 as a function of α. Three values of the coefficient of variation of the load effect are considered: 0.08, 0.12, and 0.15. The results indicate that the prestressed beam is the

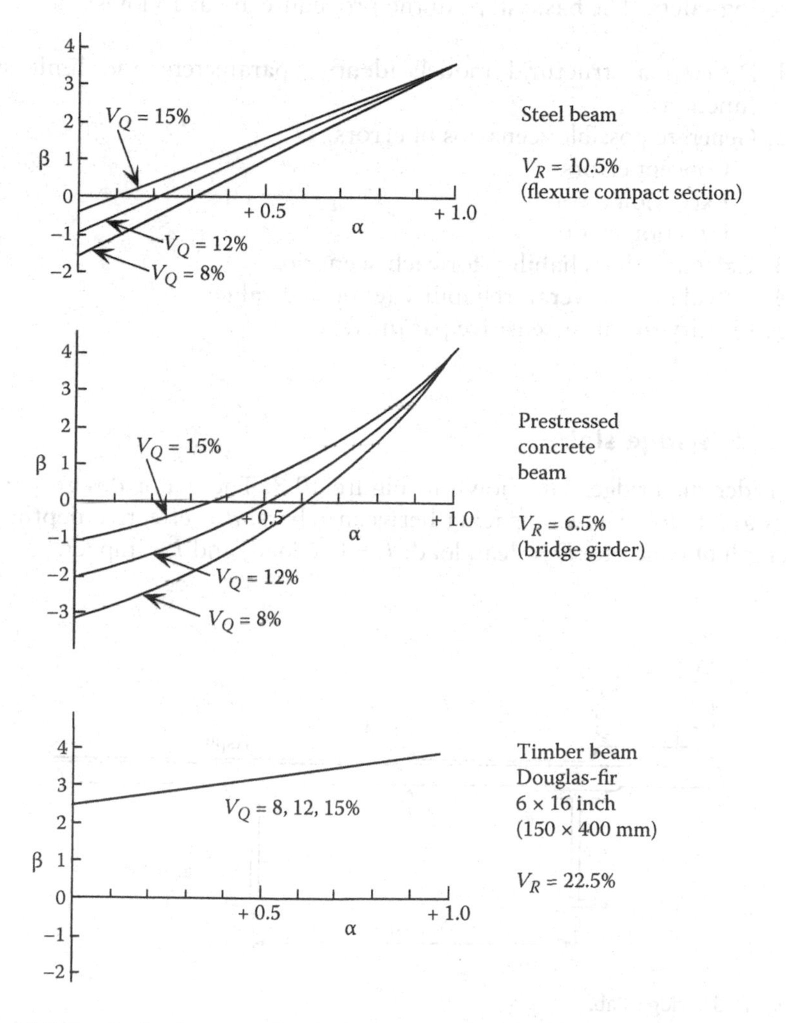

Figure 10.7 Sensitivity functions for the beam considered in Example 10.4.

most sensitive to load distribution errors. On the other hand, a wooden beam is least sensitive. The difference is attributed to the very large natural variation of wood strength compared to prestressed concrete and steel. In the case of wood, the uncertainty in material strength dominates the uncertainty in load in the reliability analysis.

10.5 SENSITIVITY ANALYSIS

10.5.1 Procedure

Sensitivity analysis is performed to identify the most important parameters affecting safety. The basic steps in the procedure are as follows:

1. Develop a structural model; identify parameters and limit state functions.
2. Generate possible scenarios of errors:
   - Concept errors
   - Execution errors
   - Intention errors
3. Calculate the reliability for each scenario.
4. Calculate the overall reliability (expected value).
5. Identify the most sensitive parameters.

10.5.2 Bridge slab

Consider the bridge slab shown in Figure 10.8. The major design parameters are as follows: s = spacing between rebars; d = effective depth; f_c' = strength of concrete; D = dead load; L = live load; and I = impact.

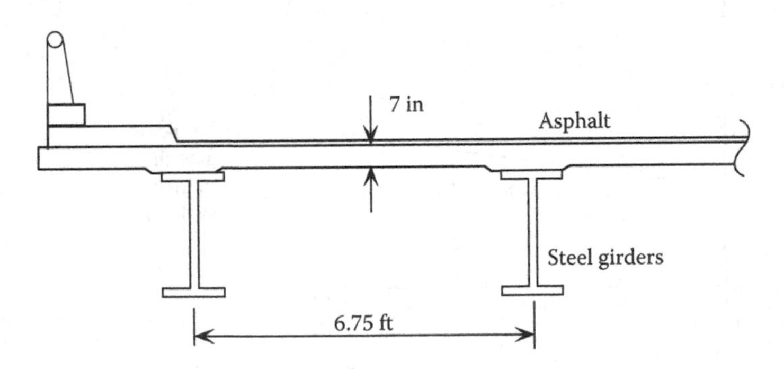

Figure 10.8 Bridge slab.

The following are the design (nominal) values of parameters:

Concrete	$f_c' = 3000$ psi
Rebars	$f_y = 40$ ksi (#6 bars at s = 7.5 in)

American Association of State Highway and Transportation Officials (AASHTO, 2012) moment-carrying capacity = 10.9 k-ft

Effective depth, d	4.7 in
Dead load moment	0.40 k-ft
Live load moment	3.30 k-ft
Impact	1.00 k-ft

Statistical data:

$\mu_D = 1.05D_n$	$V_D = 0.08$
$\mu_L = 1.58L_n$	$V_L = 0.11$
$\mu_I = 1.05D_n$	$V_I = 0.45$
$\mu_R = 1.07R_n$	$V_R = 0.11$

The reliability index is calculated using

$$\beta = \frac{\mu_R - \mu_Q}{\sqrt{\sigma_R^2 + \sigma_Q^2}} \tag{10.1}$$

The resulting sensitivity functions are shown in Figure 10.9.

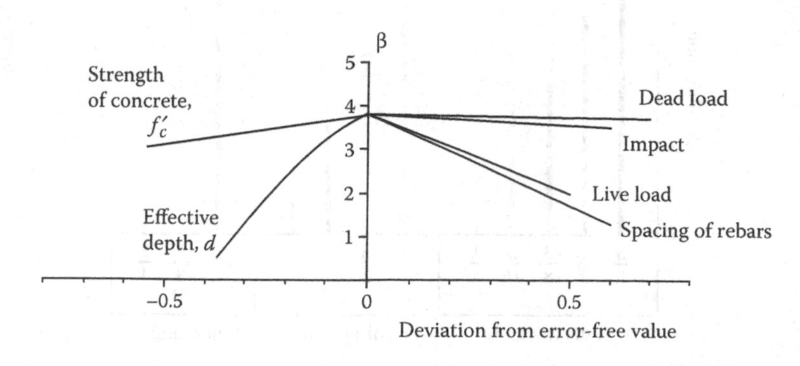

Figure 10.9 Sensitivity function for bridge slab shown in Figure 10.8.

10.5.3 Beam-to-column connection

A typical steel beam-to-column connection is considered, as shown in Figure 10.10. It is assumed that fillet welds and bolts are used. The following parameters are considered: angle thickness, number of bolts, bolt diameter, shear strength of the bolt, shear strength of the weld, dead load, and live load.

Statistical data:

$D = L$	
$\lambda_D = 1.0$	$V_D = 0.10$
$\lambda_L = 0.85$	$V_L = 0.20$
$\mu_R = 2.93\ (D + L)$	$V_R = 0.185$ for fillet weld
$\mu_R = 3.00\ (D + L)$	$V_R = 0.10$ for A325 bolts
$\mu_R = 2.51\ (D + L)$	$V_R = 0.07$ for A490 bolts

The sensitivity functions are presented in Figures 10.11 and 10.12.

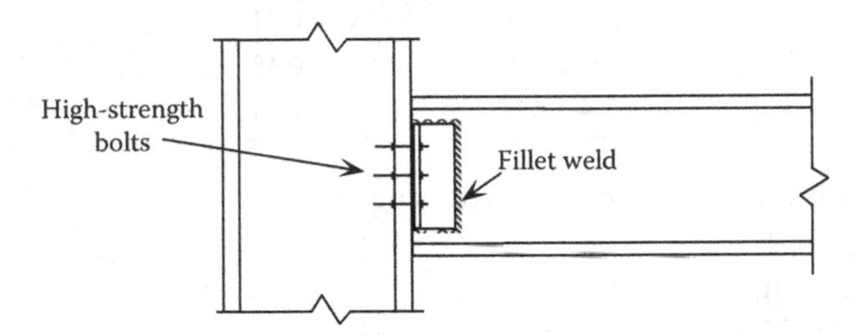

Figure 10.10 Steel beam-to-column connection.

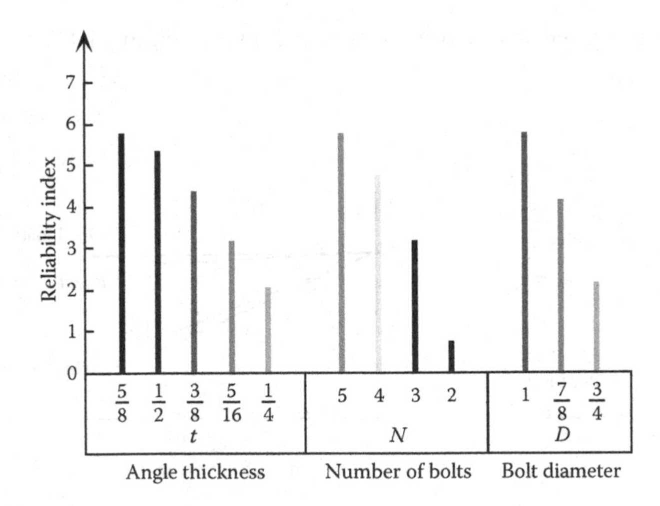

Figure 10.11 Sensitivity functions for beam-to-column connection. Sensitivity to information on bolts is shown.

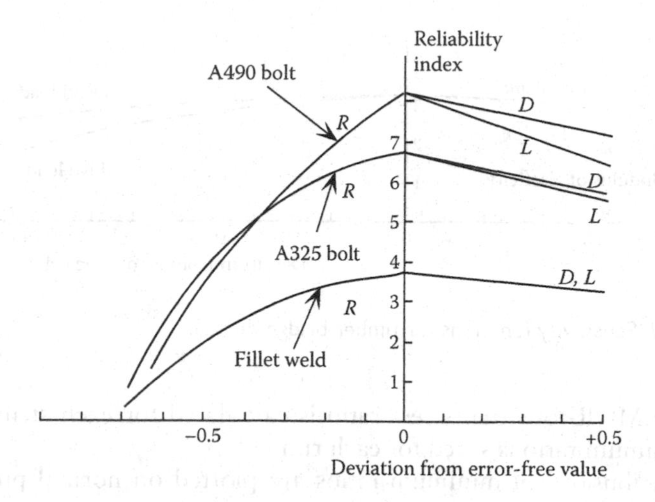

Figure 10.12 Sensitivity functions for beam-to-column connection considering both bolts and welds.

10.5.4 Timber bridge deck

A typical stringer deck is considered as shown in Figure 10.13. It is assumed that Hem Fir select structural timber is used. For the dimensions and stringer spacing given in Figure 10.13, the maximum span allowed by the AASHTO Specifications (1992) is 12.5 ft.

The major parameters considered include the following: modulus of rupture (MOR), MOE, dead load, and live load. It is assumed that MOR and MOE are partially correlated. Reliability analysis is performed using Monte Carlo simulations. The basic procedure is as follows:

- MOR and MOE are generated for each stringer.
- Stress in each stringer is calculated using the Finite Strip Method.

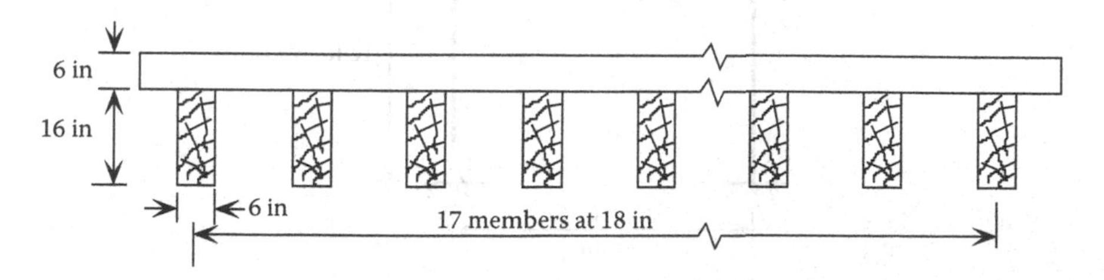

Figure 10.13 Timber bridge deck.

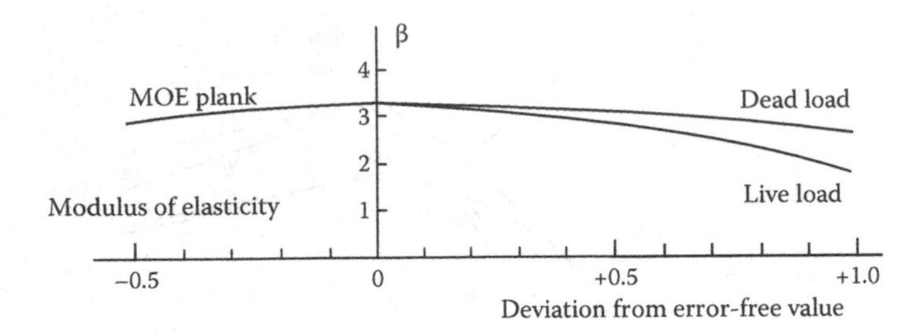

Figure 10.14 Sensitivity functions for timber bridge deck.

- The MOR-to-actual stress ratio is calculated for each stringer. The minimum ratio is saved for each run.
- Distributions of minimum ratios are plotted on normal probability paper and β is determined.

The MOE of the deck planks affects the stiffness of these planks and hence influences the lateral load distribution.

The sensitivity functions are shown in Figure 10.14.

10.5.5 Partially rigid frame structure

A partially rigid frame structure is shown in Figure 10.15. The effect of a change in support condition at A is considered for (a) a fully fixed support and (b) a partially fixed support.

Reliability indices are calculated for both cases. For a fully fixed support at A, $\beta = 2.7$; for a partially fixed support at A, $\beta = 2.0$.

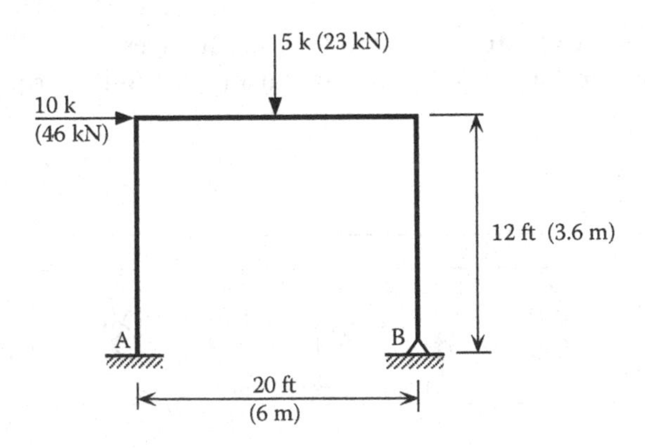

Figure 10.15 Partially rigid frame structure.

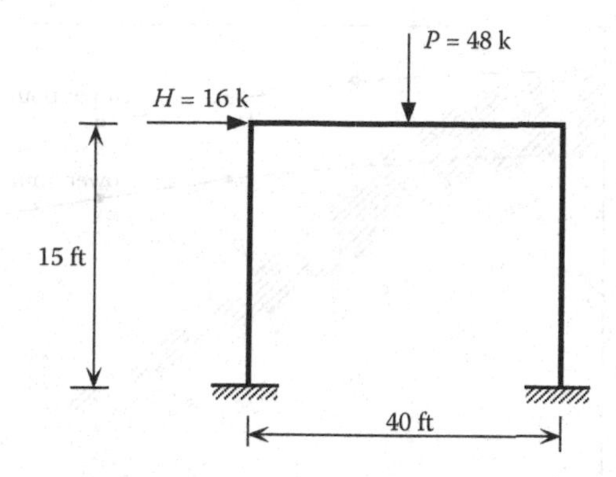

Figure 10.16 Simple rigid frame structure.

10.5.6 Rigid frame structure

The rigid frame structure shown in Figure 10.16 is considered. Three parameters are considered: vertical force P (gravity load), horizontal force H (wind), and plastic moment M_P (resistance). The statistical parameters of load and resistance are summarized in Table 10.1.

The sensitivity functions are presented in Figure 10.17. Various degrees of correlation are considered between the calculated upper and lower bounds for the reliability index.

10.5.7 Noncomposite steel bridge girder

A typical noncomposite steel bridge girder is considered. The parameters are as follows: span = 18 m, girders are W36 × 210 spaced at 2.4 m, F_y = yield strength = 250 MPa, slab thickness = 180 mm, concrete strength = f_c' = 21 MPa.

The sensitivity functions are presented in Figure 10.18.

Table 10.1 Statistical parameters of load and resistance

Parameters	*Bias factor*	*Coefficient of variation*	*Probability density function*
Material properties, M_P	1.1	0.11	Lognormal
Gravity load, P	0.6	0.20	Normal
Horizontal load, H	0.8	0.25	Extreme Type I

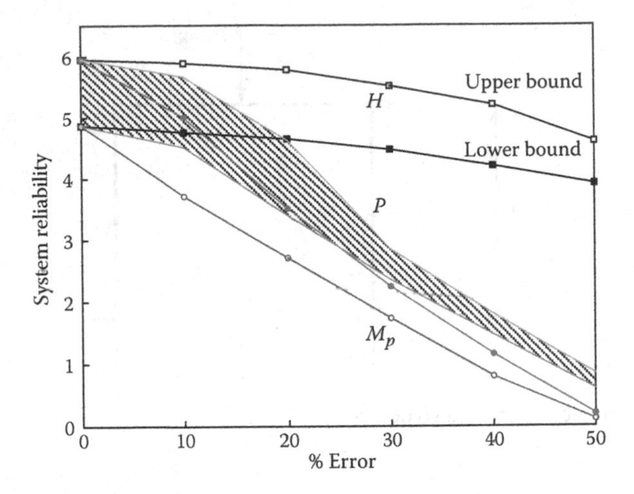

Figure 10.17 Sensitivity of system reliability of frame structure due to errors in load and resistance parameters.

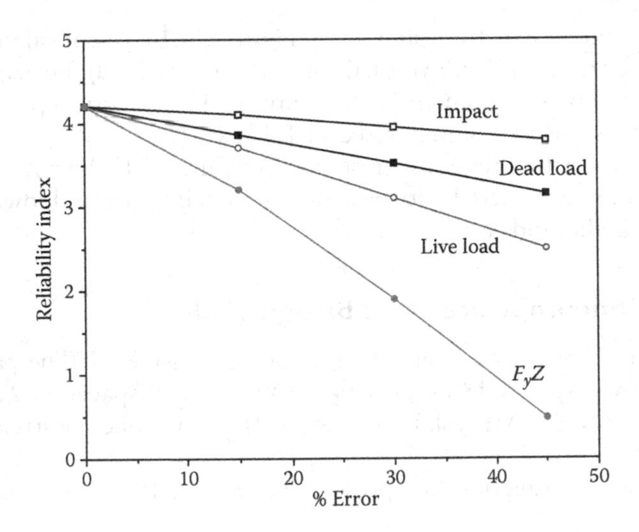

Figure 10.18 Sensitivity functions for noncomposite steel girder.

10.5.8 Composite steel bridge girder

A typical composite steel bridge girder is considered. The parameters are as follows: span = 18 m, girders are W33 × 130 spaced at 2.4 m, F_y = yield strength = 250 MPa, slab thickness = 180 mm, concrete strength = f_c' = 21 MPa.

The sensitivity functions are presented in Figure 10.19.

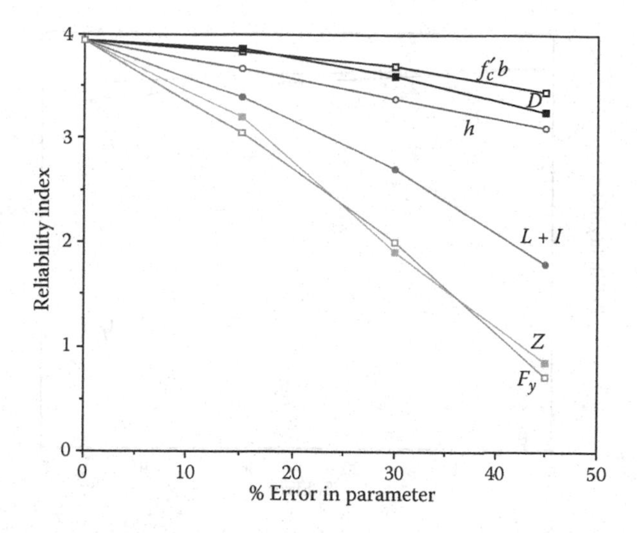

Figure 10.19 Sensitivity functions for a composite steel girder.

10.5.9 Reinforced concrete T-beam

A typical reinforced concrete T-beam is considered. The parameters are as follows: span = 18 m, beam effective depth = 915 mm, beams are spaced at 2.4 m, F_y = 275 MPa, concrete slab strength f_c' = 21 MPa.

The sensitivity functions are presented in Figure 10.20.

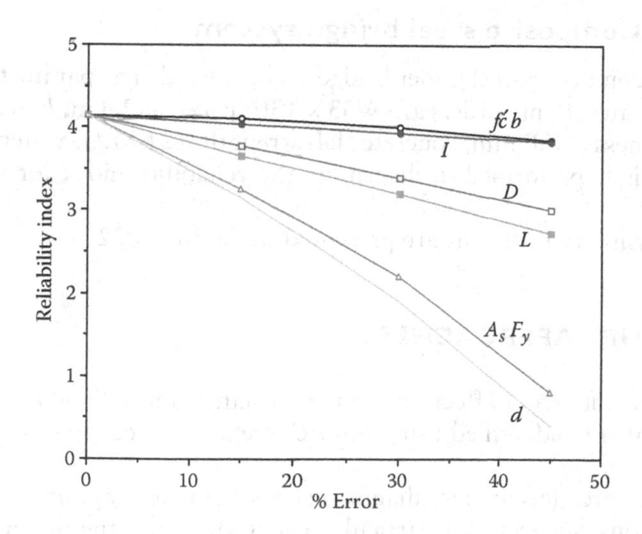

Figure 10.20 Sensitivity functions for reinforced concrete T-beam.

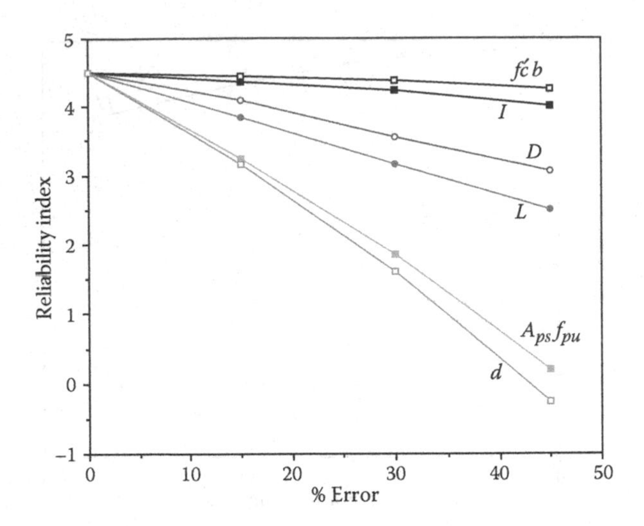

Figure 10.21 Sensitivity functions for prestressed concrete girder.

10.5.10 Prestressed concrete bridge girder

A typical prestressed concrete bridge girder (AASHTO type) is considered. The parameters are as follows: span = 18 m, f_{pu} = 1860 MPa, slab thickness = 180 mm, concrete strength for girder = f_c' = 28 MPa, concrete strength for slab = f_c' = 21 MPa.

The sensitivity functions are presented in Figure 10.21.

10.5.11 Composite steel bridge system

A typical composite steel girder bridge is considered. The parameters are as follows: span = 18 m, girders are W33 × 130 spaced at 2.4 m, F_y = 250 MPa, slab thickness = 180 mm, concrete slab strength = 21 MPa. System reliability analysis is performed to determine the reliability index for the whole system.

The sensitivity functions are presented in Figure 10.22.

10.6 OTHER APPROACHES

The important errors affecting the performance and reliability of a structure can also be identified using failure tree (event tree) analysis and fault tree analysis.

A failure tree (event tree) diagram is a schematic diagram showing the possible consequences of a particular event known as the initiating event. For any consequence to occur, usually one or more intermediate events

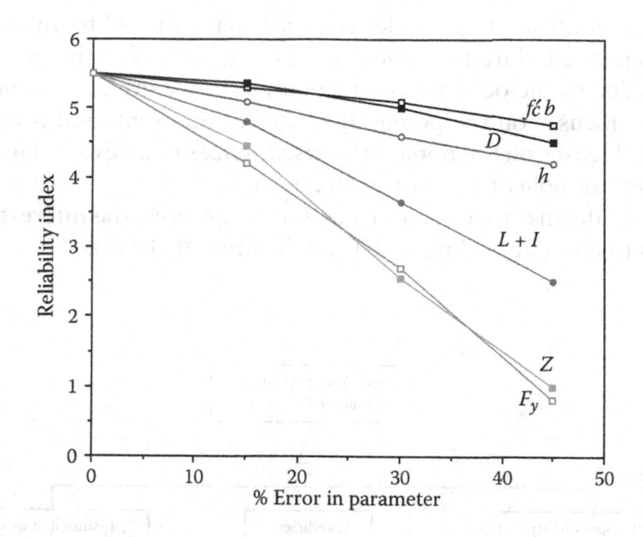

Figure 10.22 Sensitivity functions for composite steel girder bridge.

must occur between the initiating event and the consequence being considered. Therefore, each path of the tree diagram represents one possible scenario of events leading to a particular consequence. This idea is illustrated in Figure 10.23.

A fault tree diagram is a schematic diagram showing how a particularly significant event, commonly referred to as the top event, can occur given different scenarios of faults (errors) that may occur. The distinction between a fault tree diagram and a failure tree diagram is as follows: a

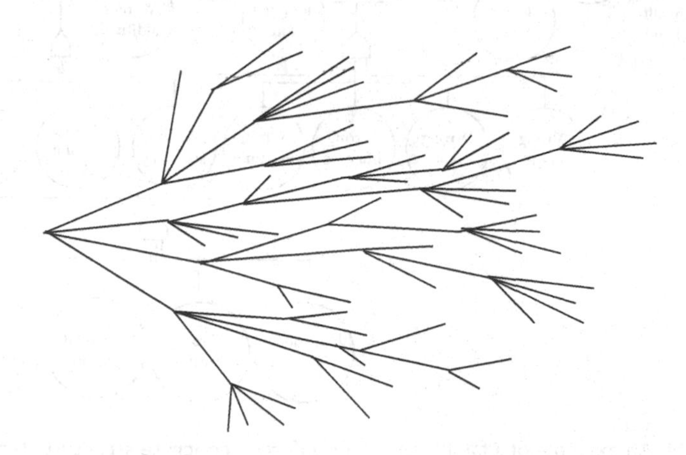

Figure 10.23 A schematic diagram of a failure tree.

fault tree diagram shows possible scenarios that can lead to an undesirable event, whereas a failure tree (event tree) diagram shows the possible consequences due to the occurrence of the undesirable event. In other words, a fault tree focuses on the potential *causes* of an event, and a failure tree (event tree) focuses on the potential *consequences* of an event. Figure 10.24 provides an example of a fault tree diagram.

For more information on fault trees and event trees, the interested reader is referred to the text by Ang and Tang (Volume II, 1984).

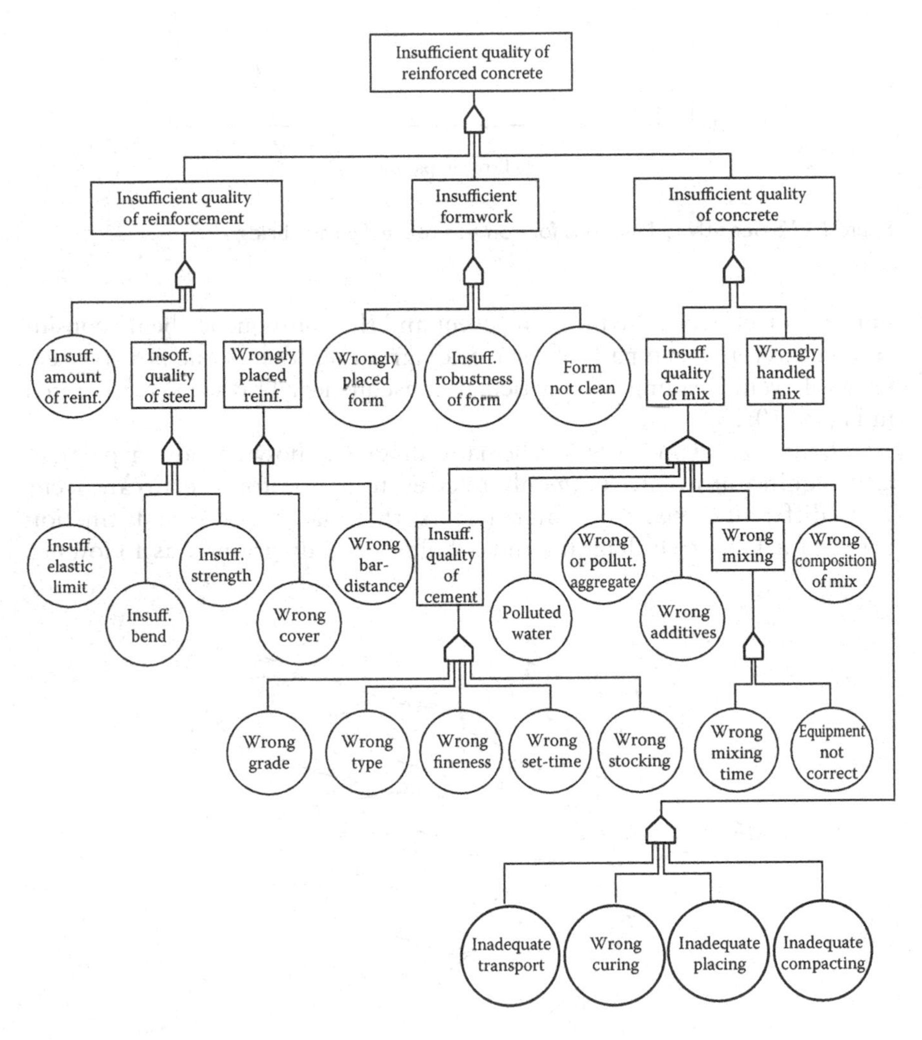

Figure 10.24 An example of a fault tree for a reinforced concrete structure. (From Task Group I, CEB, 1983. With permission.)

10.7 CONCLUSIONS

Human error is the major cause of structural failure. The reliability of structures depends, to a large degree, on controlling the causes of errors and minimizing the consequences. Optimization of error control requires identifying the most frequent errors with their resultant consequences. Error causes and frequency can be identified by surveys. Sensitivity analysis is an efficient method to identify the consequences of the errors.

Error control measures can be selected based on the sensitivity functions and the frequencies of errors. This may serve as a basis for the development of a strategy in a design office, giving priority attention to those errors that have severe outcomes. The approach may have a great effect on the basic organizational structure of the office, as well as the selection of design and construction procedures.

The probability of occurrence of errors can be controlled by checking, inspection, monitoring, foolproof design, and proof loading. The extent of damage can be controlled by safety factors, fail-safe design, and performance monitoring.

Appendix A: Acronyms

AASHTO	American Association of State Highway and Transportation Officials
ACI	American Concrete Institute
ADTT	Average daily truck traffic
AISC	American Institute of Steel Construction
AISI	American Iron and Steel Institute
ANSI	American National Standards Institute
API	American Petroleum Institute
ASCE	American Society of Civil Engineers
BOCA	Building Officials and Code Administrators
BSSC	Building Seismic Safety Council
CDF	Cumulative distribution function
CEC	Commission of the European Communities
CHBDC	Canadian Highway Bridge Design Code
CISC	Canadian Institute of Steel Construction
Cov	Covariance
COV	Coefficient of variation
CSA	Canadian Standard Association
FEMA	Federal Emergency Management Agency
FHWA	Federal Highway Administration
FLS	Fatigue limit state
GDF	Girder distribution function
IABSE	International Association for Bridge and Structural Engineering
ICBO	International Conference of Building Officials
LRFD	Load and resistance factor design
MCE	Maximum considered earthquake
MOE	Modulus of elasticity

MOR	Modulus of rupture
NBS	National Bureau of Standards
NCHRP	National Cooperative Highway Research Program
NDS®	National Design Specification® for Wood Construction
NEHRP	National Earthquake Hazard Reduction Program
OHBDC	Ontario Highway Bridge Design Code
PDF	Probability density function
PMF	Probability mass function
SEAOC	Structural Engineers Association of California
SLS	Serviceability limit state
UBC	Uniform Building Code
ULS	Ultimate limit state

Appendix B: Values of the CDF $\Phi(z)$ for the standard normal probability distribution

Values of the CDF are given for *negative* values of the standard normal variate ranging from 0 to −8.99. To find the value of $\Phi(z)$ for positive z values, use

$$\Phi(+z) = 1 - \Phi(-z)$$

Example

To find $\Phi(-1.13)$, read down the first column to the row labeled "−1.1" and then go across to the column labeled "0.03." The corresponding value of the CDF is 1.29E-01 = 0.129. This value is highlighted in Table B.1. The value of $\Phi(1.13)$ is

$$\Phi(+1.13) = 1 - \Phi(-1.13) = 1 - 0.129 = 0.871$$

Table B.1 Values of the cumulative distribution function for the standard normal variable

	0	0.01	0.02	0.03	0.04	0.05	0.06	0.07	0.08	0.09
0	5.00E-01	4.96E-01	4.92E-01	4.88E-01	4.84E-01	4.80E-01	4.76E-01	4.72E-01	4.68E-01	4.64E-01
–0.1	4.60E-01	4.56E-01	4.52E-01	4.48E-01	4.44E-01	4.40E-01	4.36E-01	4.33E-01	4.29E-01	4.25E-01
–0.2	4.21E-01	4.17E-01	4.13E-01	4.09E-01	4.05E-01	4.01E-01	3.97E-01	3.94E-01	3.90E-01	3.86E-01
–0.3	3.82E-01	3.78E-01	3.74E-01	3.71E-01	3.67E-01	3.63E-01	3.59E-01	3.56E-01	3.52E-01	3.48E-01
–0.4	3.45E-01	3.41E-01	3.37E-01	3.34E-01	3.30E-01	3.26E-01	3.23E-01	3.19E-01	3.16E-01	3.12E-01
–0.5	3.09E-01	3.05E-01	3.02E-01	2.98E-01	2.95E-01	2.91E-01	2.88E-01	2.84E-01	2.81E-01	2.78E-01
–0.6	2.74E-01	2.71E-01	2.68E-01	2.64E-01	2.61E-01	2.58E-01	2.55E-01	2.51E-01	2.48E-01	2.45E-01
–0.7	2.42E-01	2.39E-01	2.36E-01	2.33E-01	2.30E-01	2.27E-01	2.24E-01	2.21E-01	2.18E-01	2.15E-01
–0.8	2.12E-01	2.09E-01	2.06E-01	2.03E-01	2.00E-01	1.98E-01	1.95E-01	1.92E-01	1.89E-01	1.87E-01
–0.9	1.84E-01	1.81E-01	1.79E-01	1.76E-01	1.74E-01	1.71E-01	1.69E-01	1.66E-01	1.64E-01	1.61E-01
–1	1.59E-01	1.56E-01	1.54E-01	1.52E-01	1.49E-01	1.47E-01	1.45E-01	1.42E-01	1.40E-01	1.38E-01
–1.1	1.36E-01	1.33E-01	1.31E-01	1.29E-01	1.27E-01	1.25E-01	1.23E-01	1.21E-01	1.19E-01	1.17E-01
–1.2	1.15E-01	1.13E-01	1.11E-01	1.09E-01	1.07E-01	1.06E-01	1.04E-01	1.02E-01	1.00E-01	9.85E-02
–1.3	9.68E-02	9.51E-02	9.34E-02	9.18E-02	9.01E-02	8.85E-02	8.69E-02	8.53E-02	8.38E-02	8.23E-02
–1.4	8.08E-02	7.93E-02	7.78E-02	7.64E-02	7.49E-02	7.35E-02	7.21E-02	7.08E-02	6.94E-02	6.81E-02
–1.5	6.68E-02	6.55E-02	6.43E-02	6.30E-02	6.18E-02	6.06E-02	5.94E-02	5.82E-02	5.71E-02	5.59E-02
–1.6	5.48E-02	5.37E-02	5.26E-02	5.16E-02	5.05E-02	4.95E-02	4.85E-02	4.75E-02	4.65E-02	4.55E-02
–1.7	4.46E-02	4.36E-02	4.27E-02	4.18E-02	4.09E-02	4.01E-02	3.92E-02	3.84E-02	3.75E-02	3.67E-02
–1.8	3.59E-02	3.51E-02	3.44E-02	3.36E-02	3.29E-02	3.22E-02	3.14E-02	3.07E-02	3.01E-02	2.94E-02
–1.9	2.87E-02	2.81E-02	2.74E-02	2.68E-02	2.62E-02	2.56E-02	2.50E-02	2.44E-02	2.39E-02	2.33E-02
–2	2.28E-02	2.22E-02	2.17E-02	2.12E-02	2.07E-02	2.02E-02	1.97E-02	1.92E-02	1.88E-02	1.83E-02
–2.1	1.79E-02	1.74E-02	1.70E-02	1.66E-02	1.62E-02	1.58E-02	1.54E-02	1.50E-02	1.46E-02	1.43E-02

–2.2	1.39E-02	1.36E-02	1.32E-02	1.29E-02	1.25E-02	1.22E-02	1.19E-02	1.16E-02	1.13E-02	1.10E-02
–2.3	1.07E-02	1.04E-02	1.02E-02	9.90E-03	9.64E-03	9.39E-03	9.14E-03	8.89E-03	8.66E-03	8.42E-03
–2.4	8.20E-03	7.98E-03	7.76E-03	7.55E-03	7.34E-03	7.14E-03	6.95E-03	6.76E-03	6.57E-03	6.39E-03
–2.5	6.21E-03	6.04E-03	5.87E-03	5.70E-03	5.54E-03	5.39E-03	5.23E-03	5.08E-03	4.94E-03	4.80E-03
–2.6	4.66E-03	4.53E-03	4.40E-03	4.27E-03	4.15E-03	4.02E-03	3.91E-03	3.79E-03	3.68E-03	3.57E-03
–2.7	3.47E-03	3.36E-03	3.26E-03	3.17E-03	3.07E-03	2.98E-03	2.89E-03	2.80E-03	2.72E-03	2.64E-03
–2.8	2.56E-03	2.48E-03	2.40E-03	2.33E-03	2.26E-03	2.19E-03	2.12E-03	2.05E-03	1.99E-03	1.93E-03
–2.9	1.87E-03	1.81E-03	1.75E-03	1.69E-03	1.64E-03	1.59E-03	1.54E-03	1.49E-03	1.44E-03	1.39E-03
–3	1.35E-03	1.31E-03	1.26E-03	1.22E-03	1.18E-03	1.14E-03	1.11E-03	1.07E-03	1.04E-03	1.00E-03
–3.1	9.68E-04	9.35E-04	9.04E-04	8.74E-04	8.45E-04	8.16E-04	7.89E-04	7.62E-04	7.36E-04	7.11E-04
–3.2	6.87E-04	6.64E-04	6.41E-04	6.19E-04	5.98E-04	5.77E-04	5.57E-04	5.38E-04	5.19E-04	5.01E-04
–3.3	4.83E-04	4.66E-04	4.50E-04	4.34E-04	4.19E-04	4.04E-04	3.90E-04	3.76E-04	3.62E-04	3.49E-04
–3.4	3.37E-04	3.25E-04	3.13E-04	3.02E-04	2.91E-04	2.80E-04	2.70E-04	2.60E-04	2.51E-04	2.42E-04
–3.5	2.33E-04	2.24E-04	2.16E-04	2.08E-04	2.00E-04	1.93E-04	1.85E-04	1.78E-04	1.72E-04	1.65E-04
–3.6	1.59E-04	1.53E-04	1.47E-04	1.42E-04	1.36E-04	1.31E-04	1.26E-04	1.21E-04	1.17E-04	1.12E-04
–3.7	1.08E-04	1.04E-04	9.96E-05	9.57E-05	9.20E-05	8.84E-05	8.50E-05	8.16E-05	7.84E-05	7.53E-05
–3.8	7.23E-05	6.95E-05	6.67E-05	6.41E-05	6.15E-05	5.91E-05	5.67E-05	5.44E-05	5.22E-05	5.01E-05
–3.9	4.81E-05	4.61E-05	4.43E-05	4.25E-05	4.07E-05	3.91E-05	3.75E-05	3.59E-05	3.45E-05	3.30E-05
–4	3.17E-05	3.04E-05	2.91E-05	2.79E-05	2.67E-05	2.56E-05	2.45E-05	2.35E-05	2.25E-05	2.16E-05
–4.1	2.07E-05	1.98E-05	1.89E-05	1.81E-05	1.74E-05	1.66E-05	1.59E-05	1.52E-05	1.46E-05	1.39E-05
–4.2	1.33E-05	1.28E-05	1.22E-05	1.17E-05	1.12E-05	1.07E-05	1.02E-05	9.77E-06	9.34E-06	8.93E-06
–4.3	8.54E-06	8.16E-06	7.80E-06	7.46E-06	7.12E-06	6.81E-06	6.50E-06	6.21E-06	5.93E-06	5.67E-06
–4.4	5.41E-06	5.17E-06	4.94E-06	4.71E-06	4.50E-06	4.29E-06	4.10E-06	3.91E-06	3.73E-06	3.56E-06

(continued)

Table B.1 Values of the cumulative distribution function for the standard normal variable (Continued)

	0	*0.01*	*0.02*	*0.03*	*0.04*	*0.05*	*0.06*	*0.07*	*0.08*	*0.09*
–4.5	3.40E-06	3.24E-06	3.09E-06	2.95E-06	2.81E-06	2.68E-06	2.56E-06	2.44E-06	2.32E-06	2.22E-06
–4.6	2.11E-06	2.01E-06	1.92E-06	1.83E-06	1.74E-06	1.66E-06	1.58E-06	1.51E-06	1.43E-06	1.37E-06
–4.7	1.30E-06	1.24E-06	1.18E-06	1.12E-06	1.07E-06	1.02E-06	9.68E-07	9.21E-07	8.76E-07	8.34E-07
–4.8	7.93E-07	7.55E-07	7.18E-07	6.83E-07	6.49E-07	6.17E-07	5.87E-07	5.58E-07	5.30E-07	5.04E-07
–4.9	4.79E-07	4.55E-07	4.33E-07	4.11E-07	3.91E-07	3.71E-07	3.52E-07	3.35E-07	3.18E-07	3.02E-07
–5	2.87E-07	2.72E-07	2.58E-07	2.45E-07	2.33E-07	2.21E-07	2.10E-07	1.99E-07	1.89E-07	1.79E-07
–5.1	1.70E-07	1.61E-07	1.53E-07	1.45E-07	1.37E-07	1.30E-07	1.23E-07	1.17E-07	1.11E-07	1.05E-07
–5.2	9.96E-08	9.44E-08	8.95E-08	8.48E-08	8.03E-08	7.60E-08	7.20E-08	6.82E-08	6.46E-08	6.12E-08
–5.3	5.79E-08	5.48E-08	5.19E-08	4.91E-08	4.65E-08	4.40E-08	4.16E-08	3.94E-08	3.72E-08	3.52E-08
–5.4	3.33E-08	3.15E-08	2.98E-08	2.82E-08	2.66E-08	2.52E-08	2.38E-08	2.25E-08	2.13E-08	2.01E-08
–5.5	1.90E-08	1.79E-08	1.69E-08	1.60E-08	1.51E-08	1.43E-08	1.35E-08	1.27E-08	1.20E-08	1.14E-08
–5.6	1.07E-08	1.01E-08	9.55E-09	9.01E-09	8.50E-09	8.02E-09	7.57E-09	7.14E-09	6.73E-09	6.35E-09
–5.7	5.99E-09	5.65E-09	5.33E-09	5.02E-09	4.73E-09	4.46E-09	4.21E-09	3.96E-09	3.74E-09	3.52E-09
–5.8	3.32E-09	3.12E-09	2.94E-09	2.77E-09	2.61E-09	2.46E-09	2.31E-09	2.18E-09	2.05E-09	1.93E-09
–5.9	1.82E-09	1.71E-09	1.61E-09	1.51E-09	1.43E-09	1.34E-09	1.26E-09	1.19E-09	1.12E-09	1.05E-09
–6	9.87E-10	9.28E-10	8.72E-10	8.20E-10	7.71E-10	7.24E-10	6.81E-10	6.40E-10	6.01E-10	5.65E-10
–6.1	5.30E-10	4.98E-10	4.68E-10	4.39E-10	4.13E-10	3.87E-10	3.64E-10	3.41E-10	3.21E-10	3.01E-10
–6.2	2.82E-10	2.65E-10	2.49E-10	2.33E-10	2.19E-10	2.05E-10	1.92E-10	1.81E-10	1.69E-10	1.59E-10
–6.3	1.49E-10	1.40E-10	1.31E-10	1.23E-10	1.15E-10	1.08E-10	1.01E-10	9.45E-11	8.85E-11	8.29E-11
–6.4	7.77E-11	7.28E-11	6.81E-11	6.38E-11	5.97E-11	5.59E-11	5.24E-11	4.90E-11	4.59E-11	4.29E-11
–6.5	4.02E-11	3.76E-11	3.52E-11	3.29E-11	3.08E-11	2.88E-11	2.69E-11	2.52E-11	2.35E-11	2.20E-11
–6.6	2.06E-11	1.92E-11	1.80E-11	1.68E-11	1.57E-11	1.47E-11	1.37E-11	1.28E-11	1.19E-11	1.12E-11

–6.7	1.04E-11	9.73E-12	9.09E-12	8.48E-12	7.92E-12	7.39E-12	6.90E-12	6.44E-12	6.01E-12	5.61E-12
–6.8	5.23E-12	4.88E-12	4.55E-12	4.25E-12	3.96E-12	3.69E-12	3.44E-12	3.21E-12	2.99E-12	2.79E-12
–6.9	2.60E-12	2.42E-12	2.26E-12	2.10E-12	1.96E-12	1.83E-12	1.70E-12	1.58E-12	1.48E-12	1.37E-12
–7	1.28E-12	1.19E-12	1.11E-12	1.03E-12	9.61E-13	8.95E-13	8.33E-13	7.75E-13	7.21E-13	6.71E-13
–7.1	6.24E-13	5.80E-13	5.40E-13	5.02E-13	4.67E-13	4.34E-13	4.03E-13	3.75E-13	3.49E-13	3.24E-13
–7.2	3.01E-13	2.80E-13	2.60E-13	2.41E-13	2.24E-13	2.08E-13	1.94E-13	1.80E-13	1.67E-13	1.55E-13
–7.3	1.44E-13	1.34E-13	1.24E-13	1.15E-13	1.07E-13	9.91E-14	9.20E-14	8.53E-14	7.91E-14	7.34E-14
–7.4	6.81E-14	6.31E-14	5.86E-14	5.43E-14	5.03E-14	4.67E-14	4.33E-14	4.01E-14	3.72E-14	3.44E-14
–7.5	3.19E-14	2.96E-14	2.74E-14	2.54E-14	2.35E-14	2.18E-14	2.02E-14	1.87E-14	1.73E-14	1.60E-14
–7.6	1.48E-14	1.37E-14	1.27E-14	1.17E-14	1.09E-14	1.00E-14	9.30E-15	8.60E-15	7.95E-15	7.36E-15
–7.7	6.80E-15	6.29E-15	5.82E-15	5.38E-15	4.97E-15	4.59E-15	4.25E-15	3.92E-15	3.63E-15	3.35E-15
–7.8	3.10E-15	2.86E-15	2.64E-15	2.44E-15	2.25E-15	2.08E-15	1.92E-15	1.77E-15	1.64E-15	1.51E-15
–7.9	1.39E-15	1.29E-15	1.19E-15	1.10E-15	1.01E-15	9.33E-16	8.60E-16	7.93E-16	7.32E-16	6.75E-16
–8	6.22E-16	5.74E-16	5.29E-16	4.87E-16	4.49E-16	4.14E-16	3.81E-16	3.51E-16	3.24E-16	2.98E-16
–8.1	2.75E-16	2.53E-16	2.33E-16	2.15E-16	1.98E-16	1.82E-16	1.68E-16	1.54E-16	1.42E-16	1.31E-16
–8.2	1.20E-16	1.11E-16	1.02E-16	9.36E-17	8.61E-17	7.92E-17	7.28E-17	6.70E-17	6.16E-17	5.66E-17
–8.3	5.21E-17	4.78E-17	4.40E-17	4.04E-17	3.71E-17	3.41E-17	3.14E-17	2.88E-17	2.65E-17	2.43E-17
–8.4	2.23E-17	2.05E-17	1.88E-17	1.73E-17	1.59E-17	1.46E-17	1.34E-17	1.23E-17	1.12E-17	1.03E-17
–8.5	9.49E-18	8.70E-18	7.97E-18	7.32E-18	6.72E-18	6.15E-18	5.64E-18	5.18E-18	4.74E-18	4.34E-18
–8.6	3.98E-18	3.66E-18	3.36E-18	3.06E-18	2.82E-18	2.57E-18	2.36E-18	2.17E-18	1.98E-18	1.82E-18
–8.7	1.65E-18	1.52E-18	1.38E-18	1.27E-18	1.17E-18	1.06E-18	9.76E-19	8.94E-19	8.13E-19	7.59E-19
–8.8	6.78E-19	6.23E-19	5.69E-19	5.15E-19	4.88E-19	4.34E-19	4.07E-19	3.52E-19	3.25E-19	2.98E-19
–8.9	2.71E-19	2.44E-19	2.44E-19	2.17E-19	1.90E-19	1.90E-19	1.63E-19	1.36E-19	1.36E-19	1.36E-19

Appendix C: Values of the gamma function $\Gamma(k)$ for $1 \leq k \leq 2$

Table C.1 Values of the gamma function G(*k*) for 1 = *k* = 2

k	Γ(k)	k	Γ(k)	k	Γ(k)	k	Γ(k)
1.00	1.00000	1.26	0.90440	1.51	0.88659	1.76	0.92137
1.01	0.99433	1.27	0.90250	1.52	0.88704	1.77	0.92376
1.02	0.98884	1.28	0.90072	1.53	0.88757	1.78	0.92623
1.03	0.98355	1.29	0.89904	1.54	0.88818	1.79	0.92877
1.04	0.97844	1.30	0.89747	1.55	0.88887	1.80	0.93138
1.05	0.97350	1.31	0.89600	1.56	0.88964	1.81	0.93408
1.06	0.96874	1.32	0.89464	1.57	0.89049	1.82	0.93685
1.07	0.96415	1.33	0.89338	1.58	0.89142	1.83	0.93969
1.08	0.95973	1.34	0.89222	1.59	0.89243	1.84	0.94261
1.09	0.95546	1.35	0.89115	1.60	0.89352	1.85	0.94561
1.10	0.95135	1.36	0.89018	1.61	0.89468	1.86	0.94869
1.11	0.94740	1.37	0.88931	1.62	0.89592	1.87	0.95184
1.12	0.94359	1.38	0.88854	1.63	0.89724	1.88	0.95507
1.13	0.93993	1.39	0.88785	1.64	0.89864	1.89	0.95838
1.14	0.93642	1.40	0.88726	1.65	0.90012	1.90	0.96177
1.15	0.93304	1.41	0.88676	1.66	0.90167	1.91	0.96523
1.16	0.92980	1.42	0.88636	1.67	0.90330	1.92	0.96877
1.17	0.92670	1.43	0.88604	1.68	0.90500	1.93	0.97240
1.18	0.92373	1.44	0.88581	1.69	0.90678	1.94	0.97610
1.19	0.92089	1.45	0.88566	1.70	0.90864	1.95	0.97988
1.20	0.91817	1.46	0.88560	1.71	0.91057	1.96	0.98374
1.21	0.91558	1.47	0.88563	1.72	0.91258	1.97	0.98768
1.22	0.91311	1.48	0.88575	1.73	0.91467	1.98	0.99171
1.23	0.91075	1.49	0.88595	1.74	0.91683	1.99	0.99581
1.24	0.90852	1.50	0.88623	1.75	0.91906	2.00	1.00000
1.25	0.90640						

Bibliography

ACI 318-12, "Building Code Requirements for Structural Concrete," American Concrete Institute, Farmington Hills, Michigan, 2012.

Agarwal, A.C. and M. Wolkowicz, *Interim Report on 1975 Commercial Vehicle Survey*, Research and Development Division, Ministry of Transportation, Downsview, Ontario, Canada, 1976.

Algermissen, S.T. and E.V. Leyendecker, "Technique for Uniform Hazard Spectra Estimation in the US," *Proceedings of the Tenth World Conference on Earthquake Engineering*, July 19–24, 1992, Madrid, Spain, pp. 391–397.

Allen, D.E., "ACI Error Survey Canadian Data," Building Research Note No. 123, National Research Council of Canada, Ottawa, July 1977.

American Association of State Highway and Transportation Officials (AASHTO), *Standard Specifications for Highway Bridges*, Washington, DC, 17th Edition, 2002.

American Association of State Highway and Transportation Officials (AASHTO), *Load and Resistance Factor Design (LRFD) Bridge Design Specifications*, 6th Edition, Washington, DC, 2012.

American Concrete Institute (ACI), *Building Code Requirements for Structural Concrete and Commentary* (ACI 318), Detroit, 2008.

American Forest and Paper Association, *National Design Specification for Wood Construction*, 2005.

American Institute of Steel Construction (AISC), *Steel Construction Manual*, 13th edition, Chicago, 2006.

American Institute of Steel Construction (AISC), *Steel Construction Manual*, 14th edition, American Institute of Steel Construction Inc., Chicago, 2011.

American Iron and Steel Institute (AISI), *Cold-Formed Steel Design Manual*, 2008.

American Petroleum Institute (API), *Recommended Practice to A-LRFD (draft)*, Dallas, 1989.

American Society of Civil Engineers (ASCE), *Load and Resistance Factor Design: Specification for Engineered Wood Construction*, New York, 1992.

ASCE/SEI Standard 7–10, *Minimum Design Loads for Buildings and Other Structures*, American Society of Civil Engineers, Reston, VA, 2010.

Ang, A.H.-S. and W.H. Tang, *Probability Concepts in Engineering: Emphasis on Applications to Civil and Environmental Engineering*, Wiley, 2006.

Ang, A.H.-S. and W.H. Tang, *Probability Concepts in Engineering Planning and Design: Volume I—Basic Principles*, John Wiley, New York, 1975.

Ang, A.H.-S. and W.H. Tang, *Probability Concepts in Engineering Planning and Design: Volume II—Decision, Risk, and Reliability*, John Wiley, New York, 1984.

Augusti, G., A. Baratta, and F. Casciati, *Probabilistic Methods in Structural Engineering*, Chapman and Hall, London, 1984.

Ayyub, B.M., *Risk Analysis in Engineering and Economics*, Chapman and Hall/CRC, 2003.

Ayyub, B.M. and G.J. Klir, *Uncertainty Modeling and Analysis in Engineering and the Sciences*, Chapman and Hall/CRC, 2006.

Ayyub, B.M. and R.H. McCuen, *Probability, Statistics, and Reliability for Engineers and Scientists*, 2nd edition, Chapman and Hall/CRC Press, 2002.

Baecher, G. and J. Christian, *Reliability and Statistics in Geotechnical Engineering*, Wiley, 2003.

Bakht, B., and L.G. Jaeger, *Bridge Analysis Simplified*, McGraw-Hill, 1985.

Benjamin, J.R. and C.A. Cornell, *Probability, Statistics and Decision for Civil Engineers*, McGraw-Hill, New York, 1970.

Bjorhovde, R., "A Probabilistic Approach to Maximum Column Strength," *Safety and Reliability of Metal Structures*, American Society of Civil Engineers, New York, 1972.

Borges, J. and M. Castanheta, *Structural Safety*, 2nd edition, Laboratorio Nacional de Engenharia Civil, Lisbon, Portugal, 1971.

Building Officials and Code Administrators, International (BOCA), *BOCA National Building Code*, Country Club Hills, Illinois, 1999.

Boyd, D.W., "Maximum Snow Depths and Snow Loads on Roofs in Canada," National Research Council of Canada, Division of Building Research, *Research Paper No. 142*, December 1961.

Brown, C.B., "A Fuzzy Safety Measure," *Journal of Engineering Mechanics*, ASCE, October 1979, pp. 855–872.

Building Seismic Safety Council (BSSC), *NEHRP Recommended Seismic Provisions for New Buildings and Other Structures (FEMA P-750)*, 2009 edition, Washington, DC, 2009.

Canadian Highway Bridge Design Code (CHBDC), Standard CAN/CSA-S6-06, 2006.

Canadian Institute of Steel Construction (CISC), *Handbook of Steel Construction*, 10th edition, 2011.

Chalk, P.L. and R.B. Corotis, "Probability Models for Design Live Loads," *Journal of the Structures Division*, ASCE, Vol. 106, No. 10, 1980, pp. 2017–2033.

Collins, K.R., Y.K. Wen, and D.A. Foutch, "Dual-Level Seismic Design: A Reliability-Based Methodology," *Earthquake Engineering and Structural Dynamics*, Vol. 25, 1996, pp. 1433–1467.

Collins, K.R., "Performance-Based Seismic Design of Building Structures," *Proceedings of the Seventh Specialty Conference on Probabilistic Mechanics and Structural Reliability*, Worcester, Massachusetts, August 1996, pp. 792–795.

Collins, K.R., "Simulation of Earthquake Ground Motions and Generation of Uniform Hazard Spectra for Linear and Nonlinear Response," *Proceedings of US–Japan Workshop, Stochastic Simulation for Civil Infrastructural Systems*, Kyoto, Japan, November 1997.

Collins, M.P., and D. Kuchma, "How Safe are Our Large Lightly Reinforced Concrete Beams, Slabs, and Footings," *ACI Structural Journal*, Vol. 96, 1999, pp. 482–490.

Cornell, C.A., "Bounds on the Reliability of Structural Systems," *Journal of Structural Division*, ASCE, Vol. 93, No. ST1, February 1967, pp. 171–200.

Cornell, C.A., "Engineering Seismic Risk Analysis," *Bulletin of the Seismological Society of America*, Vol. 58, No. 5, 1968, pp. 1583–1606.

Cornell, C.A., "A Probability-Based Structural Code," *ACI Journal*, Title No. 66-85, December 1969, pp. 974–985.

Corotis, R.B. and V.A. Doshi, "Probability Models for Live Load Survey Results," *Journal of the Structural Division*, ASCE, Vol. 103, No. ST6, June 1977, pp. 1257–1274.

Csagoly, P.F. and R.J. Taylor, *A Development Program for Wood Highway Bridges*, Ministry of Transportation and Communications, 79-SRR-7, Downsview, ON, Canada, 1979, 57 pp.

Ditlevsen, O., "Narrow Reliability Bounds for Structural Systems," *Journal of Structural Mechanics*, Vol. 7, No. 4, 1979, pp. 453–472.

Ditlevsen, O., *Structural Reliability Methods*, John Wiley & Sons, New York, 1996.

Dunn, O.J. and V.A. Clark, *Applied Statistics: Analysis of Variance and Regression*, Wiley, New York, 1974.

Ellingwood, B., T.V. Galambos, J.G. MacGregor, and C.A. Cornell, *Development of a Probability Based Load Criterion for American National Standard A58*, National Bureau of Standards, NBS Special Publication 577, Washington, DC, 1980.

Ellingwood, B., "Wind and Snow Load Statistics for Probabilistic Design," *Journal of the Structural Division*, ASCE, Vol. 107, No. 7, July 1981, pp. 1345–1349.

Ellingwood, B., J.G. MacGregor, T.V. Galambos, and C.A. Cornell, "Probability Based Load Criteria: Load Factors and Load Combinations," *Journal of the Structural Division*, ASCE, Vol. 108, No. ST5, May 1982, pp. 978–997.

Ellingwood, B. and R. Redfield, "Ground Snow Loads for Structural Design," *Journal of Structural Engineering*, ASCE, Vol. 109, No. 4, April 1983, pp. 950–964.

Ellingwood, B. and M. O'Rourke, "Probabilistic Models of Snow Loads on Structures," *Structural Safety*, Vol. 2, 1985, pp. 291–299.

Ellingwood, B. and D. Rosowsky, "Combining Snow and Earthquake Loads for Limit States Design," *Journal of Structural Engineering*, Vol. 122, No. 11, November 1996, pp. 1364–1368.

Ellingwood, B. and P.B. Tekie, "Wind Load Statistics for Probability-Based Structural Design," *Journal of Structural Engineering*, ASCE, Vol. 125, No. 4, April 1999, pp. 453–463.

EN EUROCODES, n.d., http://eurocodes.jrc.ec.europa.eu/home.php.

Federal Emergency Management Agency (FEMA), *Performance-Based Seismic Design of Buildings*, FEMA Report 283, September 1996.

Feller, W., *An Introduction to Probability Theory and Its Applications*, John Wiley, New York, 1971.

Fraczek, J., "ACI Survey of Concrete Structure Errors," *Concrete International*, December 1979, pp. 14–20.

Freudenthal, A.M., "The Safety of Structures," *ASCE Transactions*, Vol. 112, 1947, pp. 125–159.

Freudenthal, A.M., "Safety and the Probability of Structural Failure," *ASCE Transactions*, Vol. 121, 1956, pp. 1337–1397.

Galambos, T.V., B. Ellingwood, J.G. MacGregor, and C.A. Cornell, "Probability Based Load Criteria: Assessment of Current Design Practice," *Journal of the Structural Division*, ASCE, Vol. 108, No. ST5, May 1982, pp. 959–977.

Gardiner, R.A. and D.S. Hatcher, "Material and Dimensional Properties of an Eleven-Story Reinforced Concrete Building," *Structural Division Research Report No. 52*, Washington University, St. Louis, August 1970.

Ghosn, M. and F. Moses, "A Markov Renewal Model for Maximum Bridge Loading," *Journal of Engineering Mechanics*, ASCE, Vol. 111, No. 9, September 1985.

Gorman, M.R., *Reliability of Structural Systems*, Report No. 79-2, Civil Engineering Dept., Case Western Reserve University, Cleveland, 1979.

Grigoriu, M., *Risk, Structural Engineering, and Human Error*, University of Waterloo Press, 1984.

Grouni, H.N. and A.S. Nowak, "Calibration of the Ontario Highway Bridge Design Code," *The Canadian Journal of Civil Engineering*, Vol. 11, No. 4, December 1984, pp. 760–770.

Gumbel, E.J., "Statistical Theory of Extreme Values and Some Practical Applications," *Applied Mathematics Series 33*, National Bureau of Standards, Washington, DC, February 1954.

Haldar, A. and S. Mahadevan, *Probability, Reliability, and Statistical Methods in Engineering Design*, Wiley, 1999.

Han, S.W. and Y.K. Wen, *Method of Reliability-Based Calibration of Seismic Structural Design Parameters*, Structural Research Series Report No. 595, Department of Civil Engineering, University of Illinois, November 1994.

Harper, R.F., *The Code of Hammurabi King of Babylon*, The University of Chicago Press, Chicago, 1904.

Hart, G.C., *Uncertainty Analysis of Loads and Safety in Structural Engineering*, Prentice Hall, Englewood Cliffs, 1982.

Hasofer, A.M. and N. Lind, "An Exact and Invariant First-Order Reliability Format," *Journal of Engineering Mechanics*, ASCE, Vol. 100, No. EM1, February 1974, pp. 111–121.

Helander, M. (editor), *Human Factors/Ergonomics for Building and Construction*, Wiley, New York, 1981.

Hwang, E.-S. and A.S. Nowak, "Simulation of Dynamic Load for Bridges," *Journal of Structural Engineering*, ASCE, Vol. 117, No. 5, May 1991, pp. 1413–1434.

Iman, R.L. and W.J. Conover, "Small Sample Sensitivity Analysis Techniques for Computer Models, with an Application to Risk Assessment," *Communications in Statistics: Theory and Methods*, Vol. A9, No. 17, 1980, pp. 1749–1842.

International Conference of Building Officials (ICBO), *Uniform Building Code*, California, 1997.

Johnston, B.G. (editor), *Guide to Stability Design Criteria for Metal Structures*, 3rd edition, Wiley-Interscience, New York, 1976.

Kaplan, W., *Advanced Calculus*, 4th edition, Addison-Wesley, 1991.

Kennedy, D.J.L., *Study of Performance Factors for Section 10 Ontario Highway Bridge Design Code*, Morrison, Hershfield, Burgess & Huggins, Ltd., Report No. 2805331, Toronto, Canada, 1982.

Knoll, F., "Safety, Building Codes and Human Reality," Introductory Report to 11th Congress, IABSE, Vienna, 1979, pp. 247–258.

Kurata, M. and H. Shodo, "The Plastic Design of Composite Girder Bridges, Comparison with Elastic Design," DobokaGokkai-Shi, Japan (1967), translated by Arao, S., Civil Engineering Department, University of Ottawa, Canada.

Lind, N.C. and A.G. Davenport, "Towards Practical Application of Structural Reliability Theory," *Probabilistic Design of Reinforced Concrete Buildings*, ACI SP-31, 1972, pp. 63–110.

Lind, N.C., "Models of Human Error in Structural Reliability," *Structural Safety*, Vol. 1, 1983, pp. 167–175.

Lind, N.C. and A.S. Nowak, "Pooling Expert Opinions on Probability Distributions," *The Journal of Engineering Mechanics*, ASCE, Vol. 114, No. 2, Feb. 1988, pp. 328–341.

Luco, N., B.R. Ellingwood, R.O. Hamburger, J.D. Hooper, J.K. Kimball, and C.A. Kircher, "Risk-Targeted versus Current Seismic Design Maps for the Conterminous United States," *Proceedings of the SEAOC 76th Annual Convention*, Structural Engineers Association of California, Sacramento, California, 2007.

Madsen, B. and P.C. Nielsen, *In Grade Testing of Beams and Stringers*, Department of Civil Engineering, University of British Columbia, Vancouver, Canada, 1978.

Madsen, H.O., S. Krenk, and N.C. Lind, *Methods of Structural Safety*, Prentice-Hall, Englewood Cliffs, NJ, 1986.

Madsen, H.O., S. Krenk, and N.C. Lind, *Methods of Structural Safety*, Dover Publications, 2006.

Marek, P., M. Gustar, and T. Anagnos, *Simulation-Based Reliability Assessment for Structural Engineers*, CRC Press, Boca Raton, 1996.

Matousek, M., "Outcomings of a Survey on 800 Construction Failures," *IABSE Colloquium on Inspection and Quality Control*, Cambridge, England, July 1977.

Mayer, M., *Die Sicherheit der Bauwerke und ihreBerechnungnachGrenzkraftenstattnachzulassigenSpannungen*, Springer-Verlag, Berlin, 1926.

McGuire, R.K., *Seismic Hazard and Risk Analysis*, Monograph MNO-10, Earthquake Engineering Research Institute, Oakland, California, 2004.

McKay, M.D., R.J. Beckman, and W.J. Conover, "A Comparison of Three Methods for Selecting Values of Input Variables in the Analysis of Output from a Computer Code," *Technometrics*, Vol. 21, No. 2, May 1979, pp. 239–245.

Melchers, R.E., *Structural Reliability Analysis and Prediction*, Ellis Horwood Limited, Chichester, England, 1987.

Melchers, R.E., *Structural Reliability Analysis and Prediction (Civil Engineering)*, Wiley, 1999.

Miller, I., J.E. Freund, and R.A. Johnson, *Probability and Statistics for Engineers*, 8th edition, Prentice Hall, Englewood Cliffs, 2010.

Milton, J.S. and J.C. Arnold, *Introduction to Probability and Statistics: Principles and Applications for Engineering and the Computing Sciences*, 4th edition, McGraw-Hill, 2002.

Montgomery, D.C. and G.C. Runger, *Applied Statistics and Probability for Engineers*, 5th edition, John Wiley, 2010.

Moses, F., R.E. Snyder, and G.E. Likins, *Loading Spectrum Experienced by Bridge Structures in the United States*, Report No. FHWA/RD-85/012, Case Western Reserve University, Cleveland, OH, 1985.

Murzewski, J., *Reliability of Engineering Structures*, Arkady, Warszaewa, 1989.

Nassif, H. and A.S. Nowak, "Dynamic Load Spectra for Girder Bridges," *Transportation Research Record*, No. 1476, 1995, pp. 69–83.

Nassif, H. and A.S. Nowak, "Dynamic Load for Girder Bridges under Normal Traffic," *Archives of Civil Engineering*, Vol. XLII, No. 4, 1996, pp. 381–400.

Nowak, A.S., "Effect of Human Errors on Structural Safety," *Journal of the American Concrete Institute*, September 1979, pp. 959–972.

Nowak, A.S., "Calibration of LRFD bridge code," *ASCE Journal of Structural Engineering*, Vol. 121, 1995, pp. 1245–1251.

Nowak, A.S. and C.D. Eamon, "Reliability Analysis of Plank Decks," *ASCE Journal of Bridge Engineering*, Vol. 13, No. 5, September 2008, pp. 540–546.

Nowak, A.S. and N.C. Lind, "Practical Bridge Code Calibration," *Journal of the Structural Division*, ASCE, December 1979, pp. 2497–2510.

Nowak, A.S., *Modelling Properties of Timber Stringers*, Report UMCE 83R1, Department of Civil Engineering, University of Michigan, Ann Arbor, 1983.

Nowak, A.S. and P.V. Regupathy, "Reliability of Spot Welds in Cold-Formed Channels," *The Journal of Structural Engineering*, ASCE, Vol. 110, No. 6, June 1984, pp. 1265–1277.

Nowak, A.S. and R.I. Carr, "Classification of Human Errors," *Proceedings, the ASCE Symposium on Structural Safety Studies*, Denver, May 1985a, pp. 1–10.

Nowak, A.S. and R.I. Carr, "Sensitivity Analysis for Structural Errors," *The Journal of Structural Engineering*, ASCE, Vol. 111, No. 8, August 1985b, pp. 1734–1746.

Nowak, A.S. (editor), *Modeling Human Error in Structural Design and Construction*, NSF Workshop Proceedings, ASCE Publication, 1986.

Nowak, A.S., J. Czernecki, J. Zhou, and R. Kayser, *Design Loads for Future Bridges*, FHWA Project, Report UMCE 87-1, University of Michigan, Ann Arbor, July 1987.

Nowak, A.S. and T.V. Galambos (editors), *Making Buildings Safer for People During Hurricanes, Earthquakes, and Fires*, Van Nostrand Reinhold, New York, 1990.

Nowak, A.S. and Y.-K. Hong, "Bridge Live Load Models," *Journal of Structural Engineering*, ASCE, Vol. 117, No. 9, 1991, pp. 2757–2767.

Nowak, A.S. and H. Nassif, *Effect of Truck Loading on Bridges*, Report UMCE 91-11, Department of Civil Engineering, University of Michigan, Ann Arbor, MI, 1991.

Nowak, A.S., A.M. Rakoczy, and E. Szeliga, "Revised Statistical Resistance Models for R/C Structural Components," ACI SP-284-6, American Concrete Institute, March 2012.

Nowak, A.S., *Calibration of LRFD Bridge Design Code*, NCHRP Report 368, Transportation Research Board, Washington, DC, 1999.

Nowak, A.S., *Calibration of LRFD Bridge Design Code*, Report UMCE 93-22, Department of Civil and Environmental Engineering, University of Michigan, Ann Arbor, MI, December 1993.

Nowak, A.S., "Calibration of the OHBDC 1991," *Canadian Journal of Civil Engineering*, 21, 1994, pp. 25–35.

Nowak, A.S., A.S. Yamani, and S.W. Tabsh, "Probabilistic Models for Resistance of Concrete Bridge Girders," *ACI Structural Journal*, Vol. 91, No. 3, 1994, pp. 269–276.

Nowak, A.S., A.M. Rakoczy, and E. Szeliga, "Revised Statistical Resistance Models for R/C Structural Components," ACI SP-284-6, March 2012, pp. 1–16.

OHBDC, *Ontario Highway Bridge Design Code*, Ministry of Transportation, Downsview, Ontario, Canada, 1st edition 1979, 2nd edition 1983, 3rd edition 1991.

Perlis, S., *Theory of Matrices*, Addison-Wesley Publishing Company, 1952.

Peterka, J.A., "Improved Extreme Wind Prediction for the United States," *Journal of Wind Engineering and Industrial Aerodynamics*, Elsevier, Amsterdam, the Netherlands, Vol. 41, 1992, pp. 533–541.

Peterka, J.A. and S. Shahid, "Design Gust Wind Speeds inthe United States," *Journal of Structural Engineering*, Vol. 124, 1998, pp. 207–214.

Press, W.H., S.A. Teukolsky, W.T. Vetterling, and B.P. Flannery, *Numerical Recipes in Fortran*, 2nd edition, Cambridge University Press, 1992.

Pugachev, V.S., *Probability Theory and Mathematical Statistics for Engineers*, Pergamon Press, Oxford, England, 1984.

Pugsley, A.G., "The Prediction of the Proneness to Structural Accidents," *Structural Engineer*, No. 6, June 1973, pp. 195–196.

Rackwitz, R., "Note on the Treatment of Errors in Structural Reliability," Technische Universitat Muenchen, Heft 21, 1977, pp. 23–35.

Rackwitz, R. and B. Fiessler, "Structural Reliability under Combined Random Load Sequences," *Computers and Structures*, Vol. 9, 1978, pp. 489–494.

Rakoczy, A.M. and A.S. Nowak, "Statistical Parameters for Ice Thickness," Progress Report for NCHRP 10-80, Report No. UNL-CE-8-2011, August 2011.

Rakoczy, A.M. and A.S. Nowak, "Statistical Parameters for Wind Speed," Progress Report for NCHRP 10-80, Report No. UNL-CE-3-2012, March 2012.

Rakoczy, A.M. and A.S. Nowak, "Resistance Model of Lightweight Concrete Members," ACI Materials Journal, American Concrete Institute (accepted).

Rakoczy, A.M. and A.S. Nowak, "Resistance Factors for Lightweight Concrete Members," ACI Materials Journal, American Concrete Institute (under review).

RAND Corporation, *A Million Random Digits with 1,000,000 Normal Deviates*, Free Press, Glencoe, Illinois, 1955.

Rao, S.S., *Reliability-Based Design*, McGraw-Hill, New York, 1992.

Reineck, K.-H., D. Kuchma, K.S. Kim, and S. Marx, "Shear Database for Reinforced Concrete Members without Shear Reinforcement," *ACI Structural Journal*, Vol. 100, No. 2, 2003, pp. 240–249.

Rosenblueth, E., "Point Estimates for Probability Moments," *Proceedings of the National Academy of Sciences*, Vol. 72, No. 10, 1975, pp. 3812–3814.

Rosenblueth, E., "Two-Point Estimates in Probabilities," *Applied Mathematical Modeling*, Vol. 5, No. 5, 1981, pp. 329–335.

Ross, S.M., *Simulation*, 2nd edition, Academic Press, San Diego, 1997.

Ross, S., *A First Course in Probability*, 8th edition, Prentice-Hall, Upper Saddle River, New Jersey, 2009.

Rubinstein, R.Y., *Simulation and the Monte Carlo Method*, John Wiley, New York, 1981.

Schneider, J. (editor), *Quality Assurance Within the Building Process*, Proceedings, Vol. 47, IABSE Workshop, Rigi, Switzerland, June 1983.

Schneider, J., *Introduction to Safety and Reliability of Structures*, International Association for Bridge and Structural Engineering (IABSE), Zurich, Switzerland, 1997.

Sexsmith, R.G., P.D. Boyle, B. Rovner, and R.A. Abbott, *Load Sharing in Vertically Laminated Post-Tensioned Bridge Decking*, Technical Report 6, Western Forest Products Laboratory, Vancouver, British Columbia, Canada, 1979.

Simiu, E., "Extreme Wind Speeds at 129 Stations in the Contiguous United States," *Series No. 118*, National Bureau of Standards Building Science, Washington, DC, 1979.

Siriaksorn, A. and A.E. Naaman, *Reliability of Partially Prestressed Beams at Serviceability Limit States*, Report No. 80-1, University of Illinois at Chicago Circle, Chicago, Illinois, 1980.

Soong, T.T. and M. Grigoriu, *Random Vibration of Mechanical and Structural Systems*, Prentice-Hall, 1993.

Spiegel, M.R. and L. Stephens, *Schaums Outline of Statistics*, 4th edition, McGraw-Hill, New York, 2011.

Stankiewicz, P.R. and A.S. Nowak, "Material Testing for Wood Plank Decks," UMCEE 97-10, Report submitted to US Forest Service, April 1997.

Streletskii, N.S., *Foundations of Statistical Account of Factor of Safety of Structural Strength*, State Publishing House for Buildings, Moscow, 1947.

Stuart, A.J., "Equally Correlated Variates and the Multinormal Integral," *Journal of the Royal Statistical Society, Series B*, Vol. 20, 1958, pp. 373–378.

Swain, A.D. and H.E. Guttmann, *Handbook of Human Reliability Analysis with Emphasis on Nuclear Power Plant Applications*, NUREG/CR-1278, US Nuclear Regulatory Commission, Washington, DC, August 1983.

Tabsh, S.W., *Reliability-Based Sensitivity Analysis of Girder Bridges*, PhD Dissertation, Department of Civil Engineering, University of Michigan, Ann Arbor, MI, 1990.

Tabsh, S.W. and A.S. Nowak, "Reliability of Highway Girder Bridges," *Journal of Structural Engineering*, ASCE, Vol. 117, No. 8, August 1991, pp. 2373–2388.

Tantawi, H.M., *Ultimate Strength of Highway Girder Bridges*, PhD Dissertation, Department of Civil Engineering, University of Michigan, 1986.

Tantawi, H.M., A.S. Nowak, and N.C. Lind, "Point Distribution Methods for Bridge Reliability Analysis," *Journal of Forensic Engineering*, Vol. 3, No. 2/3, 1991, pp. 137–145.

Teng, S., H.K. Cheong, K.L. Kuang, and J.Z. Geng, "Punching Shear Strength of Slabs with Openings and Supported on Rectangular Columns," *ACI Structural Journal*, Vol. 101, No. 5, 2004, pp. 678–687.

Thoft-Christensen, P. and M.J. Baker, *Structural Reliability Theory and Its Applications*, Springer-Verlag, Berlin, 1982.

Thoft-Christensen, P. and Y. Murotsu, *Application of Structural Systems Reliability Theory*, Springer-Verlag, Berlin, 1986.

Thom, H.C.S., "Distribution of Maximum Annual Water Equivalent of Snow on the Ground," *Monthly Weather Review*, Vol. 94, No. 4, April 1966, pp. 265–271.

Ting, S.-C., *The Effects of Corrosion on the Reliability of Concrete Bridge Girders*, PhD Dissertation, Department of Civil Engineering, University of Michigan, Ann Arbor, MI, 1989.

Tobias, P.A. and D.C. Trindade, *Applied Reliability*, 3rd edition, Chapman and Hall/CRC, 2010.

Turkstra, C.J., "Theory of Structural Design Decisions," Ph.D. Dissertation, University of Waterloo, 1970.

Turkstra, C.J., "Theory of Structural Design Decisions," *Solid Mechanics Study No.* 2, University of Waterloo, Waterloo, Canada, 1972.

Turkstra, C.J. and H. Madsen, "Load Combinations for Codified Structural Design," *Journal of the Structures Division*, ASCE, Vol. 106, No. 12, December 1980, pp. 2527–2543.

US Geological Service, National Seismic Hazard Maps, 2008, http://earthquake.usgs.gov/designmaps.

Vecchio, F.J. and M.P. Collins, *The Response of Reinforced Concrete to In-Plane Shear and Normal Stresses*, Research Final Report, University of Toronto, Department of Civil Engineering, Toronto, Canada, 1982.

Vecchio, F.J. and M.P. Collins, "The Modified Compression Field Theory for Reinforced Concrete Elements Subjected to Shear," *ACI Journal*, Vol. 83, No. 2, 1986.

Vickery, P.J., and D. Wadhera, "Development of design wind speed maps for the Caribbean for application with the wind load provisions of ASCE 7," ARA Rep. No. 18108-1, Pan American Health Organization, Regional Office for The Americas World Health Organization, Disaster Management Programme, 525 23rd Street NW, Washington, DC, 2008.

Vickery, P.J., D. Wadhera, M.D. Powell, and Y. Chen, "A Hurricane Boundary Layer and Wind Field Model for Use in Engineering Applications," *Journal of Applied Meteorology and Climatology*, Vol. 48, Issue 2, February 2009a, p. 381.

Vickery, P.J., D. Wadhera, L.A. Twisdale, and F.M. Lavelle, "U.S. Hurricane Wind Speed Risk and Uncertainty," *Journal of Structural Engineering*, Vol. 135, March 2009b, pp. 301–320.

Vickery, P.J., D. Wadhera, J. Galsworthy, J.A. Peterka, P.A. Irwin, and L.A. Griffis, "Ultimate Wind Load Design Gust Wind Speeds in the United States for Use in ASCE-7," *Journal of Structural Engineering*, Vol. 136, 2010, pp. 613–625.

Wen, Y.K., "Statistical Combination of Extreme Loads," *Journal of the Structural Division*, ASCE, Vol. 103, No. ST5, May 1977, pp. 1079–1092.

Wen, Y.K., "Wind Direction and Structural Reliability," *Journal of Structural Engineering*, ASCE, Vol. 109, No. 4, April 1983, pp. 1028–1041.

Wen, Y.K., *Structural Load Modeling and Combination for Performance and Safety Evaluation*, Elsevier, Amsterdam, 1990.

Wen, Y.K., S.W. Han, and K.R. Collins, "Building Reliability and Codified Designs under Seismic Loads," *Proceedings of the ICASP 7 Conference*, Paris, France, July 1995, pp. 939–945.

Wen, Y.K., K.R. Collins, S.W. Han, and K.J. Elwood, "Dual-Level Designs of Buildings under Seismic Loads," *Structural Safety*, Vol. 18, No. 2/3, 1996, pp. 195–224.

Wierzbicki, W., *Safety of Structures as a Probabilistic Problem*, Przeglad, Technivzny, 1936 (in Polish).

Yamani, A.S., *Reliability Evaluation of Shear Strength in Highway Girder Bridges*, PhD Dissertation, Department of Civil Engineering, University of Michigan, Ann Arbor, MI, 1992.

Zhou, J.-H., *System Reliability Models for Highway Bridge Analysis*, PhD Dissertation, Department of Civil Engineering, University of Michigan, 1987.

Zhou, J.-H. and A.S. Nowak, "Integration Formulas for Functions of Random Variables," *Journal of Structural Safety*, Vol. 5, 1988, pp. 267–284.

Zokaie, T., T.A. Osterkamp, and R.A. Imbsen, *Distribution of Wheel Loads on Highway Bridges*, NCHRP 12-26/1, Proposed Changes in AASHTO, Imbsen and Associates, Sacramento, California, 1992.

Index

D

E

N

O

P

R

Printed in the United States
by Baker & Taylor Publisher Services

Printed in the United States
by Baker & Taylor Publisher Services